www.ingramcontent.com/pod-product-compliance
Lightning Source LLC
LaVergne TN
LVHW071341210926
848486LV00003B/201
9781908230454

Computational and Experimental Methods in Multiphase and Complex Flow X

WIT*PRESS*

WIT Press publishes leading books in Science and Technology.
Visit our website for new and current list of titles.
www.witpress.com

WIT*eLibrary*

Home of the Transactions of the Wessex Institute.
Papers published in this volume are archived in the WIT eLibrary in
volume 123 of WIT Transactions on Engineering Sciences (ISSN 1743-3533).
The WIT eLibrary provides the international scientific community with immediate and
permanent access to individual papers presented at WIT conferences.
http://library.witpress.com

Tenth International Conference on Computational and Experimental Methods in Multiphase and Complex Flow

Multiphase Flow X

Conference Chairmen

S. Hernández
University of A Coruña, Spain
Member of WIT Board of Directors

P. Vorobieff
University of New Mexico, USA

International Scientific Advisory Committee

Organised by

Wessex Institute, UK
University of New Mexico, USA

Sponsored by

WIT Transactions on Engineering Sciences
International Journal of Computational Methods and Experimental Measurements

WIT Transactions

Editorial Board

Computational and Experimental Methods in Multiphase and Complex Flow X

Editors

S. Hernández
University of A Coruña, Spain
Member of WIT Board of Directors

P. Vorobieff
University of New Mexico, USA

WIT*PRESS* **Southampton, Boston**

Editors:

S. Hernández
University of A Coruña, Spain
Member of WIT Board of Directors

P. Vorobieff
University of New Mexico, USA

Published by

WIT Press
Ashurst Lodge, Ashurst, Southampton, SO40 7AA, UK
Tel: 44 (0) 238 029 3223; Fax: 44 (0) 238 029 2853
E-Mail: witpress@witpress.com
http://www.witpress.com

For USA, Canada and Mexico

Computational Mechanics International Inc
25 Bridge Street, Billerica, MA 01821, USA
Tel: 978 667 5841; Fax: 978 667 7582
E-Mail: infousa@witpress.com
http://www.witpress.com

British Library Cataloguing-in-Publication Data

A Catalogue record for this book is available
from the British Library

ISBN: 978-1-78466-329-2
eISBN: 978-1-78466-330-8
ISSN: 1746-4471 (print)
ISSN: 1743-3533 (on-line)

The texts of the papers in this volume were set individually by the authors or under their supervision. Only minor corrections to the text may have been carried out by the publisher.

Preface

This volume contains papers presented at the 10th International Conference on Computational and Experimental Methods in Multiphase and Complex Flow (Multiphase Flow 2019) which took place in Lisbon, Portugal, organized by the Wessex Institute of Technology. Previous meetings were held in Orlando, Florida (2001); Santa Fe, New Mexico (2003); Portland, Maine (2005); Bologna, Italy (2007); New Forest, UK (2009); Kos, Greece (2011), A Coruña, Spain (2013), Valencia, Spain (2015), and Tallinn, Estonia (2017).

As before, both the conference participation and the scope of studies presented in Lisbon were truly global, as one can see from these proceedings, which cover a wide range of theoretical, experimental, and numerical studies related to many diverse applications involving multiphase flow.

It would be probably fair to say that while the field of fluid dynamics, especially classical, is dominated by studies of single-phase media – fluid, gas, or plasma, the real world around us is actually fundamentally multiphase on all relevant scales. Earth's atmosphere carries liquid droplets, bioagents, respirable and non-respirable solids, all of which have practical importance disproportionate to their volume fraction. Perhaps the most conspicuous reminder of the role of particles in the atmosphere is the color of sunset: the striking orange/red palette one often sees is due to shorter wavelengths of light (violet and blue) being scattered by particles in the atmosphere. Atmospheric particulates and aerosols affect both air quality and climate, with many effects still poorly understood. Likewise, ocean waters are permeated with plankton, natural sediment, and lately and most disturbingly – plastic particulates, whose ubiquity and large-scale effects on marine life still need a thorough study to develop appropriate mitigation strategies. On astronomical scales, recent studies (of particular – pun intended – interest here is the Infrared Astronomical Satellite survey, IRAS) reveal that vast volumes of space are filled with clouds of dusty gas whose temperature (15-30K) is largely determined by the heating of dust grains by distant starlight. Dusty cosmic matter remains the target of choice for the Spitzer Space Telescope, perhaps better known among the general public for exoplanet discoveries.

As in the natural world, multiphase-flow phenomena are nearly omnipresent in engineering applications. A good example here is the oil-gas-water flow, characterized both by considerable complexity and great practical importance. Multiphase effects are critically important for carbon dioxide sequestration, prediction of the operation of catalytic and fluidized-bed chemical reactors, and contamination of groundwater with drilling chemicals. Multiphase effects can both aid and impede drag reduction in flows near surfaces. They also play a dominant role in many areas of biomechanics, from hemodynamics to airflow and particle deposition in respiratory airways.

It was mentioned above that, while multiphase flows are ubiquitous, classical fluid mechanics mostly deals with single-phase flows. The reason for that is simple: single-phase formulation is just easier to work with – both in terms of theory and of experimental characterization. Moreover, until relatively recent times, proper numerical consideration of multiphase effects was often computationally prohibitive. However, recent important advances both in experimental and numerical studies of multiphase phenomena provide valuable insights into the relevant physics and lead to a better theoretical understanding. A good example here is the identification and characterization in 2011–2019 of a class of hydrodynamic instabilities peculiar to multiphase flows, such as shock-driven multiphase instability (SDMI) and multiphase gravity-driven instability analogous to, but in many important aspects distinct from, the single-phase Rayleigh-Taylor instability. It is worth mentioning that the first suggestion that a new class of such instabilities of multiphase media may exist appeared in an article published in 2011 in the sixth volume of this series.

Advances in numerical studies of multiphase phenomena are driven by several factors. One is availability of better validation benchmarks made possible by progress in experimental techniques, allowing to produce spatially and temporally resolved portraits of phase behavior in multiphase flow. Advancements in computing power play a role that would be difficult to overlook, but just as importantly, computational methods are improving their fidelity as well. Paraphrasing William J. Rider, a Distinguished Member of Technical Staff at Sandia National Laboratories, numerical modelers must always ask themselves two questions. “Are we modeling the physics correctly?” This question largely pertains to the accuracy of the chosen numerical scheme. The second question is “Are we modeling the correct physics?” The answer to the second question may be less obvious, and it requires thoughtful contemplation of the numerical model in combination with theory and experiment. Accordingly, different computational approaches must be applied to different multiphase flow regimes, depending on the dominant physics. In some cases, Lagrangian particle tracking can be applied to individual particles (or groups of particles) in the flow. In other situations, both phases can be regarded as fluids, using Eulerian approach. Finally, situations when phase boundaries must be accounted for explicitly are also quite common. One of the most powerful tools available to the modeler for such situations, when a regular finite-difference grid is ill-suited to reproduce the complex and time-dependent geometry, is the boundary element method (BEM) and its derivatives. This method was developed by Carlos A. Brebbia, the founder of the Wessex Institute and the original editor of this Transaction series, whose work was a key influence in shaping the field of modern computational mechanics and specifically of multiphase flow studies.

The papers included in this book address several ranges of topics relevant to advancement of the state of the art in multiphase flow. The present volume is part of the WIT Transactions on Engineering Sciences. It is widely distributed throughout the world in digital as well as paper format. The electronic versions of all the papers are archived in the e.library of the Wessex Institute (http://www.witpress.com/elibrary) where they are easily available to the scientific community.

We express our gratitude to the members of the Board and our colleagues who provided careful reviews of the contributions to this volume. The rigorous review process was essential for maintaining the high quality of this publication. The editors also thank the authors featured in this volume for submitting many excellent and innovative papers.

The Editors
Lisbon, Portugal
2019

Contents

SECTION 1
EXPERIMENTAL MEASUREMENTS

DRAG OF SHOCK-ACCELERATED MICROPARTICLES

KATHY PRESTRIDGE
Los Alamos National Laboratory, USA

ABSTRACT
Experiments of shock-accelerated microparticles show high drag coefficients when the particles are tracked from initial acceleration through the relaxation times. An eight-pulse particle tracking diagnostic measures individual particle positions, and a schlieren system measures shock location, with pressure transducers providing shock speed at the test section. These diagnostics give us detailed measurements of particle positions versus time for Mach 1.2, 1.3 and 1.4 experiments, allowing us to calculate accelerations and drag. Findings show that early-time drag is underestimated by current quasi-steady drag models due to several factors. The exact physical mechanisms causing the large drag are likely due to body pressure forces caused by the particle wake disturbing the carrier phase.
Keywords: unsteady drag, microparticles, shocks, shock tube.

1 INTRODUCTION

The passage of a shock wave past microparticles causes them to accelerate behind the shock wave then relax into mechanical and thermal equilibrium with the post-shock gas [1]. Experimental studies of accelerated particles have been performed, but usually the experiments and drag models are limited to the acoustic time scale of the flow, when the shock is just passing across the particle [2]. The total forces on a rigid sphere are generally modelled using quasi-steady, inviscid unsteady, viscous unsteady, gravity, and lift forces [3]. Theoretical and empirical unsteady drag models are developed using linearized asymptotic analyses [4]–[6] for small oscillations about a mean flow around a sphere or empirical corrections [7]–[9] such as adding compressibility effects to the linearized Basset–Boussinesq–Oseen (BBO) equation in the limit of zero particle Mach and Reynolds numbers. For shock-particle dynamics, these models do very well predicting drag on acoustic time scales, just when the shock is interacting with the particle. Unfortunately, these linearized models have limited applicability to real flows, where the particle Mach and Reynolds numbers are non-zero.

High drag coefficients have been observed by some experiments [10]–[14], but the experiments of unsteady drag behind a shock wave have mostly been limited to tracking large clouds of particles. High drag coefficients have been observed in several different experiments, but few have looked at individual particles to reduce uncertainty and study disperse flow conditions. Although high drag has been observed in post-shock flows, the physical mechanisms causing the high drag are suspected to be a time-dependent, nonlinear combination of inviscid drag forces. With these questions in mind, we have begun a series of experiments [15], [16] to observe microparticle behaviour in shock tubes and identify the physical mechanisms driving the high drag.

2 EXPERIMENTAL FACILITY AND DIAGNOSTICS

The experiments were performed at the Los Alamos Horizontal Shock Tube (HST) facility. The facility was designed to use a pneumatic, membraneless driver to produce shocks of Ma<2. The set up for these experiments is described in [15], and the HST is pictured in Fig. 1. The cross-section of the HST is 76x76 mm and the pneumatic driver is described in [17].

WIT Transactions on Engineering Sciences, Vol 123, © 2019 WIT Press
www.witpress.com, ISSN 1743-3533 (on-line)
doi:10.2495/MPF190011

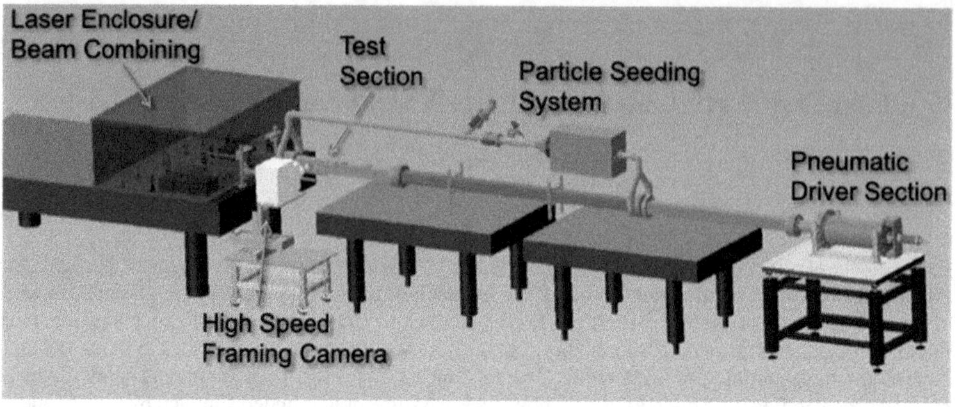

Figure 1: Horizontal shock tube with 8-pulse laser sheet system, high speed camera, and driver. The tube is 6.5 m long. The schlieren optics, not shown, are directed into the high speed framing camera from the opposite side of the test section. Pressure transducers run along the length of the driven section of the HST, all the way to the test section.

Experiments are performed using nylon particles (TSI 10084) with mean diameter of 4 μm and density of 1140 kg/m^3. Particles are illuminated by four dual-head Nd:YAG 532 nm lasers, and each of the eight pulses is timed independently. The exact timing of the laser pulses is confirmed using a high speed photodetector. To resolve the small particles, and to ensure that the centroids of the particles can be determined in each image, significant magnification of the flow is necessary. The field of view of the framing camera is 2.74x2.05 mm, with a resolution of 2.14 μm/pixel.

3 EXPERIMENTAL RESULTS

Fig. 2 shows measured particle positions for many different experiments, with four different particle tracks highlighted in colour. The particle drag, velocities and accelerations are calculated using a piecewise fitting method that is described in [18].

The position versus time measurements show that the particles are moving very quickly, and the calculated velocities and accelerations show that the early-time part of the flow, after the passage of the shock, is dominated by very high accelerations. This part of the flow is unsteady, gradually leading to particle relaxation to the post-shock flow speed at late times. Drag coefficients for the particles are calculated using the expression:

$$\ddot{x} = \frac{3\rho_g\left(u_g-\dot{x}_p\right)^2}{4\rho_p d_p} C_D(t),$$

where the dynamic properties are all post-shock. If $C_D(t)$ is known, this equation can be integrated numerically or analytically to determine an expression for the particle position versus time. The full description of the piecewise fitting method to determine drag coefficient from the particle position, including uncertainty analysis, can be found in Bordoloi et al. [18].

Particle drag is significantly higher than Clift–Gauvin, as shown in Fig. 3 for two different Mach number studies with the nylon particles. As the particles reach relaxation with the main flow, the drag gets closer to standard drag. Sets of experiments focusing on the initial

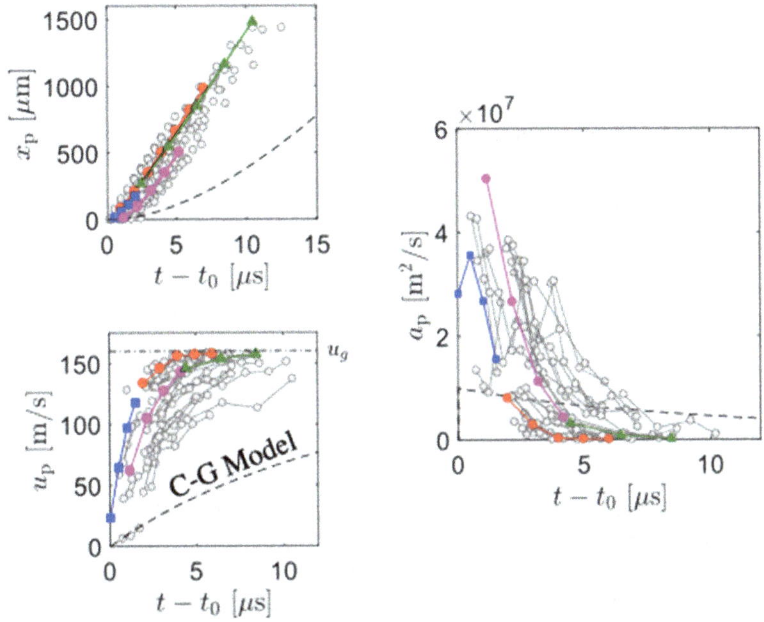

Figure 2: Particle positions, velocities and accelerations versus time after shock passage show much higher accelerations compared to those predicted by the Clift–Gauvin [19] model.

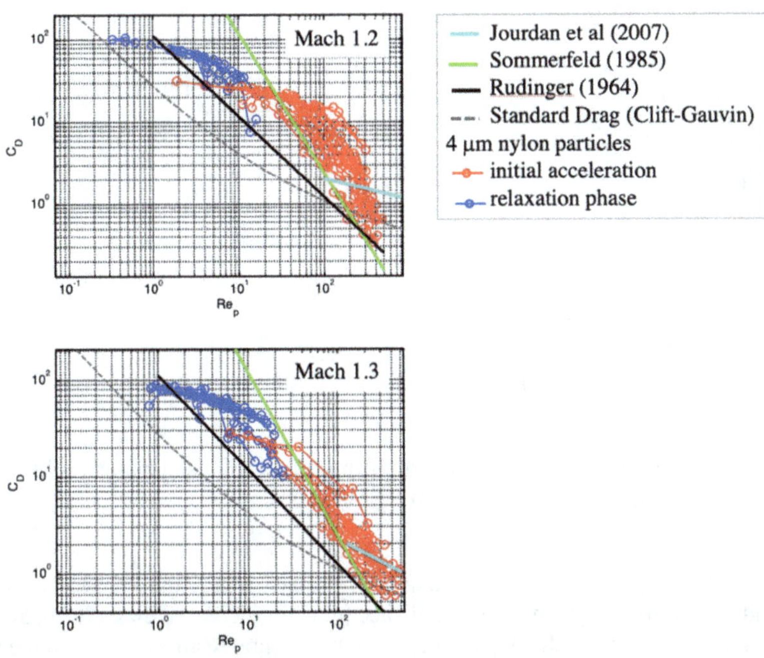

Figure 3: Drag coefficients from past and current shock-accelerated flow experiments.

acceleration phase were performed (in red), but to capture the last part of the acceleration, experiments at later times had to be performed separately (in blue). This is because the limited field of view, for high resolution, did not allow the full range of particle acceleration, from rest to the freestream velocity, to be captured in one experiment. The later-time measurements (mostly in blue) are where the Reynolds number, defined as $Re_p(t) = d_p[u_g - u_p(t)]/\nu_g$ is approaching zero because the particle slip velocity ($u_g - u_p$) is approaching zero, but there is no distinct separation between the early and late time experiments.

4 ANALYSIS

To better quantify drag as a function of the time scales of this unsteady flow, we look at the convective (t^*) and acoustic (τ_s) time scales,

$$t^* = {d_p}/{|u_g - u_p(t)|} \quad \text{and} \quad \tau_s = {d_p}/{u_s}.$$

The convective time scale is dependent upon the particle slip velocity, that is decreasing with time. The ratio of the convective and acoustic time scales can be written as a modified Knudsen number for the flow, where $Kn = M_s/Re_p$. The relationship between the Knudsen number and particle drag coefficient is shown in Fig. 4, where the drag coefficient increases sharply and becomes very high early on in the flow, and then it levels off.

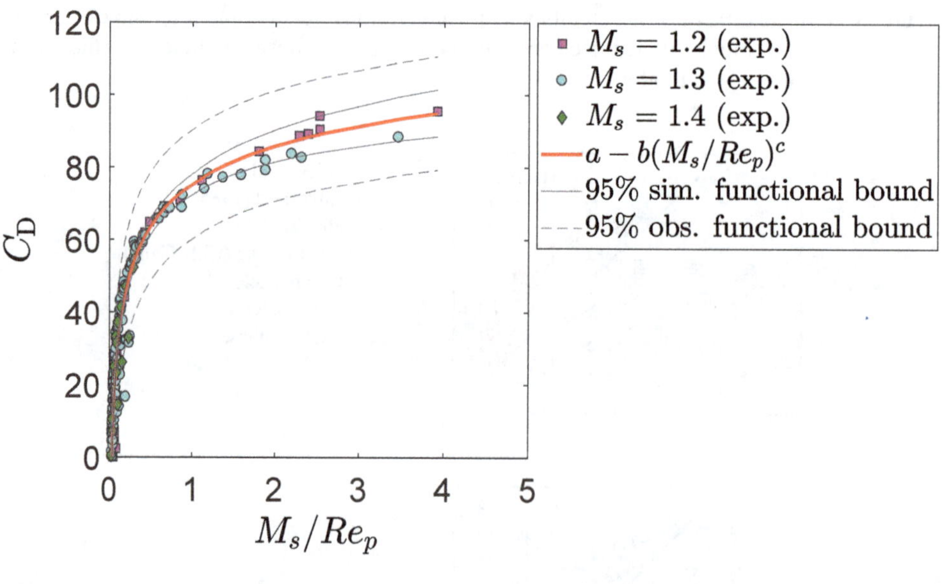

Figure 4: Drag coefficients as a function of the modified Knudsen number of the flow. The functional fit, in red, is $C_D = 167.5 - 92.2\ Kn^{-0.18}$.

This result is compared to the Clift–Gauvin [19] quasi-steady drag model in Bordoloi et al. [16], and the quasi-steady drag model only accounts for increased drag during the acoustic time scale. At the time of the shock passage, the experimental results agree with the unsteady drag models proposed by Parmar et al. [9], including: Clift–Gauvin (quasi-steady), Inviscid unsteady, and viscous unsteady (Bassett history). As soon as the particle starts accelerating, the experiments diverge from the model, and drag is dominated by an unsteady force for which there is no model currently.

The quasi-steady model of Clift–Gauvin is applicable over the full range of the experiment, since the constraints of Kn<<1 and M_p<0.6 are met. For these experiments, $Kn \approx 10^{-2}$ and 0.002<M_p<0.5. The inviscid and viscous unsteady models proposed by Parmar et al. [9] have limited range of applicability. Namely, their inviscid unsteady model is applicable only when the times of the flow are smaller than the acoustic time scales, or $t < \tau_s$. In our experiments, $t \approx 1000\tau_s$. Their viscous unsteady model is applicable only when the convective times of the flow are much smaller than the inverse of the product of the particle Reynolds number and shock Mach number, or $\tau_c Re_p M_s \ll 1$. In our experiments, $210.6 < \tau_c Re_p M_s < 436.8$ [16].

In unsteady flows, the classical Basset term is also not uniformly valid for long times, even at low particle Reynolds numbers. When a particle accelerates from rest, its approach to steady state scales as t^{-2} instead of $t^{-1/2}$ [20].

We are left with a large unsteady force that is causing the particle to have very high drag. This means that the particle will accelerate much faster than the quasi-steady model predicts over times greater than the acoustic time scale, and this has implications for simulations of shock-driven multiphase flows.

Given the current limitations of unsteady drag models, we must rely on calculations to help us understand the forces on the particle. Current work at Los Alamos is to run simulations of a single particle that can move freely after shock passage. A possible reason for past simulations not seeing this high drag over long time scales is that in most cases, simulations hold the particle stationary. So, the conditions that the particle sees are quiescent flow and then a uniform constant velocity.

Fig. 5 shows a possible explanation for the experimental results. Initially, the particle is stationary in a quiescent flow (Initial Condition). A Mach 1.5, for example, shock then passes over the particle (Acoustic Time Scale). At that time, the particle has not begun to move, and

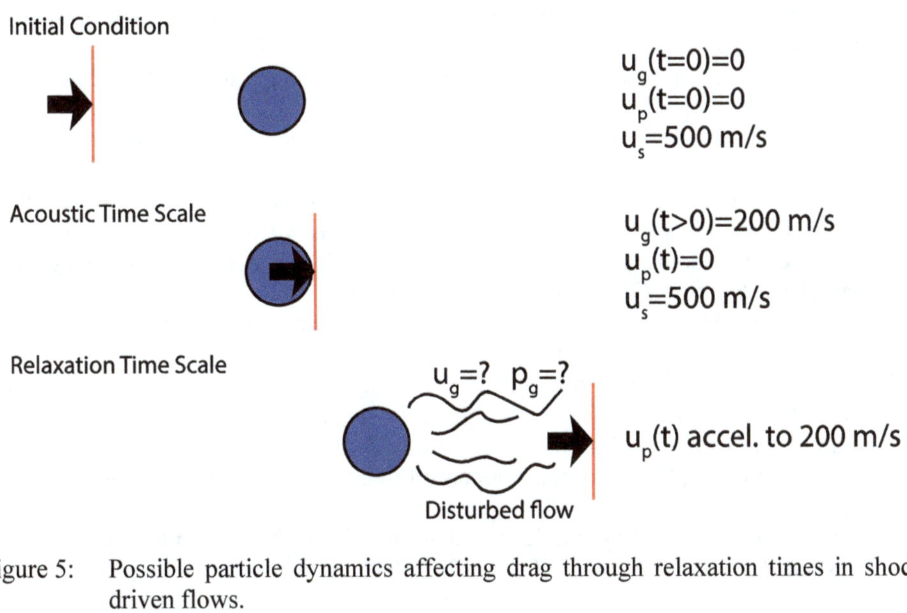

Figure 5: Possible particle dynamics affecting drag through relaxation times in shock-driven flows.

the gas behind the shock is moving at about 200 m/s. The shock then moves away from the particle. This sets up a condition where the particle is starting to move, but it is moving into a *disturbed flow*, not a quiescent or constant velocity flow. The disturbance is caused by the wake of the particle after the shock passes. This will create lower pressure conditions, and the particle is getting accelerated into those conditions. Because of the impulsive force on the particle, it will not convect with its wake, but rather get accelerated into the wake. It is possible that this imbalance causes the very high initial accelerations seen in the experiments.

To better understand the mechanisms driving these microparticles, calculations are beginning at Los Alamos that will allow the particle to move freely in the flow field. Additionally, experiments are continuing that will examine drag on liquid droplets. We are also working on diagnostic improvements to measure the carrier phase velocities and potentially visualize the particle wake dynamics using molecular tagging velocimetry.

ACKNOWLEDGEMENTS

The experiments were carefully performed and analyzed by Greg Orlicz and Adam Martinez.

REFERENCES

[1] Rudinger, G., Some properties of shock relaxation in gas flows carrying small particles. *Physics of Fluids*, **7**(5), pp. 658–663, 1964.

[2] Sun, M., Saito, T., Takayama, K. & Tanno, H., Unsteady drag on a sphere by shock wave loading. *Shock Waves*, **14**(1), pp. 3–9, 2005.

[3] Maxey, M.R. & Riley, J.J., Equation of motion for a small rigid sphere in a nonuniform flow. *Physics of Fluids*, 1983.

[4] Miles, J.W., On virtual mass and transient motion in subsonic compressible flow. *Q. J. Mech. Appl. Math.*, **4**, pp. 388–400, 1951.

[5] Longhorn, A.L., The unsteady, subsonic motion of a sphere in a compressible inviscid fluid. *Q. J. Mech. Appl. Math.*, **5**, pp. 64–81, 1952.

[6] Saito, T., Saba, M., Sun, M. & Takayama, K., The effect of an unsteady drag force on the structure of a non-equilibrium region behind a shock wave in a gas-particle mixture. *Shock Waves*, **17**, pp. 255–262, 2007.

[7] Henderson, C., Drag coefficients of spheres in continuum and rarefied flows. *AIAA Journal*, **14**(6), pp. 707–708, 1976.

[8] Loth, E., Compressibility and rarefaction effects on drag of a spherical particle. *AIAA Journal*, **46**(9), pp. 2219–2228, 2008.

[9] Parmar, M., Haselbacher, A. & Balachandar, S., Improved drag correlation for spheres and application to shock-tube experiments. *AIAA Journal*, **48**, pp. 1273–1276, 2010.

[10] Rudinger, G., Some properties of shock relaxation in gas flows carrying small particles. *Physics of Fluids*, **7**(5), pp. 658–663, 1964.

[11] Sommerfeld, M., The unsteadiness of shock waves propagating through gas particle mixtures. *Exp. Fluids*, **3**, pp. 197–206, 1985.

[12] Igra, O. & Takayama, K., Shock tube study of the drag coefficient of a sphere in a non-stationary flow. *Proc. R. Soc. A*, **442**, pp. 231–247, 1993.

[13] Jourdan, G., Houas, L., Igra, O., Estivalezes, J.L., Devals, C. & Meshkov, E.E., Drag coefficient of a sphere in a non-stationary flow: new results. *Proc. R. Soc.*, **463**, pp. 3323–3345, 2007.

[14] Sun, M., Saito, T., Takayama, K. & Tanno, H., Unsteady drag on a sphere by shock wave loading. *Shock Waves*, **14**(1), pp. 3–9, 2005.

[15] Martinez, A.A., Orlicz, G.C. & Prestridge, K.P., A new experiment to measure shocked particle drag using multi-pulse particle image velocimetry and particle tracking. *Exp. Fluids*, **56**, p. 1854, 2015.
[16] Bordoloi, A., Martinez, A.A. & Prestridge, K., Relaxation drag history of shock accelerated microparticles. *J. Fluid Mech.*, **823**, R4, 2017.
[17] Mejia-Alvarez, R., Wilson, B.M., Leftwich, M.C., Martinez, A.A. & Prestridge, K.P., Design of a fast diaphragmless shock tube driver. *Shock Waves*, **25**(6), pp. 635–650, 2015.
[18] Bordoloi, A., Ding, L., Martinez, A.A., Prestridge, K. & Adrian, R.J., A new method to calculate unsteady particle kinematics and drag coefficient in a subsonic post-shock flow. *Meas. Sci. Tech.*, **29**(7), 074001, 2018.
[19] Clift, R. & Gauvin, W.H., The motion of particles in turbulent gas streams. *Proc. Chemeca*, **70**(1), pp. 14–28, 1970.
[20] Mei, R. & Adrian, R.J., Flow past a sphere with an oscillation in the free-stream velocity and unsteady drag at finite Reynolds number. *J. Fluid Mech.*, **237**, 1992.

DIRECT MEASUREMENT OF SOLID DRAG FORCE IN FLUID–PARTICLE FLOW

ANA MOSQUERA GOMEZ, MARKKU NIKKU & PAYMAN JALALI
LUT Energy, Lappeenranta University of Technology, Finland

ABSTRACT
Many industrial applications involve fluidization systems in which it is possible to distinguish solid particles acting as a separate phase. The fluid–particle drag forces are well known and have been widely described in the literature to account for the momentum transfer between the phases whereas the study of particle–particle interaction still has several challenges especially for dense flow regimes. In this work, a novel method for the measurement of the particle–particle drag force was developed, using a cantilever beam with four strain gauges. For validation purposes, fluid–particle drag was investigated, the measurement device was evaluated considering two systems, in the first case, a spherical object was exposed to an air flow in a vertical tube where different velocities were used and for the second one, the forces resulting from impinging jet flows on the sphere were considered. The results show good agreement with the theoretical values. The results show good agreement with the theoretical values. The average magnitude of drag force upon a larger spherical object immersed on a flow of small particles was measured under different streams conditions, for different impinging height, size, and density of the particles.
Keywords: particle–particle drag force, fluid–particle drag force, strain gauge, impinging jet.

1 INTRODUCTION

Fluidized bed processes are utilized in a variety of industrial applications due to their high heat and mass transfer rates between the fluid and solid phase. For many decades, significant effort has been dedicated to increasing the understanding of the multiphase flow in these fluid–particle systems to improve on the design and scale-up of fluidized beds. The advances in computational performance have increased the popularity of numerical modelling in studies of fluidized beds, but due to the complexity of the fluid–particle and particle–particle interactions, active research is ongoing in this area.

Two approaches can be utilized in simulations of fluid–particle flows, an Eulerian–Lagrangian formulation, which treats the particles as a discrete phase and the fluid as a continuous phase, or an Eulerian–Eulerian approach, where both phases are considered continuous. Both approaches have their advantages and disadvantages, namely the Eulerian–Lagrangian approach can be computationally more expensive when the number of particles increases, though this can be somewhat remedied with grouping methods such as multiphase particle-in-cell method. For many large scale fluidized beds, the Eulerian–Eulerian approach is still more practical.

The description of particle as a continuous phase is based on the Kinetic Theory of Granular Flows [1], [2], which makes simplifications and models all the particle level phenomena, such as the particle collisions. For high solids concentrations, the collisional and frictional interactions between the particles increases and their effect is significant on the system behaviour, and in these conditions, the Eulerian models may fail to correctly predict the particle interactions [3]. As the Eulerian approach cannot consider momentum transfer and movements of individual particles, the momentum transfer is modelled as a drag force similarly to fluid–solid momentum transfer.

Literature presents several studies for fluid–particle drag force but for particle–particle drag force only a few studies can be found [4], [5]. These models are developed for or assume

WIT Transactions on Engineering Sciences, Vol 123, © 2019 WIT Press
www.witpress.com, ISSN 1743-3533 (on-line)
doi:10.2495/MPF190021

that all the particles are monosized spheres, and in the Eulerian approach the particle are distributed uniformly inside every computational cell. Both assumptions may affect the model results, especially when the size ratio between the particles is large.

Measuring the collisional force between two individual particles is possible but when the number of particles and collisions increase, is not any more feasible. In the present study, collisional forces from falling particles are measured using a force sensor based on strain gauges. The aim of the method is to be able to measure the net momentum transfer from the collisions, i.e. the drag force from the particle phase, in real process conditions, for example inside fluidized beds, to enable particle–particle drag force model development. Strain gauge systems have been previously used for direct measurement of forces and moments [6], [7]. The measurement method, sensor layout and calibration are explained in the next section. The validation measurements of fluid–particle drag force and experimental results of particle forces generated by a cloud of particles colliding with a large sphere are presented and discussed in Section 3. Finally, the concluding remarks and application perspective of the developed force measurement system are presented in Section 4.

2 FORCE SENSOR DEVELOPMENT AND CALIBRATION

When strain gauges pairs are arranged in a Wheatstone bridge circuit, the electric resistance of them can be affected by external forces and the output voltage reading is directly related to those applied forces. Four strain gauges (Kyowa KFG-1b-120-c1) were mounted on a rectangular steel beam (SAE 1074, 3 mm x 0.5 mm) to measure bending moment applied to the beam, and a spherical plastic particle of 40 mm diameter (table tennis ball) was attached to the tip of the steel beam. The elasticity modulus of the used steel was $2.10 \cdot 10^9$ N/m^2. The other end of the beam was fixed with a rigid support. A full Wheatstone bridge was used to increase the sensitivity and signal to noise ratio of the measurement signal. The sensor was connected to an analog data acquisition board and data were recorded at a frequency of 200 Hz using LabVIEW and processed with a low-pass filter. The sensor was calibrated with a set of known weights in order to correlate the voltage signal of the sensor to known gravitational forces. A schematic of the measurement system and the calibration curve are presented in Fig. 1.

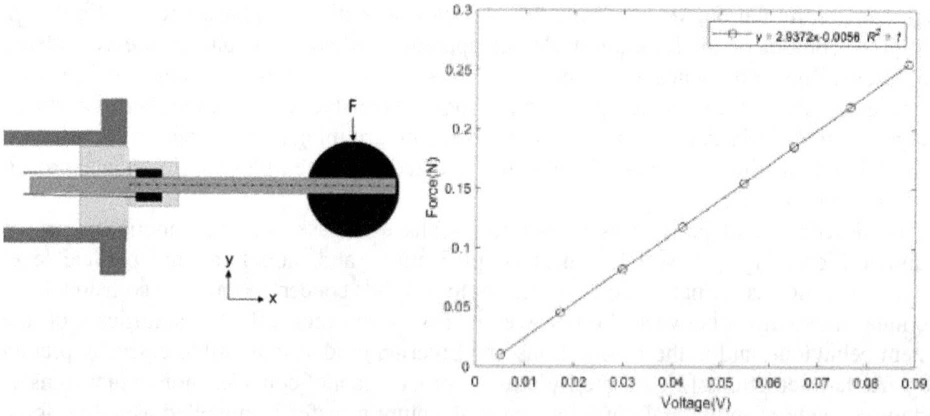

(a) (b)

Figure 1: Measurement system. (a) Schematic representation; and (b) Calibration chart.

3 VALIDATION METHODS, MEASUREMENTS AND RESULTS

For validation purposes, the sensor was initially tested with a confined flow inside a tube and later effects of an impinging jet were studied. Finally, the sensor used to measure force from particle flows with different conditions; with varying impact height, size and density of the particles.

3.1 Aerodynamic drag acting on a sphere

A stationary object immersed in a fluid flow experiences a drag force from the flow, which varies with fluid velocity and properties as well as the object characteristics, such as shape and size. For this study, the sensor was immersed in an upward airflow in ambient conditions inside a tube with 11 cm in diameter and flow velocities up to 5 m/s. The force measured by the sensor consisted of the drag force exerted on the beam and the ball, and to separate these, the force measurements were conducted without the ball and subtracted from all the collected data. The drag force can be written as:

$$F_D = \frac{1}{2} C_D A \rho V^2, \tag{1}$$

where C_D is the drag coefficient, A the cross sectional area to the flow, ρ and V are the density and velocity of the fluid, respectively. The drag coefficient equations for a sphere are presented as an empirical relations to the Reynolds number of a particle, given as:

$$Re_s = \frac{\rho_f |v_s - v_f| d_p}{\mu_f}, \tag{2}$$

where ρ_f is the density of the surrounding fluid, v_s is the velocity of the solid, which in this case is equal to zero since the sensor is fixed, v_f is the superficial velocity of the fluid, d_p is the particle diameter and μ_f is the dynamic viscosity of the fluid. In the calculation, the air was considered as an ideal gas.

The measured forces are compared with theoretical drag forces calculated with three correlations for the drag coefficient:

- Clift et al. [8]:

$$C_D = \begin{cases} 24/Re_s + 3/16 \text{ for } Re_s < 0.01 \\ (24/Re_s)\left(1 + 0.1315 Re_s^{0.82 - 0.05 log Re}\right) \text{ for } 0.01 < Re_s \le 20 \\ (24/Re_s)\left(1 + 0.1935 Re_s^{0.6305}\right) \text{ for } 20 < Re_s \le 260 \\ 10^{1.6435 - 1.1242\, log\, Re_s + 0.1558\, log^2 Re_s} \\ \text{for } 260 < Re_s \le 1500 \\ 10^{-2.4571 + 2.5558\, log Re_s - 0.9295\, log^2 Re_s + 0.1049\, log^3 Re_s} \\ \text{for } 1500 < Re_s \le 1.2 \times 10^4 \\ 10^{-1.9181 + 0.637\, \log Re_s - 0.0636 \log^2 Re_s} \\ \text{for } 1.2 \times 10^4 < Re_s \le 4.4 \times 10^4 \\ 10^{-4.339 + 1.5809\, log Re_s - 0.1546\, log^2 Re_s} \\ \text{for } 4.4 \times 10^4 < Re_s \le 3.38 \times 10^5 \\ 29.78 - 5.3\, log Re_s \text{ for } 3.38 \times 10^5 < Re_s \le 4 \times 10^5 \\ 0.1 log Re_s - 0.49 \text{ for } 4 \times 10^5 < Re_s \le 4 \times 10^6 \\ 0.19 - 8 \times 10^4 / Re_s \text{ for } Re_s > 10^6 \end{cases} \tag{3}$$

- Schiller and Neumann [8]:

$$C_D = \begin{cases} \frac{24}{Re_s}(1 + 0.15Re_s^{0.687}), \text{ for } Re_s < 1000 \\ 0.44, \text{ for } Re_s \geq 1000. \end{cases} \quad (4)$$

- Clift and Gauvin [9], [10]:

$$C_D = \frac{24}{Re_s}(1 + 0.15Re_s^{0.687}) + \frac{0.42}{1+42500Re^{-1.16}}. \quad (5)$$

The results of the measured and theoretical drag forces are presented in Fig. 2. The Mean Square Error varies from 12.67% to 28.5% between the three evaluated cases. The initial calibration can affect the values and to reduce the inaccuracy of the measurement, a frequently calibration of the system is made during the samples time between experiments.

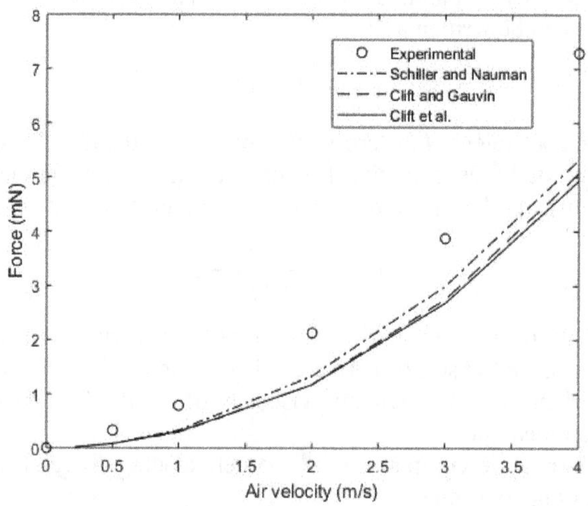

Figure 2: Drag force on a sphere immersed in an air flow.

3.2 Jet impingement on a sphere

To expand the measurements to another type of momentum transfer, impingement tests were performed with a water jet. A syringe with a nozzle diameter of 3.98 mm was used to produce a controlled water jet, with a weighted object pressing the piston with a constant force. Compared to particles, it is more straightforward to determine the momentum impacting the sensor, as the jet is continuous and the determination of the velocity is simpler. Normal tap water was used as the working fluid with a temperature of 20°C, viscosity of 0.001 Pa s and density of 998.2 kg/m^3. Each test was repeated 5 times for three different distances between the nozzle and the impingement surface, the data were processed according the description formulated in Section 2. The impingement test setup is described in Fig. 3.

As the distance in height between the nozzle and the surface of the sphere increases, the larger the forces obtained as the gravitational acceleration affects the jet. The force can also be controlled by changing the weight of the pressing object. The velocity of the water jet before the impact is calculated to estimate the force over the surface, which is expressed as:

$$F = ma = \rho q_v v_i, \quad (6)$$

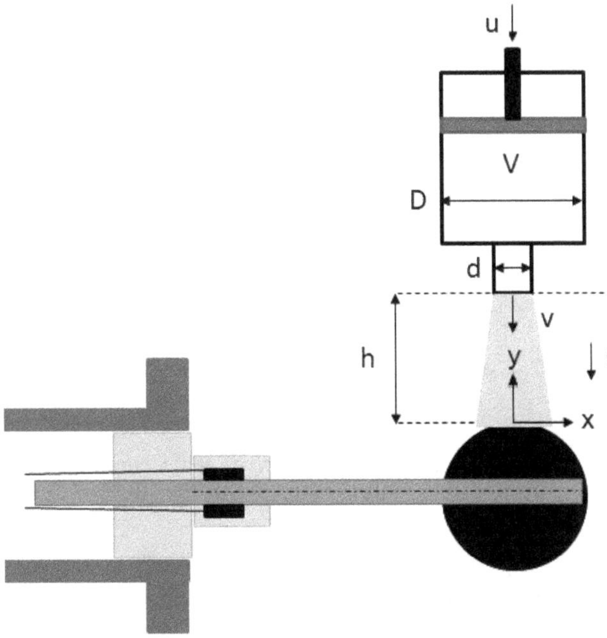

Figure 3: Water impinging jet.

where m is mass, a acceleration, ρ density of water, q_v is volumetric flow rate and v_i is velocity before the impact. Considering the water as an incompresible fluid, the volumetric flow rate remains constant throughout the syringe and the nozzle:

$$q_v = uA_1 = vA_2, \tag{7}$$

where u represents the velocity of the piston, v is the velocity at the nozzle, A_1 and A_2 are the sectional area of syringe cylinder and the nozzle, respectively. The velocity of the piston was determined with a Phantom Miro M310 High Speed Camera with a frame rate of 100 fps. The information of the location of the piston over time was processed in MatLab using component analysis of the binary images. The velocity of the jet at the nozzle was then computed with eqn (7). The effect of gravity on the stream was acknowledged in computation of the velocity before the impact with the equations of motion:

$$v_i = v + gt, \tag{8}$$

$$h = vt + 1/2\, gt^2, \tag{9}$$

in which the distance between the nozzle and the surface of the sphere is h and g is the gravitational acceleraton.

To illustrate the experimental measures and the theoretical calculated trend, an example of results for an inpinging height of 1.5 cm are shown in Fig. 4, in which is visible that velocity and the force of the jet increase during the press, with both the theoretical and measured forces presenting the same slope but the theoretical values are on average 16% larger than the force measured by the sensor.

A comparision between experimental and theoretical forces excerted by the water jet for different heights is presented in Fig. 5.

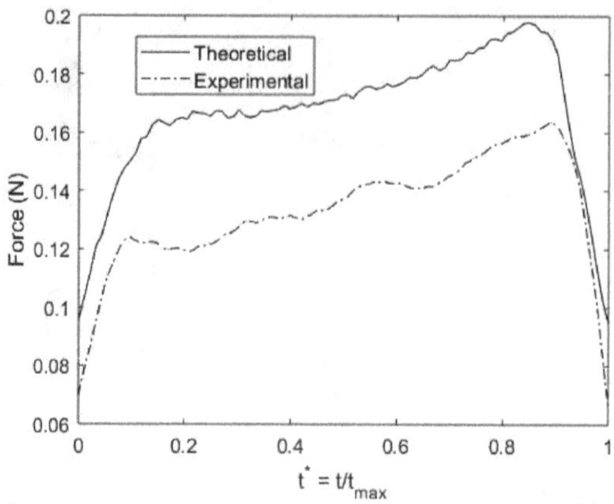

Figure 4: Experimental and theoretical forces at 1.5 cm impinging height.

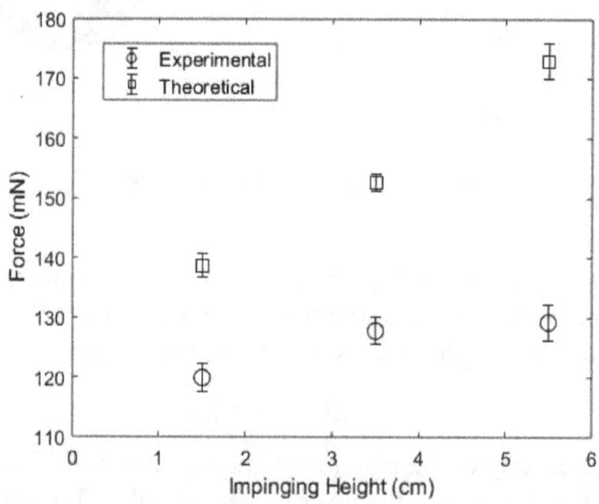

Figure 5: Experimental and theoretical forces of the impinging jet.

As seen from the results, there are clear differences between the theoretical, maximum force and the force measured with the sensor. The discrepancies between the theoretical and the experimental values increase with the height, reaching a 37.6% difference for the higher level. This is expected, as the momentum transfer from the jet to the sensor is limited. Initially, the jet hits the curved surface of the sensor, and due to the curvature, part of the momentum is redirected. After the initial impact, there remains a layer of water on top of the sensor and the incoming jet hits this layer, transferring parts of its momentum on the water layer rather than the sensor. Thus, the presented results should not be directly compared to the theoretical values, but rather they indicate the behaviour to be expected from the measurements. In this light, the obtained results are logical and reasonable.

3.3 Particle stream on a sphere

The sensor was used to measure the forces exerted on the sphere by a stream of falling particles with different properties. Unlike with the water jet, the momentum impacting the sensor could not be determined as, unfortunately for the current research, the particle velocities or concentrations at the impact could not be measured. The effect of the falling height, the size and density of the particles were studied. The falling height relates to the impact velocity of the particles, as they are accelerated more by gravity with larger height. A total of 3 repetitions were done for each configuration, in order to have statistical representative values. The output signal was processed according to the description in Section 2. A constant mass of 300 g of particles was used for all the cases. The particles were fed through a funnel with a nozzle diameter of 3 cm and mesh grids of 2, 1 and 0.5 mm were placed at the nozzle of the funnel, sizes large enough to avoid the obstruction of the mesh, to distribute the particles. The list of materials used in the experiments are summarize in Table 1. A schematic visualization of the setup is shown in Fig. 6.

Table 1: Material properties description.

Material	Material density (kg/m^3)	Particle diameter (μm)	Falling height (cm)
Glass beads	2500	700–1000	2.2,4.2,6.2
		500–560	2.2
		180–250	2.2
Zirconium oxide	4135	180–250	2.2
Iron beads	7645	180–250	2.2

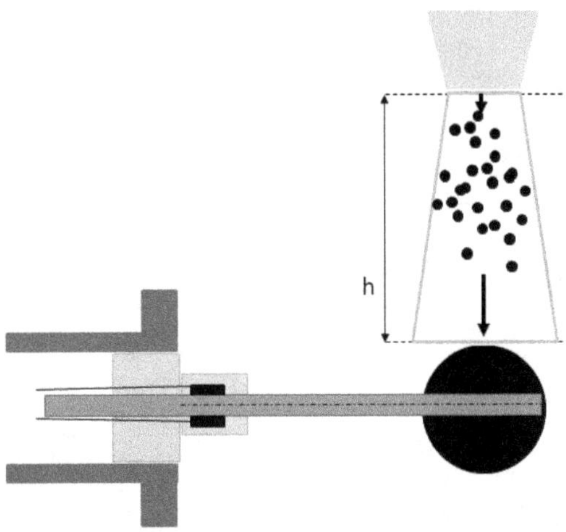

Figure 6: Schematic view of the particle–particle force measurement system.

As an example, the force measured using glass particles 500–560 μm released from a height of 2.2 cm from the surface is shown in Fig. 7, in which a relatively constant force can be observed.

The average forces for the different cases are presented in Fig. 8. The largest glass particles were used for testing the effect of different heights (impact velocities). As the distance from the surface increases, the impact velocities increase and higher forces are observed, as seen in Fig 8(a). The impact force is also a function of density of the particles,

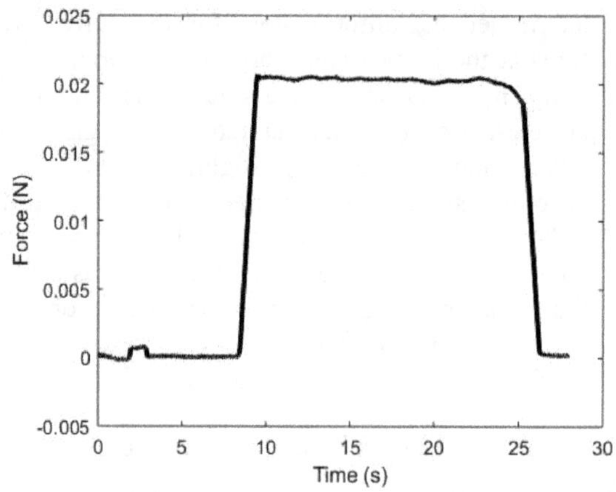

Figure 7: Force from glass beads 500–560 μm from height of 2.2 cm.

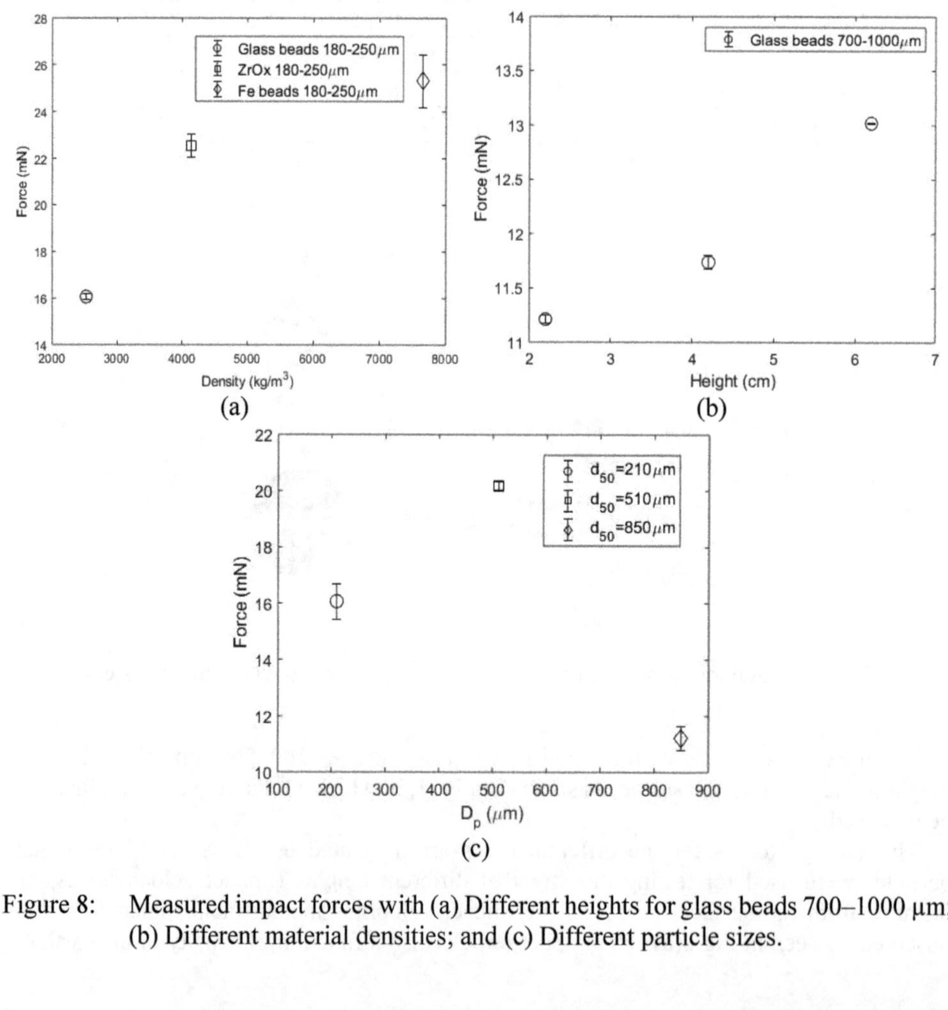

Figure 8: Measured impact forces with (a) Different heights for glass beads 700–1000 μm; (b) Different material densities; and (c) Different particle sizes.

and this was tested with glass beads, zirconium oxide and iron beads sieved to have the same particle size range. As the density of the particles increases, the resulting forces are larger as shown in Fig. 8(b). Similar analysis can be made for particles of different size, higher forces should be observed with larger particle size. This was investigated with glass bead particles of different size, with results presented in Fig. 8(c). While the forces for the small and middle size particles shown an ascendant trend, the measured forces for the large particles present a sudden drop. Unfortunately the particle velocities and concentrations could not be measured in this study, to confirm the reason for this behaviour. Most likely this is related with the particle distribution system, it is not feasible to have the same number of particle hitting the sensor with different materials, thus there were less particles large size impacting the sensor at a given time compared to the smaller particles. Additionally, a fixed mass of particles was used, which reduces the number of the larger particles compared to the smaller particles.

4 CONCLUSION

A force sensor using strain gages was developed in order to investigate momentum transfer from the particle phase in the fluidized beds. The sensor was tested in air flow to measure the gas–solid drag force on a single ball. In addition, momentum transfer to the ball was measured from a water jet and falling particle stream. These initial measurement results were reasonable which shows the reliability of the sensor and the measurement method. In future, the study will be continued by determining the velocities and concentration of particles before the impact with the object to better link the particle properties and flow conditions with the momentum transfer. Also, the setup will be used to measure particle collisional forces inside a laboratory scale fluidized bed reactor.

ACKNOWLEDGEMENT

The authors would like to acknowledge the financial support for this research from the Academy of Finland under the Grant No. 311138.

REFERENCES

[1] Gidaspow, D., Bezburuah, R. & Ding, J., Hydrodynamics of circulating fluidized beds: Kinetic theory approach. *Proceedings of the 7th Engineering Foundation Conference on Fluidization*, pp. 75–82, 1992.

[2] Lun, C.K.K., Savage, S.B., Jeffrey, D.J. & Chepurniy, N., Kinetic theories for granular flow: Inelastic particles in Couette flow and slightly elastic particles in a general flow field. *Journal of Fluid Mechanics,* **140**, pp. 223–256, 1984.

[3] Payman, J. & Hyppänen, T., Momentum transport between two granular phases of spherical particles with large size ratio: Two-fluid model versus discrete element method. *Powder Technology*, **273**, pp. 13–18, 2015.

[4] Syamlal, M., The Particle–Particle Drag Term in a Multiparticle Model of Fluidization. Topical Report DOE/MC/21353--2373, 1987.

[5] Bell, R.A., Numerical Modelling of Multi-Particle Flows in Bubbling Gas–Solid Fluidized Beds. PhD thesis, Swinburne University of Technology, 2000.

[6] Hideyuki, T., Katsuhiro, I., Tomoyuki, K., Kazuo, S. & Syuichi, U., Design and evaluation of strain gauge force balance with short test duration. *Transactions of the Japan Society for Aeronautical and Space Sciences*, **48**(159), pp. 1–6, 2005.

[7] Maruyama, T., Optimization of roughness parameters for staggered arrayed cubic blocks using experimental data. *J. Wind. Eng. Ind. Aerodyn.*, **46–47**, pp. 165–171, 1993.

[8] Clift, R., Grace, J. & Weber, M., *Bubbles, Drops and Particles*, Dover Publications: New York, 2005.
[9] Turton, R., & Levenspiel, O., A short note on the drag correlation for spheres. *Powder Technology*, **47**(1), pp. 83–86, 1986.
[10] Flemmer, R.L.C. & Banks, C.L., On the drag coefficient of a sphere. *Powder Technology*, **48**(3), pp. 217–221, 1986.

EXPERIMENTAL STUDY OF THE INTERFACIAL WAVES PRODUCED IN UPWARD VERTICAL ANNULAR FLOWS WHEN VARYING THE LIQUID-PHASE SURFACE TENSION

YAGO RIVERA, JOSÉ-LUIS MUÑOZ-COBO, CÉSAR BERNA,
JOSÉ-LUIS CUADROS & ALBERTO ESCRIVÁ
Instituto de Ingeniería Energética, Universitat Politècnica de València, Spain

ABSTRACT

A flow pattern, which appears in many industrial applications, is annular two-phase flow. Its main characteristic is that the main portion of the liquid mass is located at the walls of the tube forming an annulus, while the gas phase flows through the annulus core dragging small liquid drops. These drops are largely produced by the interactions of the gas-phase flowing through the gas core with the peaks of the disturbance waves, which are produced in the liquid/gas interface. This paper describes a set of experiments performed at the Vertical Annular Film Flow facility, in the annular flow regime, with upward liquid and gas flows, at different flow conditions varying the liquid surface tension by adding small amounts of 1-butanol to the water. The main property of 1-butanol is that small amounts of this substance significantly change the surface tension while the viscosity remains practically unchanged. The set of experiments were carried out inside a vertical tube 44 mm inner diameter, and 4.5 m high, the annular flow was created by means of a porous material with known porosity coefficient. The experimental conditions range from 2000 to 3500 l/min for the gas volumetric flow rate, and from 4 to 10 l/min for the liquid volumetric flow rate at atmospheric conditions. Several experiments have been carried out on 1-butanol varying the water surface tension from 72×10^{-3} Nm^{-1}, to 45×10^{-3} Nm^{-1}, at intermediate surface tensions. To measure the height of the disturbance waves and the thickness of the film we use a conductance probe – the electrical signal collected in the receiver sensor depends on the thickness of the liquid film layer. Correlations for the amplitude of the disturbance waves, the film base thickness and other physical magnitudes have been obtained with a good value for the coefficient of determination and the root mean square error.

Keywords: annular flow, interfacial disturbance waves, film thickness, ripple waves.

1 INTRODUCTION

Annular flow is found in many industrial applications and it is one of the fundamental heat transfer modes in industrial processes involving phase change, as for instance boiling water reactors (BWR), heat exchangers, refrigeration, some solar plants among others [1], [2]. In power plants with vertical tubes heated by nuclear energy as in BWR or other energy sources as coal, gas or solar, the two-phase mixture flows upward. Beyond a given distance from the bottom of the heated tubes the two-phase mixture becomes annular, this means that the liquid phase flows upward forming an annulus in contact with the tube walls. This ring or annulus has a gaseous core, which also flows in the upward direction and that exerts a shear stress on the liquid/gas interface. Because the gas velocity is normally bigger than the liquid velocity this shear stress can drag small drops from the crest of the waves formed at the interface. Two different types of waves coexist at the interface, the first ones that have bigger amplitudes and longer lifetimes are known as disturbance waves (DW). These waves are coherent in the streamwise direction and move according to Azzopardi et al. and Belt et al. with approximately constant velocity. The second ones, which have small amplitudes and that are not coherent over large distances are known as ripples waves (RW) [3]–[7].

Many studies have been carried out on annular flow (Berna et al. [3], [4], Alekseenko et al. [5], [8], Belt et al. [6] and Azzopardi [7]) most of them focused on the characterization of the gas–liquid interfacial waves behaviour. These interfacial waves play a determinant role in the droplet entrainment processes, increasing also the pressure drop along the pipe and affecting significantly to the mass, momentum and heat transfer processes, because of these interfacial structures promote an extra drag on the gas flow like roughness [4], [6], [8]. This extra roughness can be accounted for considering the frontal area of these DWs. Due to its high relevance, in all these transfer processes, it is necessary an adequate characterization of the interfacial waves specially the DWs. From previous studies it has been found, for instance, the increase of the liquid film thickness with the liquid flow rate, its decrease with the gas flow rate, the increase of the wave velocity and frequency with the flow rate, for both, gas and liquid phases, the weak dependency of DWs characteristics with the pipe diameter, and so forth [9].

The research of the physical phenomena hidden behind annular flow are pursued throughout this work, having as a major objective, to produce correlations for the major interfacial variables and to compare it against other widely used expressions. As another important goal of the present work, we investigate the influence of the surface tension on the DWs behaviour, variable which has not been so widely studied. In previous studies it has been found that the Kapitza number plays a fundamental role in the development of the interfacial waves [10], this nondimensional number depends fundamentally of the surface tension and the viscosity. Because the height of the disturbance waves is significantly affected by the surface tension value, then it is important to address experimental studies in this direction. This paper presents an experimental study of changing the surface tension of the water by adding small amount of 1-butanol, which has the property that small amount of 1-butanol, changes the surface tension of the water without practically affecting the viscosity. Because an increase in the amount of 1-butanol added to the water produce a diminish in the surface tension with a smaller change in the viscosity then we can study how the amplitude of the disturbance waves is affected by the temperature by simulating the increases in the fluid temperature by the increments in the concentration of 1-butanol. The paper has been organized as follows: Section 2 is devoted to describing the experimental facility and the experimental method used. Section 3 is devoted to describing the experiments performed and the main experimental results. Finally, Section 4 is devoted to present the main conclusions of the paper.

2 THE EXPERIMENTAL VERTICAL ANNULAR FILM FLOW FACILITY AND THE EXPERIMENTAL METHODS

2.1 Description of the VAFF experimental facility

A schematic view of the VAFF facility (Vertical Annular Flow Facility) is displayed in Fig. 1. The VAFF facility is an air–water loop with different sensors placed along its length, in order to measure the water film thickness and to know if the flow is fully developed. The facility main dimensions for the test section are almost 5 m of useful vertical length with an inner pipe diameter of 44 mm. Its major components are the air production circuit, the water production system, the injector/mixer system, the test section and the air–water separator system. The main physical variables are measured with different sensor types, which are strategically distributed along the facility for the purpose pursued.

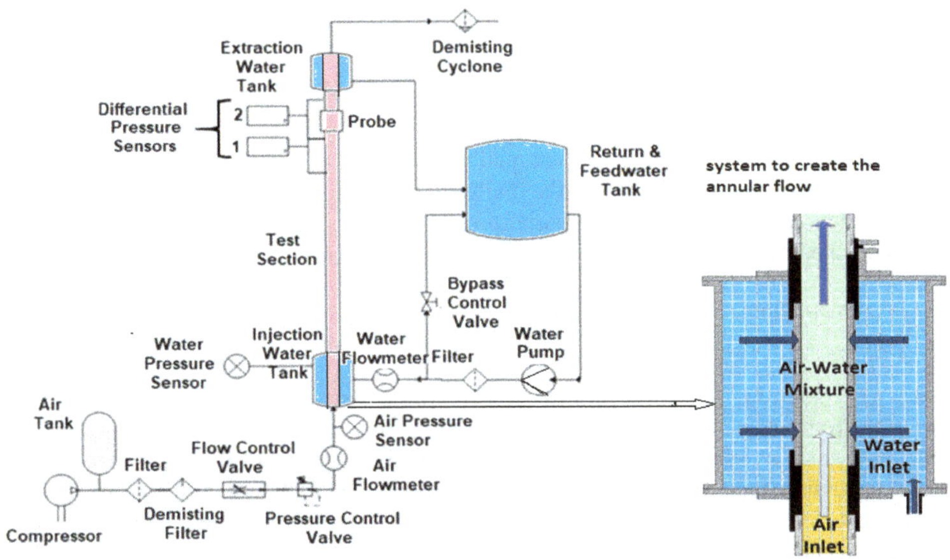

Figure 1: Schematic layout of the VAFF facility, with a detailed description of the water injector system, which creates the annular flow.

The operation and characteristics of the facility can be summarized as follows:

- Air is injected into the system through a compressor, whose maximums working pressure and volumetric flow rate are 8 bars and 3750 l/min respectively.
- Water is injected through a water pump, whose maximum discharge pressure is 4.2 bars.
- A specific designed pressurized water injector system is used to create a thin annulus liquid film just before the lower part of the test section. The water injector's operating principle is based on the pressure difference between both sides of the sintered stainless-steel tube with a given porosity value.
- The water injection system is a cylindrical tank displayed also in Fig. 1, with a hollowed central part and a sintered stainless-steel tube in it. The outer part of the cylinder is pressurized, while through its inner part circulates the air stream at almost atmospheric pressure. Due to the porosity of the sintered material the water current can pass through it, being the water volumetric flow rate function of the pressure difference between the inner and outer sides of the sintered stainless-steel material.
- The two-phase flow crosses through the test section, in which most of the experimental measurements are taken.
- The outer liquid annulus is removed through the extraction water tank in the upper part of the test section. The operating principle of the extraction tank is the same one as in the water injection system.
- The water is returned to a storage tank, from which the water is sucked by the recirculation pump and driven to the injector system once again, starting a new water cycle.
- The two-phase mixture, composed by the gas phase and entrained droplets, is separated through a centrifugal demisting cyclone.

- The air is vented to the atmosphere through the final section of the centrifugal demisting cyclone.
- Fresh air is injected back into the experimental facility by the compressor and the air cycle starts once again.

The test section consists of a vertical methacrylate tube of approximately 4 meters high and an inner diameter of 44 mm. All the major experimental measurements are taken along this test section, absolute and differential pressure sensors, thermocouples and, the most important one, the conductance probe.

2.2 Instrumentation and data acquisition

2.2.1 Facility instrumentation

The VAFF facility is equipped with two differential pressure sensors located in the upper part of the test section, as displayed in Fig. 1, in order to know if the two-phase flow is fully developed at the measuring point for the experimental conditions of the experiment. These relative pressure measurements are made with SMC PSE550 low differential pressure transducers. These sensors have a pressure range measurement between 0 to 2 kPa, with accuracy equal or higher than ±1.0% at full scale (F.S.) and a linearity error equal or less than ±0.5% F.S.

The local measurements of pressure are made with Burkert 8314 relative pressure transducers (difference in pressure between the inner part of the facility and the surrounding atmosphere). These sensors have a pressure range measurement between 0 and 1 bar, with a sum of linearity, hysteresis and repeatability errors less than ±0.3% F.S. and error adjustment accuracy zero point and full scale less than ±0.3% F.S.

Temperature measurements were performed with Type T thermocouples (copper-constantan), which are suitable in the range -200 to 350°C. Their error, for measurements above 0°C, is the greater value between 1.0°C and 0.75% of the measured temperature.

Water volumetric flow rate measurements are made with a Kroni Optiflux 4300 C Series electromagnetic flowmeter. Its volumetric flow rate measurement interval ranges from 0 to 20 l/min and has an accuracy higher than ±0.3% of span.

Measurements of the high-speed gas volumetric flow rate are made with a SMC PF2A706H magnetic flowmeter. Its volumetric flow rate measurement range covers from 300 to 6000 l/min, the flowmeter has a linearity of ±1.5 of full scale (F.S.) or less and a repeatability of ±1.0 F.S. or less.

2.2.2 Conductance probe

To estimate the characteristics of the interfacial waves we use a conductance probe. The conductance probe used in these measurements is described in reference [11]. The physical basis of the conductance sensors is the dependence of the water conductivity on the film thickness value [12]. The conductance probe design consists of three electrodes mounted flush with the wall and aligned in the same flow direction of the liquid film and the gas flow. The central electrode is connected to the ground. The transmitter electrode is excited, with a 300 KHz and 4 Vpp sinusoidal signal. The receiver electrode is responsible for collecting the electric current that the water film conduces. The total amount of current collected by the receiver is proportional to the liquid film thickness between both electrodes i.e. the transmitter and the receiver. This dependency is caused by the existing correlation between the liquid film thickness and the electric field lines, because of higher film thickness leads to higher number of electric field lines, which can be collected by the receiver. Consequently, the electric signal collected by the receiver changes when an interfacial wave crosses the sensor region, and this change allows the determination of the liquid film thickness variations with time. The receiver signal is rectified and amplified by an electronic circuit explained below and then registered in the data acquisition system. At the same time, this signal gives us the wave's form when it is plotted versus time.

To insert the sensors of the conductance probe in the test section we have used a Poly-Propylene (PPR) pipe section denoted as port. The geometrical characteristics of the port are an inner diameter of 44 mm, equal to that of the test section, and a total length of 170 mm. Slots have been created at both ends of the port, so that the measuring tube fits perfectly and does not cause any disturbance in the liquid film entering or leaving the port. The electrodes have been placed in the central part of the port.

2.2.3 Electronic circuit and data acquisition system

The internal operations carried out by the sensor electronic circuit consist on the following steps: The sinusoidal signal produced by the generator is amplified before sending it to the excitation electrode. Next, the receiver electrode captures the signal, which has been transmitted through the liquid layer. This last signal is amplified again and transformed in DC current by a rectifier. Then, the data acquisition system acquires this DC signal, whose value is related to the liquid film thickness existing between both electrodes. Then, the variations with time in the value of this DC signal will indicate the passage of waves present in the liquid film.

During the test runs all the instrumentation devices, described in the previous sections, are monitoring in real time all the major experimental conditions, which include air and water volumetric flow rates, temperatures and pressures, in addition to the recording of the experimental data along the test section, i.e. conductance probe signal and pressure test-section values. The data acquisition system has been programmed in LabView. This system distinguishes between two types of measurements: The conductance probe measurements, which suffer fast variation with time and need high sample rates, and the rest of measurements as temperature and pressure. The acquisition system for the conductance probe records 10^5 samples per second (S/s) (National Instruments PCI 6255 with 80 analogic channels and a maximum rate of 1.25 MS/s). For the rest of experimental data, which suffers smaller variations with time, the acquisition system records 100 samples per second (National Instruments cDAQ-9174 chassis compact-DAQ with a module NI 9207 of 16 analogic channels and a maximum rate of 500 S/s).

2.3 Calibration

The final step performed before placing the conductance probe in the experimental facility is the calibration of the conductance sensor. The objective is to relate the voltage signal with the water film thickness. For the measurement performed adding 1-butanol to the water, the sensor calibration must be repeated for each 1-butanol concentration, because the water conductivity changes with the 1-butanol concentration. Proceeding is this way it is possible to relate the information contained in the signals (V) with the amplitude of the waves produced in the fluid interface (mm). We remind that the conductance sensor consists of 3 electrodes with 2 mm of inner diameter, spaced 1.5 mm between them, and aligned in the direction of flow and flush in the inner wall of the pipe.

The calibration process was carried out with a calibration device, displayed in Fig. 2. This device consists of three major components: one optical precision table; a precision positioning system mounted on this table with an error of 0.002 mm; and several cylinders with known diameters and made with a dielectric material. By setting the dielectric cylinder at the desired position, in the inner part of the conductance probe port, by means of the precision positioning system, we can know the fluid thickness between the dielectric and the probe electrodes. Measuring the voltage in the probe electrode, we can find the relationship between the signal voltage and the fluid thickness.

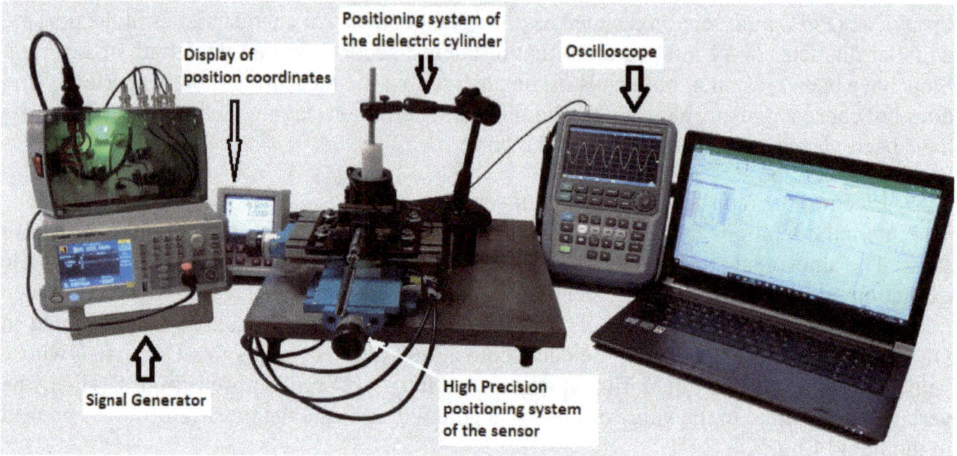

Figure 2: High precision calibration device.

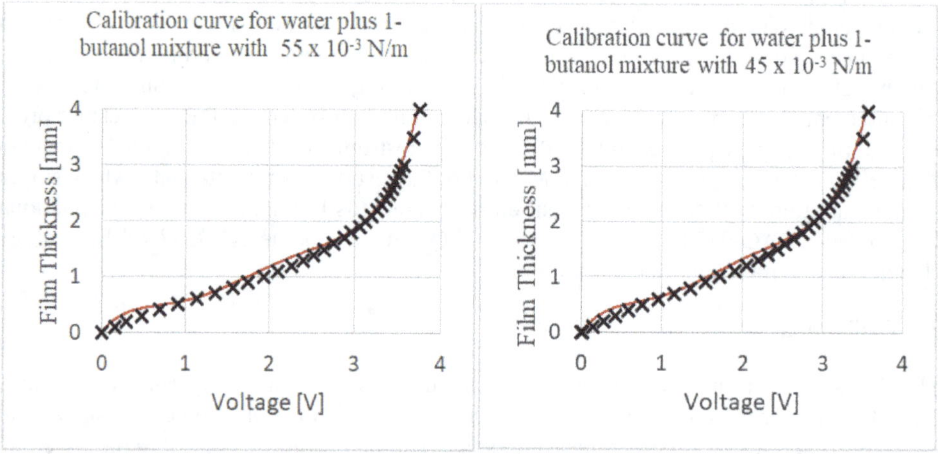

Figure 3: Calibration curve for two of the experiments performed adding 1-butanol to the water, the crosses denote the experimental points.

A total number of 33 calibration measurements were made with the calibration device for each one of the three 1-isobutanol concentrations values, which are denoted by their surface tensions at 25°C, 68.5 10^{-3} N/m, 55 10^{-3} N/m and 45 10^{-3} N/m. Two of the calibration curves obtained for the cases of 55 10^{-3} N/m and 45 10^{-3} N/m are displayed at Fig. 3. We notice that the calibration curves displayed have similar shapes although their values are not the same.

The calibration curve fitted to the 1-butanol calibration experiments was a polynomial of fifth degree given by eqn (1), with coefficients denoted by the symbols x_0 to x_5, while for the pure water was the curve given by eqn (2):

$$\delta(mm) = x_0 + x_1\,V + x_2V^2 + x_3V^3 + x_4V^4 + x_5V^5 \tag{1}$$

$$\delta(mm) = a + b_1\,V^{b_2} + c_1V^{c_2} + d_1V^{d_2} + e_1V^{e_2} \tag{2}$$

The coefficients of the curves for each one of the 1-butanol concentrations are displayed in Table 1. The uncertainty in the calibration using eqn (1) was estimated using the methods explained in Chapter 7 of Bevington and Robinson [13]. The estimated uncertainty for the 1-butanol case with $\sigma = 68.5\ 10^{-3}$ N/m, was 0.28 mm.

Table 1: Coefficients of the calibration curves for the 1-butanol plus water experiments.

$\sigma\ (N/m)$	x_5	x_4	x_3	x_2	x_1	x_0
$68.5\ 10^{-3}$	0.038	-0.384	1.388	-2.089	1.535	0.
$55\ 10^{-3}$	0.088	-0.733	2.203	-2.779	1.789	0.
$45\ 10^{-3}$	0.105	-0.833	2.403	-2.946	1.925	0.
Pure water ($\sigma = 72\ \times\ 10^{-3}$)						
a	b_1	b_2	c_1	c_2	d	d_2
0.046	-5.416	3.481	-3.032	0.958	6.242	4.381
e_1	e_2					
5.176	0.953					

3 EXPERIMENTAL MEASUREMENTS AND RESULTS

The test matrix consisted of four experimental data series, each data series was performed with a constant concentration of 1-butanol, which produced a given surface tension. Each data series consisted of several runs with different boundary conditions i.e. different volumetric flow rates of air and water respectively, which are displayed at Table 2. The number of data series was 4, one for water with a surface tension of 72×10^{-3} N/m, at 25°C and three additional ones obtained adding small amount of 1-butanol, which yields surface tensions of 68.5 10^{-3} N/m, 55. 10^{-3} N/m and 45. 10^{-3} N/m. For each solution of water-1-butanol 49 experimental runs were made, which were obtained by combination of seven gas and seven liquid volumetric flow rates. The seven liquid flow rates covered from 4 to 10 l/min, with increments of 1 l/min from run to run. While the seven air volumetric flow rates changed between 2000 l/min to 3500 l/min (i.e., 0.0333–0.0583 m^3/s), increasing 250 l/min between different runs. This procedure was repeated for the rest of 3 water-1-butanol mixtures. The major objective of changing the water surface tension is to study the strongly effects suffered by the interfacial wave's behaviour, in response to variation in the surface tension. Because diminishing the water surface tension value is quite like the effect of the temperature increase. During all the experimental measurements the pressure in the upper part of the facility was kept at atmospheric conditions, the water and air temperatures were kept approximately constant at about 30 and 25°C respectively.

The most important variables of the experiment were monitored in real time during the 49 test runs of each one of the four data series. Then main variables that were monitored and stored during the sixty seconds of each experimental run were: air and water volumetric flow rates, temperatures and pressures, in addition to the conductance probe voltage signal, which was converted in mm using the calibration curves displayed in Section 2.3.

Table 2: Boundary conditions of the 49 experimental runs performed for each surface tension, the units of Q_g and Q_l are l/min.

Boundary conditions used for the surface tension $\sigma = 72 \times 10^{-3}$ N/m						
$Q_g \times Q_l$	Q_g(l/min) × Q_l(l/min)					$Q_g \times Q_l$
2000x4	2250x4	2500x4	2750x4	3000x4	3250x4	3500x4
2000x5	2250x5	2500x5	2750x5	3000x5	3250x5	3500x5
2000x6	2250x6	2500x6	2750x6	3000x6	3250x6	3500x6
2000x7	2250x7	2500x7	2750x7	3000x7	3250x7	3500x7
2000x8	2250x8	2500x8	2750x8	3000x8	3250x8	3500x8
2000x9	2250x9	2500x9	2750x9	3000x9	3250x9	3500x9
2000x10	2250x10	2500x10	2750x10	3000x10	3250x10	3500x10

To eliminate the noise fluctuation of the signal a Savitzky–Golay moving average filtering was applied to the raw signals [14]. Fig. 4 displays the film thickness versus time obtained for the three different mixtures of water and 1-butanol and the pure water, while keeping the volumetric flow rates of air and water constants at (2250 l/min) and (7 l/min) respectively. It is observed that the height of the disturbance waves diminishes continuously when increasing the 1-butanol concentration.

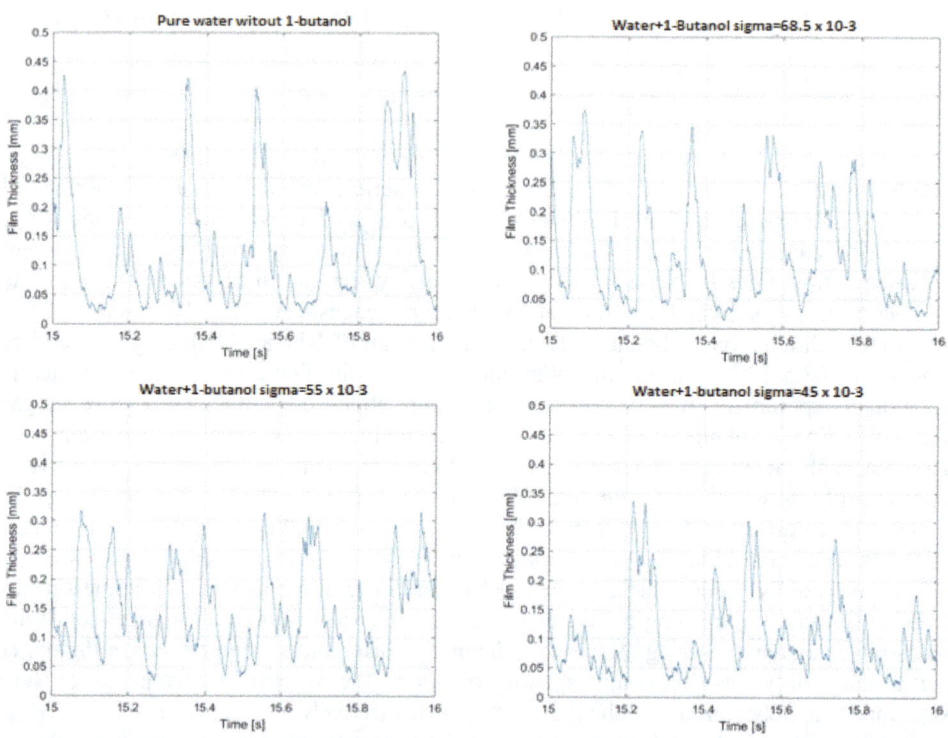

Figure 4: Film thickness evolution during one second for the different mixtures of water and 1-butanol keeping constant the volumetric flow rates of air (2250 l/min) and water (7 l/min).

Aiming to estimate the error margin of the experimental measurements for the test series a set of repeated experiments has been made. A run test has been repeated ten times. In such a way that, assuming a normal probability distribution function, then the accidental error margin of a determined variable can be estimated through the following expression with a confidence level of 95% ($\alpha = 0.05$) [15]:

$$\varepsilon_{acc} = t_{1-\alpha/2}(f)\frac{S}{\sqrt{n}}. \quad (3)$$

where $t_{1-\alpha/2}(f)$ is the Student-t $1-\frac{\alpha}{2}$ quantile for $f = n-1$ degrees of freedom, which depends on the confidence level, and that for the usual 95% confidence level has a value of 2.26, with 9 degrees of freedom. Notice that ε_{acc} is half the confidence interval for the mean given by: $\bar{\delta} \pm t_{1-\alpha/2}(f)\frac{S(\delta)}{\sqrt{n}}$, being $S(\delta)$ the square root of the variance. Table 3 displays the accidental errors computed with the data obtained repeating one of the experiments ten times.

The variables, which were obtained from the signal data were: the mean value of the amplitude of the disturbance waves, during the considered time interval for each run with given boundary conditions, denoted by A_{DW}, the mean value of the frequency of the disturbance waves denoted by ν_{DW}, the base film thickness denoted by δ_b, the mean film thickness $\bar{\delta}_m$, and the mean value of the film thickness at the disturbance waves denoted by $\bar{\delta}_{DW}$. Table 4 displays the results obtained for the runs with 1-butanol and $\sigma = 55 \times 10^{-3}$ N/m, for $Q_g = 2500$ l/m, and different values of Q_l, while Table 5 displays the same results obtained for water without adding 1-butanol, for the same cases.

From Tables 4 and 5 it is observed that when the surface tension increases the amplitude A_{DW} of the disturbance waves increases, and also increases the frequency of the disturbance waves, this last fact means that we have a bigger number of disturbance waves per second. The base film thickness is not very much affected by diminishing the surface tension by adding 1-butanol. Also, we can study from these tables the effect of increasing the volumetric flow rate of air on the results. Because of the increase of the gas flow rate, it also increases the interfacial shear stress, which diminishes the base film thickness as observed comparing

Table 3: Accidental error from the repeatability analysis results for the main variables.

	$\varepsilon_{acc}(\delta_m)$ (µm)	$\varepsilon_{acc}(\delta_b)$ (µm)	$\varepsilon_{acc}(A_{DW})$ (µm)	$\varepsilon_{acc}(\nu_{DW})$ (Hz)
ε_{acc}	29.3	5.12	53.4	0.92
$\pm\varepsilon_{rel,acc}$, %	±6.2	±1.76	±7.9	±4.5

Table 4: Experimental results for the water-1-butanol experiments with $\sigma = 55 \times 10^{-3}$ N/m. All the results are expressed in mm.

$Q_g \times Q_l$	δ_b(mm)	$\bar{\delta}_{DW}$	$\bar{\delta}_{ripples}$	$\nu_{DW}(Hz)$	$A_{ripples}$	A_{DW}	$\bar{\delta}_m$
2500x4	0.12680	0.53267	0.14103	8.966	0.00712	0.4094	0.19057
2500x5	0.13107	0.54601	0.14297	9.416	0.00595	0.4179	0.20957
2500x6	0.13405	0.54833	0.14545	10.36	0.00570	0.4171	0.22747
2500x7	0.13719	0.54720	0.14778	11.13	0.00530	0.4126	0.24316
2500x8	0.13790	0.52920	0.14683	10.88	0.00447	0.3935	0.24623
2500x9	0.14065	0.52677	0.14950	10.15	0.00443	0.3883	0.25594
2500x10	0.14269	0.50716	0.15181	10.06	0.00456	0.3667	0.26088

Table 5: Experimental results for the water experiments with $\sigma = 72 \times 10^{-3}$ N/m. All the results are expressed in mm.

$Q_g \times Q_l$	$\bar{\delta}_b$(mm)	$\bar{\delta}_{DW}$	$\bar{\delta}_{ripples}$	$\nu_{DW}(Hz)$	$A_{ripples}$	A_{DW}	$\bar{\delta}_m$
2500x4	0.12904	0.61000	0.14447	10.39	0.00771	0.48480	0.20590
2500x5	0.13365	0.62178	0.14750	11.22	0.00692	0.49158	0.22304
2500x6	0.13814	0.62931	0.15170	12.41	0.00678	0.49456	0.24312
2500x7	0.14285	0.63207	0.15614	13.52	0.00664	0.49253	0.25654
2500x8	0.13423	0.64154	0.14630	13.93	0.00603	0.51032	0.26493
2500x9	0.13559	0.64150	0.14692	14.19	0.00566	0.50874	0.26800
2500x10	0.13483	0.64811	0.14874	14.74	0.00695	0.51676	0.27814

Table 6: Experimental results for the water-1-butanol experiments with $Q_g = 3000$ l/m and $\sigma = 55\frac{N}{m}$, for a tube diameter D=44mm. All the results are expressed in mm.

$Q_g \times Q_l$	$\bar{\delta}_b$(mm)	$\bar{\delta}_{DW}$	$\bar{\delta}_{ripples}$	$\nu_{DW}(Hz)$	$A_{ripples}$	A_{DW}	$\bar{\delta}_m$
3000x4	0.09303	0.42337	0.10676	5.5	0.006867	0.33377	0.11697
3000x5	0.10095	0.45039	0.11553	6.95	0.007286	0.35308	0.13530
3000x6	0.10593	0.44924	0.12092	8.05	0.007496	0.34706	0.14798
3000x7	0.10879	0.45189	0.12457	8.68	0.007890	0.34704	0.15622
3000x8	0.11364	0.44024	0.12894	9.58	0.007649	0.33042	0.16645
3000x9	0.11762	0.43667	0.13284	10.05	0.007610	0.32285	0.17728
3000x10	0.12230	0.42726	0.13675	11.21	0.007223	0.30857	0.18825

Tables 4 and 6. Also, when increasing the gas superficial velocity diminishes the amplitude and mean thickness of the disturbance waves because the shear stress cut the peaks of the disturbance waves producing drops. As it is observed comparing Tables 4 and 6.

Finally, a total of three correlations were obtained for the base film thickness, the total film thickness, and the amplitude of the disturbance waves in terms of the liquid Reynolds number Re_{liq}, the gas Reynolds number Re_g, and the Kapitza number Ka. The correlations obtained have the following form:

$$\frac{\bar{\delta}_m}{D} = b_1 Re_{liq}^{b_2} Re_g^{b_3} Ka^{b_4}, \tag{4}$$

where the gas Reynolds number and the liquid film Reynolds number were obtained from the following expressions:

$$Re_g = \frac{\rho_g J_g D}{\mu_g}, \; Re_l = \frac{\rho_l J_l D}{\mu_l} = \frac{4\Gamma}{\mu_l}, \tag{5}$$

where J_g and J_l, are the superficial velocities of the gas and the water-1-butanol mixture respectively and Γ is the mass flow rate per unit of circumferential length. The coefficient of determination was 0.9883 and the root mean square error for $\bar{\delta}_m$ was 0.0224. The values of the coefficients were, $b_1 = 0.805, b_2 = 0.3047, b_3 = -0.9746$, and $b_4 = 0.411$.

Because of the frequency of the disturbance waves for pure water showed too much dispersion from one boundary condition to another one. Then, we repeated the experiments for all the boundary conditions three times at different days and we compute for each boundary condition the average value for the frequency, the results are displayed in the fifth column of Table 5.

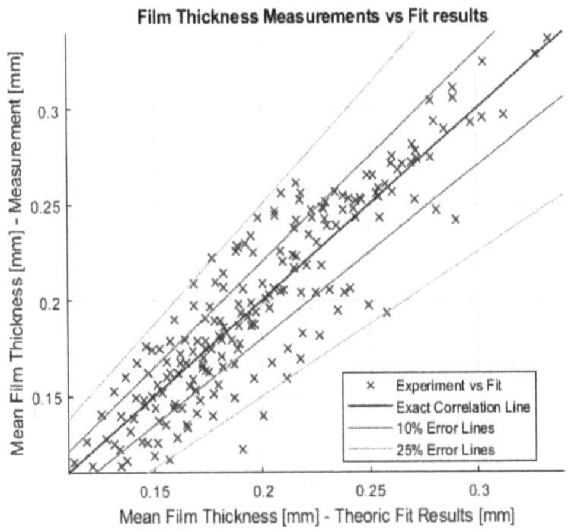

Figure 5: Experimental results versus correlation predictions for the average of the total film thickness.

4 CONCLUSIONS

The major conclusions of this paper are related to the evolution of the mean thickness $\bar{\delta}_{DW}$, and the amplitude A_{DW} of the disturbance waves and the frequency ν_{DW} of these waves when varying the surface tension, and the liquid and gas superficial velocities while maintaining constant the tube diameter. It is observed that in general the mean thickness of the disturbance waves diminishes when the surface tension diminishes maintaining constant the gas and liquid superficial velocities, also the amplitude of the disturbance waves diminishes when reducing the values of the surface tension by adding 1-butanol to the water as observed comparing Tables 4 and 5. This reduction is about 20% in the amplitude and mean thickness of the disturbance waves, for the case of Q_g =2500 l/min and different values of Q_l ranging from 4 l/min to 10 l/min when changing from normal water to water with $\sigma = 55.10^{-3}$ N/m. In addition, the frequency of the disturbance waves increases when we increase the surface tension from 55 10^{-3} to 72 10^{-3} N/m, as it is observed comparing the Tables 4 and 5. This increment in the frequency of the waves is especially noteworthy when we also increase the superficial velocity of the liquid.

Another important issue is the influence of the gas superficial velocity on the amplitude of the disturbance waves (DW), this influence can be deduced comparing Tables 4 and 6, for the same 1-butanol concentration. It is observed that A_{DW} diminish when we increase the superficial velocity of the gas from $j_g = 27.4\,\mathrm{m/s}$ to $j_g = 32.88\,\mathrm{m/s}$. This reduction can be explained on account of the drag force produced by the interfacial shear stress on the crest of the disturbance waves that is able to cut the peak of the wave producing drops. This effect reduces the average amplitude of the DW, as deduced from Tables 4 and 6.

Finally, the factor that has the bigger influence on the total film average thickness is the gas superficial velocity, this influence can be deduced from the tables but also from the correlation obtained from all the experimental data. In this correlation appears $Re_g^{-0.9746}$, this means that because of $Re_g = \frac{\rho_g J_g D}{\mu_g}$, then the mean thickness of the film diminishes as $J_g^{-0.9746}$, i.e. with the superficial velocity of the gas. The explanation is that the shear stress produced by the gas flow at the interface reduces the film thickness.

ACKNOWLEDGEMENT

The authors are indebted for the support of the Spain National Agency of Research to the project EXMOTRANSIN, ENE2016-79489-C2-1-P.

REFERENCES

[1] Collier, J.G., *Convective Boiling and Condensation,* 2nd ed., McGraw-Hill, 1981.

[2] Stephan, K., *Heat Transfer in Condensation and Boiling*, Springer Verlag, 1992.

[3] Berna, C., Escrivá, A., Muñoz-Cobo, J.L. & Herranz, L.E., Review of droplet entrainment in annular flow: Interfacial waves and onset of entrainment. *Progress in Nuclear Energy*, **74**, pp. 14–43, 2014.

[4] Berna, C., Escrivá, A., Muñoz-Cobo, J.L. & Herranz, L.E., Review of droplet entrainment in annular flow: Characterization of the entrained droplets. *Progress in Nuclear Energy*, **79**, pp. 64–86, 2015.

[5] Alekseenko, S., Antipin, V., Cherdantsev, A., Kharlamov, S. & Markovich, D., Two-wave structure of liquid film and wave interrelation in annular gas-liquid flow with and without entrainment. *Physics of Fluids*, **21**(1), 061701, 2009.

[6] Belt, R.J., Van't Westende, J.M.C., Prasser, H.M. & Portela, L.M., Time spatially resolved measurements of interfacial waves in vertical annular flow. *International Journal of Multiphase Flow*, **36**, pp. 570–587, 2010.

[7] Azzopardi, B.J., Drops in annular two-phase flow. *International Journal of Multiphase Flow*, **23**, pp. 1–53, 1997.

[8] Alekseenko, S., Antipin, V., Cherdantsev, A., Kharlamov, S. & Markovich, D., Investigation of waves interaction in annular gas-liquid flow using high-speed fluorescent visualization technique. *Microgravity Science Technology*, **20**(1), p. 271, 2008.

[9] Cuadros, J.L. et al., Experimental investigation of the gas-liquid interfacial waves in vertical upward concurrent annular flows. *Nuclear Engineering and Design*, in press, 2019.

[10] Muñoz-Cobo, J.L., Miquel, A., Berna, C. & Escrivá, A., Spatial and time evolution of non-linear waves in falling liquid films by the harmonic expansion method with spatial integration. *Proceedings of the International Conference on Heat Transfer, Fluid Mechanics and Thermodynamics HEFAT*, 2016.

[11] Muñoz-Cobo, J.L, Chiva, S., Méndez, S., Monrós, G., Escrivá, A. & Cuadros, J.L., Development of conductivity sensors for multi-phase flow local measurements at the Polytechnic University of Valencia (UPV) and University Jayme I of Castellon (UJI). *Sensors*, **17**, pp. 1–35, 2017. DOI: 10.3390/s17051077.

[12] Wayne, C., The interfacial characteristics of falling film reactors. PhD thesis, University of Nottingham, Nottingham, UK, 2001.

[13] Bevington, P.R. & Robinson, D.K., *Data Reduction and Error Analysis in the Physical Sciences*, 3rd ed., McGraw-Hill, 2003.

[14] Savitzky, A. & Golay, M.J.E., Smoothing and differentiation of data by simplified least squares procedures. *Analytical Chemistry*, **36**(8), pp. 1627–1639, 1964.

[15] Cramer, H., *Elements of Probability Theory and Some of its Applications,* 9th ed., Wiley, 1966.

INFLUENCE OF THE ORIFICE SHAPE ON MASS FLOW MEASUREMENTS OF AIR–WATER MIXTURE

BARBARA TOMASZEWSKA-WACH, MARIUSZ RZASA,
BOLESLAW DOBROWOLSKI & OREST SEREDIUK
Opole University of Technology, Faculty of Mechanical Engineering, Poland

ABSTRACT

Flow measurements using differential pressure meters are common in industrial applications. In such cases, the flow of gas is often accompanied by conditions that can lead to liquid condensation. As a consequence, flow measurements basically involve gas–liquid mixture metering. For this reason, errors occur in the metering equipment resulting from the variations in the characteristics of the continuous phase that is present in the flow. In addition, the existence of a dispersed phase leads to the development of flow disturbance and pressure pulsations. Therefore, new methods and tools are being sought to enable the measurements of gas–liquid mixture flows that will offer a suitable accuracy of measurement in the instances of flow interference in the form of a liquid phase. This paper reports the results of a study into the application of orifice plate meters for gas–liquid mixture flow metering. The analysis of the influence of the geometry of an orifice meter on the measurement of a two-phase mixture flow was carried out for this purpose. Experimental tests were carried out by application of a standard orifice and three slotted orifice meters with various designs. The experiments included the measurements of air flow containing small amounts of dispersed water in the form of droplets. The analysis also involved the level of differential pressure that is obtained as a result of applying orifice meters, and the level of the permanent pressure loss caused by the installation of an orifice plate. The results of the research were compared with the results obtained for the standard orifice.

Keywords: standard orifice, slotted orifice, two-phase flow, gas–liquid flow, experimental research, Lockhart–Martinelli parameter.

1 INTRODUCTION

A variety of orifice meters are applied in gas flow measurements in many industry branches. When gas is transported in industrial applications, condensation of liquids contained in the gas often occurs. The occurrence of small liquid droplets results in an error due to the change in the physical properties of the fluid flow when flow measurements are performed by application of orifice meters. This is due to the loss of the homogeneous structure of the gas phase, which becomes a two-phase mixture consisting of gas and liquid phases. The use of standard methods for measurement of homogeneous fluids leads to the significant levels of the measurement error. Therefore, measurements involving two-phase mixtures form some of the most challenging problems facing flows metrology. Such measurements involving the mass flow rates of gas–liquid mixtures play an important role in many areas of science and technology, e.g. in meteorology, chemistry, power engineering, as well as petrochemical and chemical industries [1]. This is due to the fact that the presence of a dispersed phase results in the variations of the physical properties of the continuous phase [2].

Another area in which the present method can be applied is associated with the transport of natural gas saturated with liquid in which the volumetric fraction of the liquid phase does not exceed 5%, i.e. wet gas [3]. Due to the current feasibility of extracting gas from such from sources of sea bed mining, desert, arctic and other inaccessible areas, it is necessary to use metering systems with high reliability, small size and built at a relatively low cost [4], [5]. Yet another group of applications are waste composting processes [6] where mainly wet gas flows occur. In many research centers and scientific institutions, research has been

WIT Transactions on Engineering Sciences, Vol 123, © 2019 WIT Press
www.witpress.com, ISSN 1743-3533 (on-line)
doi:10.2495/MPF190041

conducted over the past years with the purpose of improving existing techniques and developing new ones to be applied for measuring flows of two-phase mixtures. Often such metering applies equipment with considerable dimensions, complex design and high cost of performance. Therefore, the demand for metering equipment with a simple design and low cost of development is a current challenge for engineering. Due to their simple design, low cost of production and reliability of operation, differential pressure meters offer some of the cheapest and best performance flow measurement solutions for application in industry. In the recent reports, there are examples of the using orifice plate meters in the measurement of mixtures comprising gas and small amounts of liquids [7], [8].

2 MEASUREMENTS OF GAS–LIQUID MASS FLOW RATES APPLYING ORIFICE PLATES

The measurements of the fluid flow rates by application of the orifice plates consist in the use of a constriction of the flow cross-section. This constriction leads to an increase in the fluid velocity, and along with the change in the velocity, the static pressure decreases and a differential pressure is developed at the orifice. The resulting differential pressure ΔP forms a reliable source serving for determination of the fluid mass or volume [9]. Among the orifice meters, we can distinguish orifice plates, nozzles and Venturi tubes [10]. In this work, various types of orifice meters were applied in experimental research. The schematic of the measurement system is presented in Fig. 1.

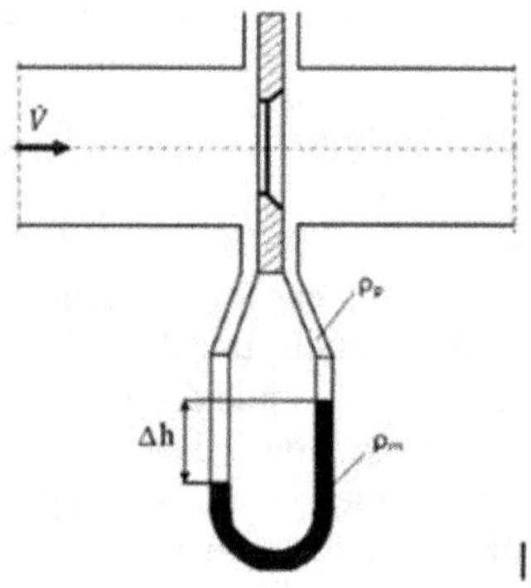

Figure 1: Operating principle of orifice plate.

The use of orifice meters specified in the norm [10] offers the possibility of calculating mass flow rates on the basis of differential pressure measurements. The mass flow rate is calculated on the basis of eqn (1):

$$\dot{m} = \frac{C}{\sqrt{1-\beta^4}}\varepsilon\frac{\pi d^2}{4}\sqrt{2\Delta P\rho}, \tag{1}$$

where C is discharge coefficient, ε is expansion number relative to fluid compressibility, β is beta ratio of the orifice, d is orifice diameter, ΔP is differential pressure at orifice plate, ρ is fluid density.

2.1 Standard orifice

Due to its simple design, the standard orifice is commonly used in fluid flow metering. The advantages of the standard orifice also include its long service life. Since such orifices have no moving parts, they can be used for measurements involving liquids, gas and suspensions. They can perform fault-free under extreme operating conditions and provide reliable metering solutions in a wide range of pressures and temperatures. Another advantage of using standard orifices includes the possibility of their easy replacement without the need to dismantle large piping sections [9], [11].

The metering using orifice plates produced to the current standard provides considerable accuracy of measurements only for the case of single-phase flows. The presence of a second phase, i.e. liquid in the gas flow leads to the variations in in the differential pressure and its pulsations, fluctuations in the density of the fluid and expansion number, etc. Therefore, the use of standard orifices is not recommended for measuring wet gases or multiphase mixtures [12].

For many years, research has focused on the development of methods that can offer the possibility of using orifices for metering wet gases and two-phase mixture flows. In addition, other solutions based on ultrasound methods, laser techniques, process tomography and other complex systems are sought so as to provide reliable measurements [1].

The reason for wet gas formation is often associated with a decrease in temperature or pressure, when condensation processes occur in the pipeline leading to the conditions in which small amounts of liquid in the form of droplets are condensed on the pipeline walls. And it is known that even small amounts of liquid in the flow of gas can lead to flow disturbance.

The Lockhart–Martinelli parameter forms one of the most commonly applied parameters defining the relative fraction of liquid in a two-phase flow. It is defined by eqn (2) [3], [4], [7].

$$X_{LM} = \frac{m_L}{m_G}\sqrt{\frac{\rho_G}{\rho_L}}, \tag{2}$$

where m_L and m_G are mass flow rates of the liquid and gas phases, respectively, and ρ_L, ρ_G are the liquid and gas densities. For the wet gas, the value of the X_{LM} parameter does not exceed 0.35.

Another way in which it is possible to determine if a given two-phase mixture is in the wet gas range is associated with learning about the parameter called Gas Volume Fraction (GVF, eqn (3)) and for this case the value should not be smaller than 95% [4]:

$$GVF = \frac{V_G}{V_{TP}}, \tag{3}$$

where V_G is the volumetric flow rate of the gas phase, and V_{TP} is the volumetric flow rate of the two-phase mixture.

2.2 Slotted orifice plate

One of the ways in which wet gas metering can be feasibly performed is based on the use and selection of an adequate design of a slotted orifice plate. The reports in the area list a variety of equipment that offers a derivative of the standard orifice. Such solutions often apply special orifices, including slotted and fractal-shaped orifices, perforated plates, etc.

(Fig. 2) as well as other designs that comprise slots laid out in various arrangements (square, triangle, oval) [12]–[16].

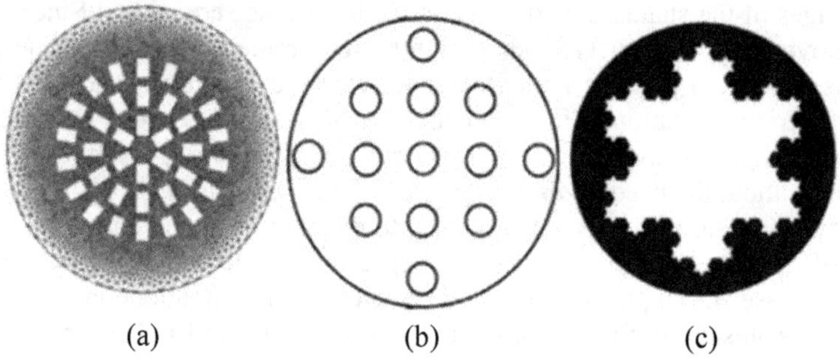

Figure 2: Examples of special orifices. (a) Slotted; (b) Perforated plate; (c) Fractal.

Morrison, among others, investigated a new type of orifice. In a study reported in Morrison et al [11], a comparison between the slotted orifice and a standard one was performed for the case of a single-phase flow. The results demonstrated that the slotted orifice generates a smaller value of the differential pressure and ensures faster pressure recovery of downstream of the orifice plate.

Geng et al. [3] performed numerical and experimental examinations of the performance of the standard and slotted orifice for single and two-phase flows. This study demonstrated that the slotted orifice performs better than the standard one in terms of the requirements regarding straight pipe sections used to stabilize the flow, stabilization of the velocity profile downstream of the orifice plate, and elimination of the effect of the liquid hold-up upstream of the orifice plate. This study also showed that orifices with a small value of β ratio tend to be more sensitive to the occurrence of the liquid phase in the flow.

The study by Kumar and Bing [17] also involved the analysis of solutions of the orifice meter. Numerical studies were conducted with regard to the optimal design of a slotted orifice in terms of the pressure recovery downstream of the orifice plate and involved the measurements of the values of the pressure drop resulting from its installation. In simulation studies, four orifice plates with various slot shapes were applied, including rectangular and circular, and the results were compared with those gained for a standard orifice. The study found that the design of the perforations does not have a major impact on the pressure recovery, however, rectangular slots with a length to width ratio equal to $l/w=3.0$ demonstrate a better static pressure recovery in comparison to orifices with lower values of this ratio.

The study by Annamalai et al. [18] focused on the homogenization of a two-phase mixture downstream of the orifice meter. The results demonstrated that it is possible to apply a slotted orifice as a stream conditioner that is capable of converting a majority of the flow patterns into a homogeneous mixture. The tests involve the analysis of the homogeneity of a two-phase gas–liquid mixture flow as a result of applying a slotted orifice in a horizontal tube for various conditions upstream of the orifice. The conditions of the tests included the bubble and slug flows. This study demonstrated that optimal homogeneity of two-phase flow occurs at a distance of 1.5 to 2.5D downstream of the slotted orifice. Therefore, it was found that the slotted orifice meter is not sensitive to the conditions of the flow downstream of the orifice.

Due to the design of the standard orifice, when the gas flow contains even a small volume ratio of the liquid phase the liquid is retained upstream of the orifice meter. This leads to flow disturbance and, as a consequence, a positive bias of Δp is obtained. In the case of a slotted orifice, the use of concentrically arranged slots in the orifice plate can limit the effects of the initial flow disturbance. By separating the flow of the mixture in the entire cross-section of the pipeline, the effect of the liquid accumulating in the obstacle was eliminated and the formation of considerable vortex structure downstream of the obstacle was prevented.

The objective of the present study involves the determination of the effect of the orifice plate geometry on the differential pressure recorded in the gas–liquid mixture flow and permanent pressure loss.

3 EXPERIMENTAL SETUP

The experimental part of the study was carried out on a setup, whose diagram is presented in Fig. 3. The air into this installation was fed via a compressor (1). The air was routed for the experiment via a throttling valve (2), which provided a constant pressure value. The flow rate of the air was regulated by a valve (3). The parameters of the air flow were controlled by a measuring system comprising a pressure sensor (4) and a temperature sensor (10) and an orifice plate (5,6). Water was supplied from the water network to a chamber (7) in which the two-phase mixture was formed, and the flow rate was regulated by a valve (9). The flow rate of the water was measured by a rotameter (8). The horizontal section of the pipeline comprised a system designed for testing slotted orifice plates (11) that could be removed and replaced. The differential pressure resulting from the installation of the slotted orifice plates was measured with a differential pressure transducer (5). The static pressure value in the pipeline was measured with a sensor (4).

A separator (12) was applied to remove the air from the liquid at the end of the measuring section in the installation. The values of signals from the measurement sensors were recorded continuously by a dedicated card on a PC throughout the duration of the experiment.

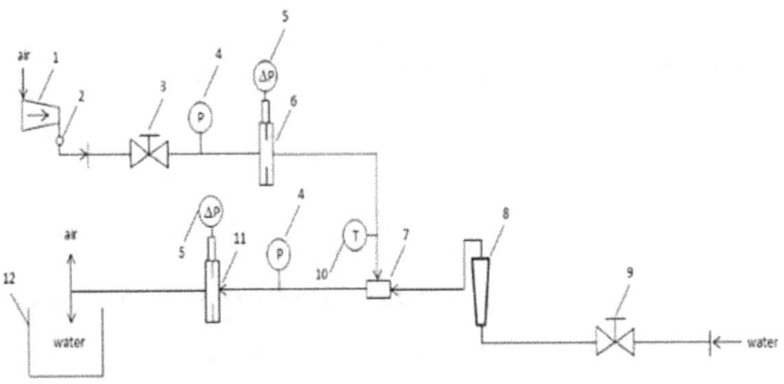

Figure 3: Setup applied for two-phase flow measurements. 1: Compressor; 2: Reducing valve; 3: Air flow regulation valve; 4: Pressure sensor; 5: Differential pressure transducer; 6: Differential pressure meter; 7: Mixing chamber; 8: Rotameter; 9: Liquid flow regulating valve; 10: Temperature sensor; 11: Slotted orifice; 12: Liquid vessel.

In order to perform measurements into the distribution of static pressures downstream and upstream of the tested orifice plate, an experimental section of the pipeline was designed and built comprising a series of pressure tapping points (Fig. 4). In this section, nonlinear distribution of static pressure tapping points was applied. The largest number of measuring points in terms of distribution was selected to occur just before and after the orifice plate.

Figure 4: View of measuring section applied for determination of the static pressure downstream and upstream of the orifice plate.

The measurements were carried out for two mass flow rates of air, which were equal to 0.06 kg/s and 0.086 kg/s, respectively. After the gas flow was stabilized, water was fed into the installation in the range of flow rates from 0.06 kg/s to 0.16 kg/s.

The experimental part of the study involved a standard orifice and three slotted orifices with various layouts of the perforations were applied (Fig. 5).

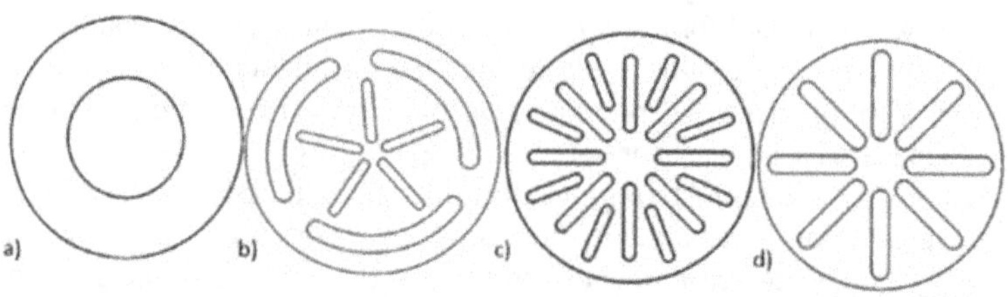

Figure 5: Orifice plates applied in the experimental research. (a) Standard orifice; (b) Slotted orifice 1; (c) Slotted orifice 2; (d) Slotted orifice 3.

The size and number of slots were selected in such a way that the beta ratio for all orifice plates and equal to $\beta = 0.5$. The value of β is derived from eqn (4):

$$\beta = \sqrt{\frac{A_{slots}}{A_{pipe}}}, \tag{4}$$

where A_{slots} is surface area of the slots and A_{pipe} is cross-sectional area of the pipe.

4 RESULTS OF EXPERIMENTAL RESEARCH

Fig. 6 contains the results of the measurements of the differential pressure as a function of the mass fraction of liquid in the constant gas flow rates, for 0.06 kg/s and 0.087 kg/s, respectively. The values in terms of the mass fraction of the liquid were determined on the basis of the measurements of gas and liquid flows by application of the following relation (eqn (5)):

$$\alpha = \frac{m_L}{m_L+m_G}, \tag{5}$$

where m_L and m_G are mass flow rates of the liquid and gas, respectively.

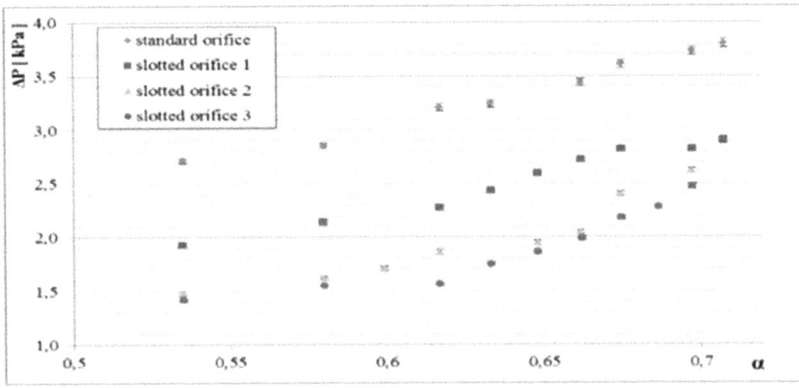

(a)

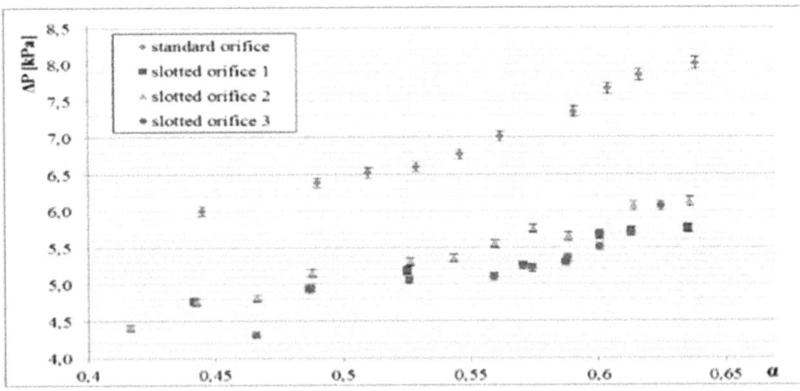

(b)

Figure 6: Dependence between the differential pressures and water mass fraction for the airflow rates equal to: (a) 0.06 kg/s; (b) 0.087 kg/s.

On the basis of the analysis of the results of the measurements presented in Fig. 6, the study found that the presence of liquid in the constant air flow leads to a change in the differential pressure developed at the orifice. The greater mass fraction of the liquid in the air flow results in a greater differential pressure. The distribution of the pressure depending on the mass fraction of the liquid is non-linear. It was observed that for the case of the air–water mixture flows, in comparison to the standard orifice, the slotted orifice generates a smaller differential pressure. The slotted orifice with a radial arrangement of slots demonstrates the smallest sensitivity to changes in the mass fraction of the liquids. This is due to the fact that throughout the entire cross-section of the pipeline there is a steady local decrease in the diameter ratio of the perforations. The liquid passes freely through the obstacle along with the gas, and only a small ratio of it is stopped by the orifice. The arrangement of the slots along the entire cross-section of the pipe leads to the liquid drop entrainment by the gas flow. In the case of a standard orifice, a large part of the liquid is held up downstream of the orifice, the flowing liquid accumulation that leads to additional, unfavorable flow disturbances.

The charts in Fig. 7 present a comparison of the mass flow rate of the gas taking into account the differential pressure measured between the single-phase gas flow and flow gas containing small amounts of liquid.

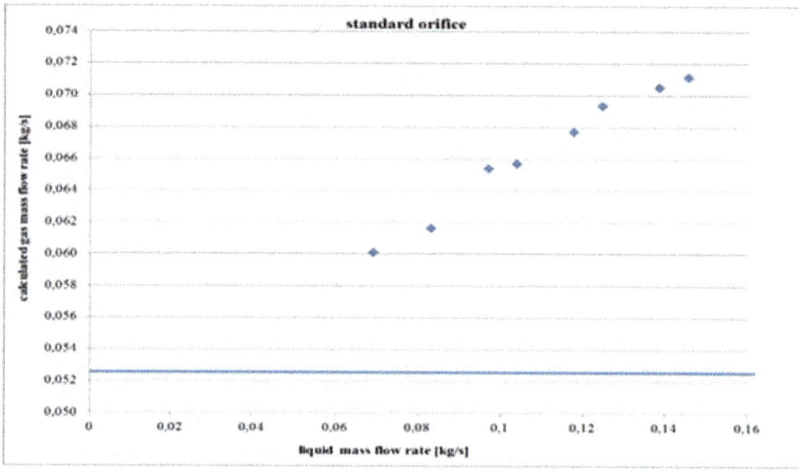

(a)

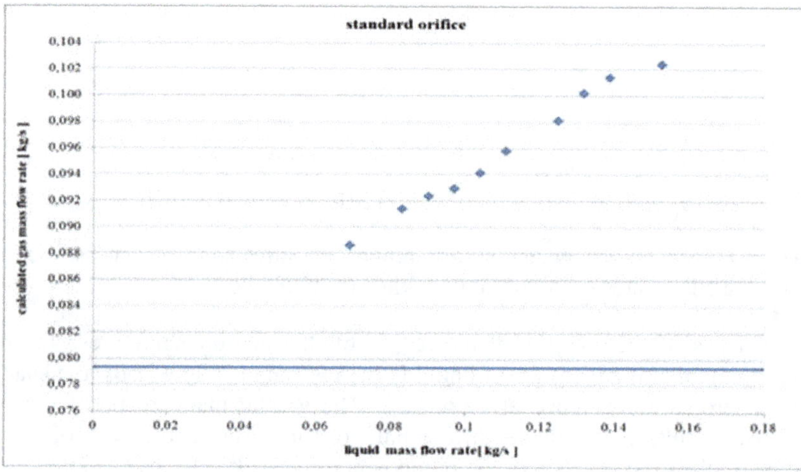

(b)

Figure 7: Dependence between calculated gas mass flow rate and liquid mass flow rate for the different inlet airflow rates.

The charts above contain the results of a comparison between the mass flow rates of gas and the mass flow rate of water that was measured in the investigated orifices plates. The gas mass flow rate was calculated on the basis of eqn (1) taking into account the differential pressure developed at the orifice plate during the gas–liquid fluid flow. The horizontal lines on the diagrams show the mass flow rates for the case of single-phase gas flow.

We can also note that along with the increase in the flow rate of water, the measurement error of the flow rate of the gas increases. As a result of the application of the standard orifice in the flow of gas with a small amount of liquid, the measured flow rate of gas accounting for the gas–liquid differential pressure is 12–28%.

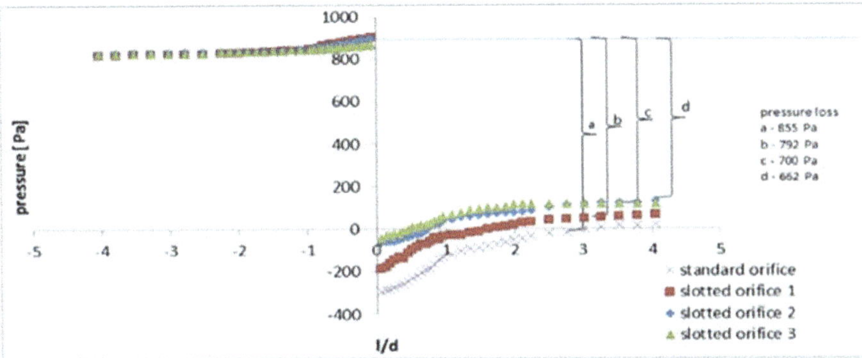

Figure 8: Distribution of pressure along the pipeline and determination of permanent pressure loss.

The results of the measurement of the static pressure distribution downstream and upstream of the orifice plate are presented in Fig. 8. Under the same flow conditions, the use of a standard orifice results in a greater differential pressure caused by its presence compared to the case of the use of a slotted orifice. In terms of metrology parameters, this forms an advantage since in the earlier case, an increase in the measurement sensitivity is gained. However, as a result of the application of slotted orifice plate, pressure recovery occurs faster than in the case of a standard orifice. Also, permanent pressure loss is smaller when slotted orifices are used. The smallest permanent pressure loss, which is equal to 662 Pa, was recorded in the slotted orifice design marked by 3, whereas the largest one was recorded the standard orifice, whose permanent pressure loss is 855 Pa.

5 CONCLUSIONS

The tests carried out in this paper demonstrate that the design and distribution of the perforations on the surface of slotted orifices affect the length of the hydraulic stabilization zone of the flow and the value of permanent pressure loss. This is due to the separation of the flux into several or a dozen or so smaller ones. In this case, a large recirculation zone is not developed, which normally occurs when a standard orifice is applied, and forms the principal reason for generation of energy losses. The hydraulic stabilization width of the flow depends on the design of the slotted orifices and it is smaller compared to the case of a standard orifice. In addition, the presence of perforations has a positive effect on flow homogenization and reduction of pulsations, which are the sources of an additional measurement error. A certain disadvantage of slotted orifices is associated with the decrease in the differential pressure, as the sensitivity of metering is limited. This requires further research related to finding a compromise between the advantages and disadvantages of slotted orifices in specific engineering applications. This applies particularly with regard to determining the optimal design, arrangement and number of slots in the specific flow conditions.

REFERENCES

[1] Bertani, C., De Salve, M., Malandrone, M., Monni, G. & Panella, B., State-of-Art and selection of techniques in multiphase flow measurement. Electrical System Research Report, RdS/2010/67, 2010.

[2] Falcone, G., Hewitt, G. & Alimonti, C., *Multiphase Flow Metering: Principles and Applications*, Developments in Petroleum Science Series, vol. 54, 2009.

[3] Geng, Y., Zheng, J. & Shi, T., Study on the metering characteristics of a slotted orifice for wet gas flow. *Flow Measurement and Instrumentation,* **17**, pp. 123–128, 2006.

[4] Hua, C. & Geng, Y., Wet gas metering technique based on slotted orifice and swirlmeter. *Flow Measurement and Instrumentation,* **30**, pp. 138–143, 2013.

[5] Anders, D. & Rząsa, M.R., The possibility of composting animal waste products. *Environment Protection Engineering*, **33**(2), 2007.

[6] Zheng, G.B., De Jin, N., Jia, X.H., Lv, P.J. & Liu, L.B., Gas–liquid two phase flow measurement method based on combination instrument of turbine flowmeter and conductance sensor. *International Journal of Multiphase Flow,* **34**, pp. 1031–1047, 2008.

[7] Steven, R. & Hall, A., Orifice plate meter wet gas flow performance. *Flow Measurement and Instrumentation,* **20**, 141–151, 2009.

[8] Fang, L., Zhang, T. & Jin, N., A comparison of correlations used for Venturi wet gas metering in oil and gas industry. *Journal of Petroleum Science and Engineering,* **57**, pp. 247–256, 2007.

[9] Taler, D., *Pomiar Ciśnienia, Prędkości i Strumienia Przepływu Płynu*, Wydawnictwo Naukowo-dydaktyczne: Kraków, 2006.

[10] ISO, Measurement of fluid flow by means of pressure-differential devices – Guidelines for the specification od orifice plates, nozzles and Venturi tubes beyond the scope of ISO 5167, 2007.

[11] Morrison, G.L. et al. Comparison of orifice and slotted plate flowmeters. *Flow Measurement and Instrumentation*, **5**(2), pp. 71–77, 1994.

[12] Morrison, G.L., Terracina, D., Brewer, C. & Hall, K.R., Response of a slotted orifice flow meter to an air/water mixture. *Flow Measurement and Instrumentation,* **12**, pp. 175–180, 2001.

[13] Morrison, G.L., Hall, K.R., Holste, J.C., DeOtte, R.E., Macek, M.L. & Ihfe, L.M., Slotted orifice flowmeter. *AIChE Journal*, **40**(10), pp. 1757–1760, 1994.

[14] Bayazit, Y., Sparrow, M. & Joseph, D., Perforated plates for liquid management: Plate geometry effects and flow regime. *International Journal of Thermal Sciences*, **85**, pp. 101–111, 2014.

[15] Huang, S., Ma, T., Wang, D. & Lin, Z., Study on discharge coefficient of perforated orifices as a new kind of flowmeter. *Experimental Thermal and Fluid Science,* **46**, 74–83, 2013.

[16] Abou El-Azm Aly, A., Chong, A., Nicolleau, F. & Beck, S., Experimental study of pressure drop after fractal-shaped orifice in turbulent pipe flow. *Experimental Thermal and Fluid Science*, **34**, pp. 104–111, 2010.

[17] Kumar, P. & Bing, M.W.M., A CFD study of low pressure wet gas metering using slotted orifice meters. *Flow Measurement and Instrumentation*, **22** pp. 33–42, 2011.

[18] Annamalai, G., Pirouzpanah, S., Gudigopuram, S.R. & Morrison, G.L., Characterization of flow homogeneity downstream of a slotted orifice plate in a two-phase flow using electrical resistance tomography. *Flow Measurement and Instrumentation*, **50**, pp. 209–215, 2016.

COMPARISON OF THE EFFECTS OF EXPLOSIONS INITIATED AT DIFFERENT LOCATIONS IN MULTIPHASE MEDIA: AIR AND STRUCTURE

PETR P. PROCHAZKA
Civil Engineering, Czech Technical University in Prague, Czech Republic

ABSTRACT
In connection with the assessment of underground structures, the question arises of the mechanical behavior of the different types of materials of the load-bearing structure exposed to the explosion and the location of its source. This means that the stresses induced by shock waves due to the explosion are in direct relation with the properties of the lining material and the location of the charge. There are a number of numerical methods that deal not only with the issue of shock waves propagation in a closed or semi-closed environment, but also nonlinear properties of the material can be described. In order to feed the numerical procedures with the relevant data, it is essential to carry out practical tests that are best in scale models, saving time and money. Recall that there are means to evaluate the actual behavior of air, structure and surrounding rocks from observation in scale models. This paper focuses on the description of concrete circular tubes simulating the tunnel in the scale of 1/20. The three-phase flow is reduced to two phases (air, structure), for simplicity. The charge is placed on the longitudinal axis and also eccentrically, for comparison. The numerical method is suggested for the completeness of the flow behavior in the multiphase medium. The overpressure exerted on the lining is monitored by sensors that are located directly above the source of the explosion, further on the distal parts of the structure and ultimately overpressure at the mouth of the tube is monitored by the so-called pencil sensor. The source of the explosion is the Semtex 1A explosive in both explosion locations.
Keywords: tunnel tubes, different location of charges, different concrete linings, scale models.

1 INTRODUCTION

An explosion is a physicochemical process leading to a sudden and very rapid release of energy, a local increase in temperature and pressure, and the degradation of the properties of the environment in which the explosion occurs. The effect of explosion strongly depends on the place, in which it arises, on the shape of structure, which is exposed to the shock waves initiated by the explosion and the material of the structure. First, the description of effects of explosion in an open air is relatively simple, although it may cause also complications if the surface of the structure is irregular (wavy, for example). In this paper, the charge is located in a semi-enclosed space. Second, the location of the charge is either at the longitudinal axis or non-symmetric. The material is always concrete, simple, or reinforced by plastic fibers, or steel fibers.

It is worth noting that the solution of the basic equations describing the flow in certain media and the propagation of the shock waves in reality is quite similar, if solid phase, air or liquid is considered. The principal goal of the most observations in the above field in civil engineering is obtaining the basis for assessment of mechanical behavior of structures, such as plastic behavior or damage. Typically, BLEVE (boiling liquid expanding vapor explosion) occurs when liquid in the vessel is at a temperature above its atmospheric boiling point [1].

Today's civilized world has a very urbanized structure – a large part of the population lives in cities and underground, generally enclosed, spaces, are one of the most important areas of modern towns. However, the structures are very vulnerable to sudden changes in load, associated with terrorist attacks, VBIED, unpredictable explosions (traffic accidents, fire, rapid economic development), etc., that can cause huge damage to structures and may

WIT Transactions on Engineering Sciences, Vol 123, © 2019 WIT Press
www.witpress.com, ISSN 1743-3533 (on-line)
doi:10.2495/MPF190051

have fatal consequences for human lives. Therefore, it is very important to address and anticipate the dynamic effects of explosions in enclosed spaces. The location of the explosive center is very important. It turns out that if its position is on the surface of the building or over the terrain, no follow-up wave and the interaction of the main shock wave with the secondary waves are expected. Another situation occurs when describing the time evolution in closed spaces, where wave interactions play an important role.

Propagation of shock waves in gas (air) that fill underground spaces is described by non-linear equations of conserving density, momentum and energy. The influence of gas pressure is usually given for adiabatic state for this type of problem. The effects of explosions, i.e. initial airborne shock waves, and then interactions with secondary waves, are calculated for time changes in mass density, motion velocity and changes in internal and kinetic energy. The states of stress and deformation in the structure are also generally described by non-linear relations.

The issue of the effects of explosions on structures in enclosed spaces is dealt with by a number of publications, [2]–[4], for example. There are a number of publications that lead, if not directly to standards, at least to useful recommendations that come from the findings of practical knowledge [5]–[8].

At present, the most commonly used material for underground structures is concrete, either classic, which is reinforced with steel rods or fiber-reinforced concrete. The latter is characterized by a cement matrix and unevenly distributed and randomly oriented fibers of various materials (synthetic-polypropylene, steel, glass, etc.); the fibers may also have different shapes and sizes. Thanks to these specific properties, such a composite has a higher tensile and shear strength. The abovementioned fibrous materials can be combined, [9], [10]. Steel fibers serve predominantly to bridge cracks in the material, while synthetic fibers increase the ductility and deformability of the concrete. They reduce or even suppress the spalling of concrete. Appropriate application of the free hexagon method may well simulate properties of different types of fiber reinforced concrete.

In order to obtain the material properties of objects subjected to a sudden change in load, tests are carried out which, for more complex problems of the propagation of shock waves in enclosed spaces and their influence on structures are often carried out on physical and scale models [12]–[15]. Compared to 1:1 or in situ models make it possible to achieve very reasonable results while saving time, material and energy.

The formulation is based on hydrodynamic nonlinear Euler equations derived from the Landau–Lifschitz theory. This can be reconstructed in cylindrical coordinates, which are introduced in this paper.

2 MODELS OF TEST SPECIMENS

Two models were made at a scale of 1/30 with regard to the explanatory value of the experiment and manipulation options. This is a group of five circular tubes which are linked on a half-groove. The tubes have diameters 1 x 1 m^2 (external diameter x length), thickness 15 mm. Therefore, the total length of the tunnel (made from tubes) was 5 m. Due to the compactness, the tubes were connected by a cement screed near the sensors (Fig. 1).

The goal of this paper is to compare the results from measurements and computations in the above said models. The first concerns the location of the charge at the axis of horizontal symmetry and is studied in details in [16]. The observations following from this paper are here taken for comparison with a new observation, which follows from explosions positioned outside the horizontal axis.

Test specimens of plain concrete are made from class of concrete C 25/30. The test site consists of the individual concrete units and the explosive items are (Fig. 2):

Figure 1: The tubes prepared for the tests and during the test.

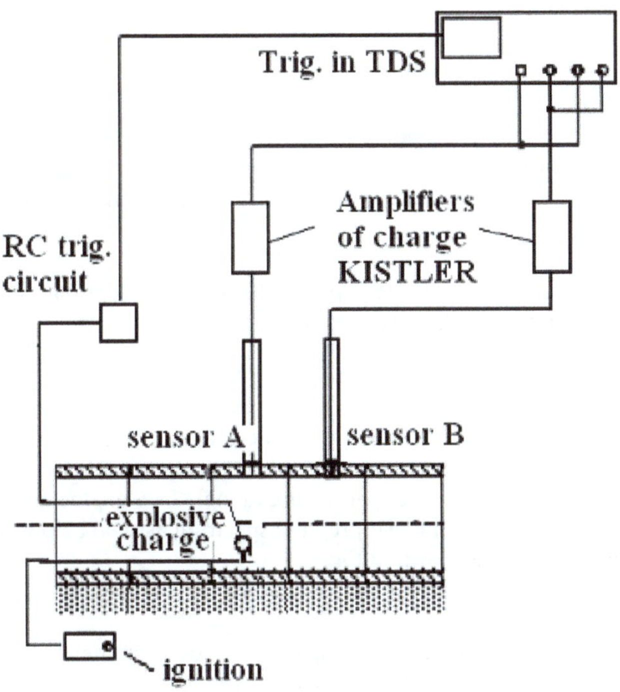

Figure 2: Adjustment of the measurement devices.

- Bulk plastic burst blasting explosives with a nominal mass $m_X = 20$ g initiated by a SO8 detonator (medium resistance). The detonator is activated by means of a capacitor firing device.

- A switching sensor is inserted into the charge, which is short-circuited by the pressure and thermal effects of the fumes, which leads to the creation of a synchronization signal in the RC circuit. This pulse triggers the running time of the TDS 3014 synchro.
- The permissible load of a pair of pressure sensors reaches 600 MPa, their sensitivity is 1.99 pC/100 kPa.
- Signals of these sensors are processed using KISTLER 5001 and 5006 amp amplifiers with an upper limit of 170 kc/s. The basic parameter of the amplifier – the ratio of the released hub to the output voltage – was controlled by the calibration unit. In this case, the sensitivity of both transmitters was set to 5 MPa/V.
- The synchro recording logs register the output signals of the transducers at two different sensitivities and at the total signal recording length of 4 ms.

In Fig. 3, locations of sensors are shown.

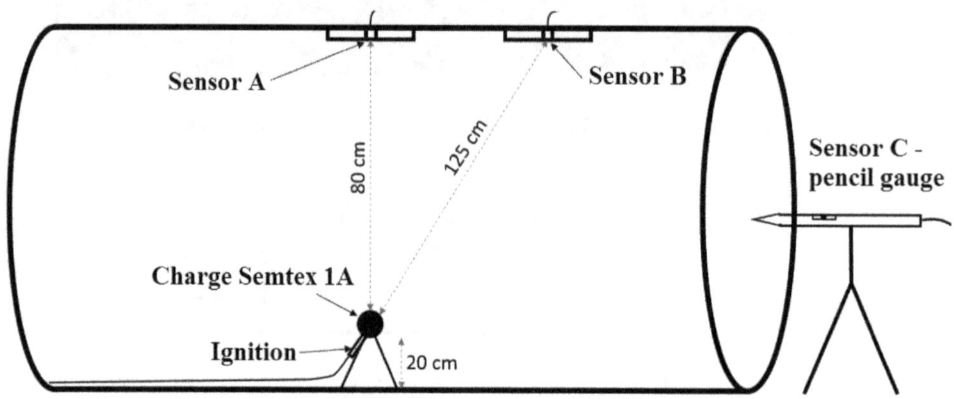

Figure 3: Scheme of the set-up of the explosion and sensors locations, charge location in the non-symmetric charge deposition.

3 PROCESSING THE EXPERIMENTS

Two locations of charges were centered inside the tunnel (2.5 m from both open parts). Vertically, the location of the charges was in the first case put in the horizontal axis, and in the second case 20 cm from the lower part of the tunnel. Two explosions were initiated in each case providing almost the same overpressure spectrum as seen below. The position and the charge itself in our case are shown in Fig. 3.

Sensors A and B were placed in a nylon box and the pencil gauge C are located as seen in Fig. 3.

Semtex 1A plastic explosive of a round shape (Fig. 4), of a nominal weight of 20 g (similar to the symmetric case [16]), was mounted above the sensor A.

4 RESULTS FROM EXPERIMENTS

In order to compare the effects of explosions on the behavior of shock waves and interaction with the structure the case when the charge was positioned symmetrically and non-symmetrically, the results in the form of overpressure spectra recorded at the sensors A and B are depicted in Figs 5 and 6 (symmetric case) and Figs 7 and 8 in non-symmetric case. In all cases two shots were made for identification of possible discrepancies.

Figure 4: The stand located at the axis of horizontal symmetry (first case) and 20 cm above the bottom of the tunnel (second case), left stabilized version for the second explosion.

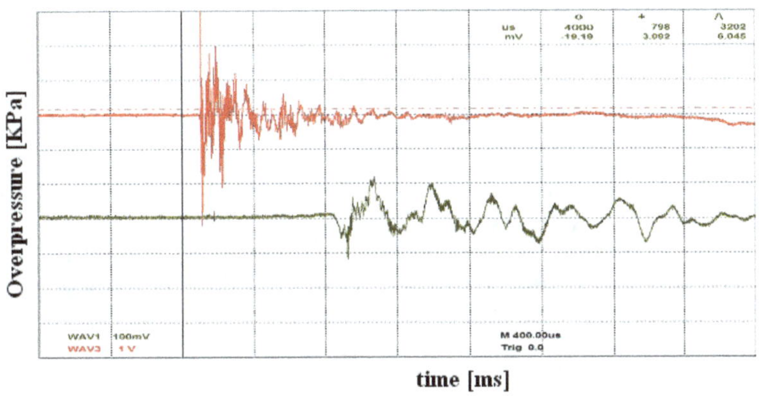

Figure 5: Symmetric position of the charge: first shot.

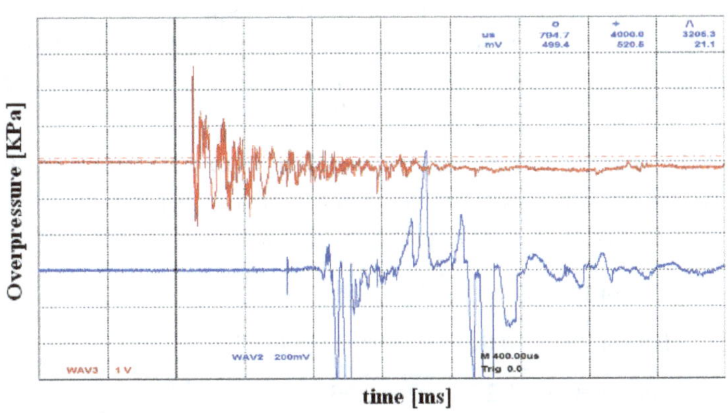

Figure 6: Symmetric position of the charge: second shot.

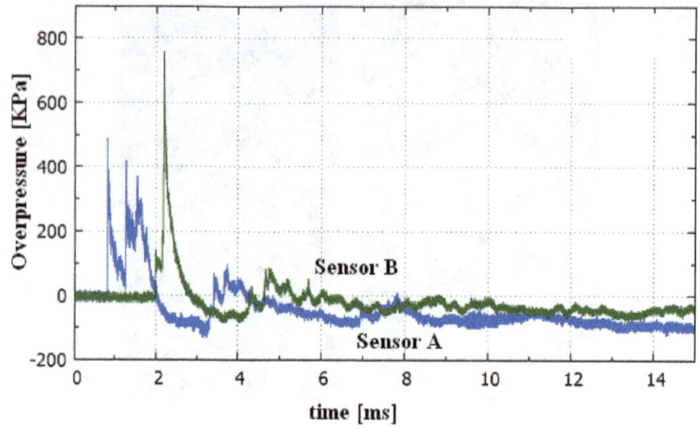

Figure 7: Non-symmetric position of the charge: first shot.

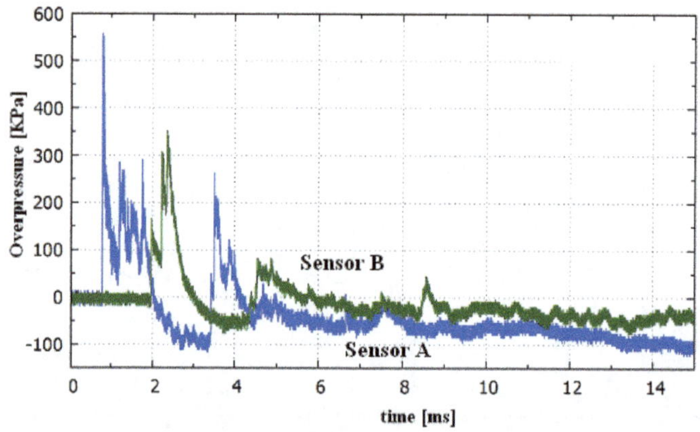

Figure 8: Non-symmetric position of the charge: second shot.

In the given 3D conditions, the distance between the explosion site and the radius of the spherical wavelength of the shock wave R is expressed by the reduced radius r as:

$$r = R / \sqrt[3]{m_x}$$

where m_x is the mass of the charge. Since the inner radius of the tunnel is 0.5 m and the nominal mass of the used charges is 0.02 kg then the reduced distance is:

1st case: Sensor A: $r = 0.5\sqrt[3]{0.02} = 0.135\ \text{m.kg}^{1/3}$,

Sensor B: $r = 1.08\sqrt[3]{0.02} = 0.292\ \text{m.kg}^{1/3}$,

2nd case: Sensor A: $r = 0.8\sqrt[3]{0.02} = 0.216\ \text{m.kg}^{1/3}$,

Sensor B: $r = 1.25\sqrt[3]{0.02} = 0.338\ \text{m.kg}^{1/3}$.

The geometric parameters are obtained from Fig. 3.

From the above overpressure spectra, the impact time of the shock wave front can be recorded on the inner surface of the tube, the first peak of the positive signal representing the maximum overpressure behind the reflected wave front and the fundamental frequency of the radial component of the oscillation. Higher frequencies are superimposed on the base frequency, which are probably caused by the mutual mechanical interaction between the sensor, its silon sleeve and the body of the tube.

Note that from Fig. 4 in [16] it is seen that the behavior of the shock waves measured at sensor A are really similar for both explosions (Fig. 4 of this paper) but records at sensor B need filtration and probably suffer from some kind of instability.

The model time course of overpressure behind the front of shock wave is, according to [8], shown in Fig. 9.

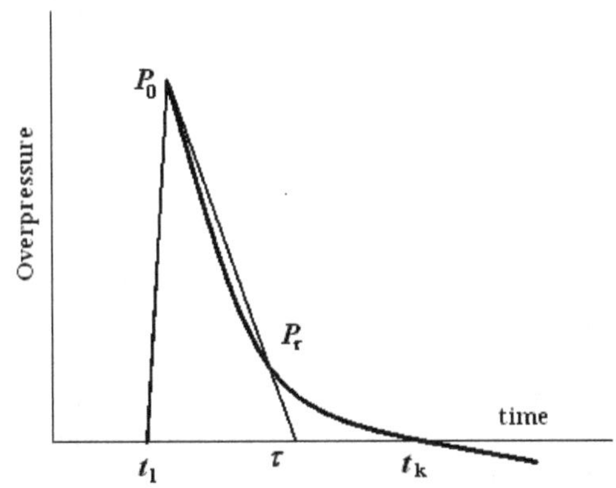

Figure 9: General time course of overpressure.

In what follows, the characteristics follow the recommendations given in [8]. In Fig. 9, the symbols $t_1, t_k, \tau = (t_k - t_1)/2.2, P_0$ and $P_t \approx 0.37 P_0$ denote successively the moment of explosion, the end of the overpressure phase, the front pressure of the shock wave and the pressure at time τ. In the situation of low values of reduced distance, i.e. $P_0 >> P_{\text{atmospheric}}$, the influence of the dilute phase on the course of the effects of explosion can be neglected. The time course of the pressure diminution on the wave front within the limits $t_1 \leq t \leq t_k$ generates with sufficient precision an envelope, Fig. 9, defined by the exponential function

$$p(t) = \exp(-t/\tau)$$

In the case of centered position of the charge, the maximum generated overpressures P_0 were measured at A in the range of 0.7 to 1.2 MPa, giving an average value of 0.92 MPa. In the second case they were 500–650 kPa with the average value of 575 kPa. The behavior of the shock waves at B was reasonable in the case of the first shot with the maximums P_0 circa three times lesser. As said above, the second shot cannot be considered as serious enough.

The time, when the pressure phase ends, is denoted as t_k. It is estimated by the value 0.86 ms. The time from which the overpressure is measured is t_1 = 0.57 ms. The average of duration of the pressure phase of plain concrete is 0.86–0.57 ms = 0.29 ms. This values are taken from [16]. In the non-symmetric case the average duration of the pressure is 1.1 ms according to the measurement at A and 0.9 ms at B.

The relevant characteristics of measured and calculated values at sensors A and B are listed in Table 1.

Table 1: Relevant characteristics of measured and calculated values at sensors A and B.

	Pressure τ (MPa)	Higher frequencies	Lower frequencies (Hz)
1st Case sensor A	0.34	8500	2200
1st Case sensor B	0.12	9000	2500
2nd Case sensor A	0.21	7200	1900
2nd Case sensor B	0.18	7000	1800

5 DISCUSSION OF EXPERIMENTS

Tests were prepared quite carefully, although some factors may have affected the results. First of all, it is the attachment of sensors to the concrete structure that causes local inhomogeneity in the area of measurement. It is also possible occurrence of instability, caused by various reasons. For example, it is the propagation of gas and other products reflected from the bottom of the tube. At the bottom there was no sense to attach the sensors, due to the placement of the tube on relatively solid bedrock; the interaction with the thill will be minimal. On the other hand, slightly above the tunnel, the largest structural damage occurs, as this part is exposed to the highest pressure. In practice, however, this is a part of the structure of platforms at the metro or walkways in automobile tunnels, where, moreover, the shape of the cut is not completely circular.

In both cases, at the location of the sensors quite standard overpressure distribution over the source of the explosion (sensor A) is registered, corresponding to the exponential course of the overpressure envelopes, see [8] and Fig. 9. This is especially true for the first case (centered location). The overpressure spectrum at the first case of the second charge position is an interesting phenomenon, resulting in a peak pressure of about two times at sensor B than at A. The apparent shock wave interaction is the reason and, moreover, as the same tubes are used for all tests, the measurement results may be affected by the partial local disturbances of the lining.

In this paper the test characteristics are among the most important, but naturally they fulfill not the total set of them. Others, however, can be derived either directly from the above-mentioned figures or from norms and recommendations. It is probably mainly important that they can be used to tune the algorithms and inputs to the algorithms for computational programs and also to design and assess real underground structures, which are exposed to possible negative effects of explosion.

ACKNOWLEDGEMENTS

Financial support by the Grant agency of the Czech Republic No. 17-04204S is greatly acknowledged. The experiments have been carried out at Pardubice, Czech Republic; Assoc/Prof. Pachman was the monitor.

REFERENCES

[1] Tijsseling, A.S., Fluid-structure interaction in liquid pipe system: A review. *J. Fluids Struct.*, pp. 109–146, 1996.

[2] Mitu, M., Giurcan, V., Razus, D., Prodan, M. & Oancea, D., Propagation indices of methane-air explosions in closed vessels. *J. Loss Prevent. Process. Ind.*, **247**, pp. 110–119, 2017
[3] Mitra, M., Brandes, E. & Hirsch, W., Mitigation effects on the explosion safety characteristic data of ethanol/air mixtures in closed vessel. *Process. Safety Environ. Prot.*, **117**, pp. 190–199, 2018.
[4] Edrl, I., Feldgun, V.R., Karinski, Y.S. & Yankelevsky, D.Z., On blast pressure analysis due to a partially confined explosion: Afterburning effect. *Int. J. Protect. Struct.*, **3**(3), pp. 311–332, 2012.
[5] Cioca, I.L. & Moraru, R.I., Explosion and/or fire risk assessment methodology: A common approach structured for underground coalmine environments. *Archives of Mining Sciences*, **57**(1), pp. 53–60, 2012.
[6] Haramy, K.Y. & Brady, B.T., A methodology to determine in situ rock mass failure. Internal Report of Bureau of Mines, Denver, CO, 1995.
[7] Haramy, K.Y., Magers, J.A. & McDonnel, J.P., Mining under strong roof. *7th International Conference on Ground Control in Mining Nureau of Mines*, Denver, USA, pp. 179–194, 1992.
[8] US Army Corps of Engineers®, Methodology Manual for the Single-Degree-of-Freedom Blast Effects Design Spreadsheets (SBEDS), PDC TR-06-01 Rev. 1, 2008.
[9] Luccioni, B.M. & Luege, M., Concrete pavement slab under blast loads. *Int. J. Impact. Eng.*, **32**(8), pp. 1248–1266, 2006.
[10] Schenkera, A. et al., Full-scale field tests of concrete slabs subjected to blast loads. *Int. J. Impact. Eng.*, **35**(3), pp. 184–198, 2008.
[11] Buchana, P.A. & Chenb, J.F., Blast resistance of FRP composites and polymer strengthened concrete and masonry structures: A state-of-the-art review. *Composites Part B: Engineering*, **38**(5–6), pp. 509–522, 2007.
[12] De, A., Zimmie, T.F., Abdoun, T. & Tessari, A., Physical modeling of explosive effects on tunnels. *Fourth International Symposium on Tunnel and Safety and Security*, Frankfurt am Main, Germany, March 17–19, 2010.
[13] Zhang, P., Wang, J., Qi, S., Li, G., Wang, S. & Yan, F.. Experimental analysis of the flame speed, brightness and zone thickness of gasoline-air explosion in a closed tunnel. *J. Loss Prevent. Process. Ind.*, **53**, pp. 129–135, 2018.
[14] Prochazka, P.P. & Trckova, J., Stress and deformation states in underground structures using coupled modeling. *Acta Geodyn. and Geomat.*, **5**(4), pp. 361–375, 2008.
[15] Bakhtar, K., Impact of joints and discontinuities on the blast-response of responding tunnels studied under physical modeling at 1-g. *Int. J. Rock Mech. and Min. Sci.*, **34**(3–4), 021, 1997.
[16] Prochazka, P. & Jandekova, D., Measuring overpressure caused by the detonation of explosives in a horizontal tube. *ICPS 2019*, Poznan, Poland, pp. 23–31, 2019.
[17] Prochazka, P. & Peskova, S., Modeling the impact of explosions in enclosed spaces using discontinuous boundary elements and finite volume element methods. *SUSI*, pp. 453–455, 2018.
[18] Prochazka, P.P., Rock bursts due to gas explosion in deep mines based on hexagonal and boundary elements. *Adv Eng. Soft.*, **72**, pp. 57–65, 2014.
[19] Prochazka, P.P. & Lok, T.S., Hereditary problems in long-wall mining by free hexagons. *Int. J. Comput. Meth.*, **8**(2), pp. 293–313, 2011.

TWO-PHASE FLOW PHENOMENA IN A HORIZONTAL PIPE WITH A SUDDEN CONTRACTION: EFFECTS OF CONTRACTION RATIO ON FILM BEHAVIOUR

TAKAHIRO FUJIMATSU, MIZUKI KITO & KUNIKAZU KONDO
Department of Mechanical Engineering, National Institute of Technology, Suzuka College, Japan

ABSTRACT

Experimental studies were conducted on the behaviour of water films in a horizontal pipe with a sudden contraction to achieve a fundamental understanding of the film mechanisms of annular flow using high-speed video imaging analysis. The sudden contraction area ratios $A_C = (d_2/d_1)^2 = 0.36$ and 0.64, where d_1 is the inlet pipe diameter (50 mm) and d_2 is the contracted pipe diameter were examined. The experiments were carried out for various values of superficial gas and liquid velocities. The superficial gas velocity after mixing (j_{G1}) ranged from 18.02 to 24.41 m/s (the calculated velocity from the cross-sectional at the sudden contraction was j_{G2} = 27.98–67.39 m/s) and the superficial liquid velocity (j_{L1}) ranged from 8.62×10^{-2} to 1.23×10^{-1} m/s ($j_{L2} = 1.35\times10^{-1}$–$3.42\times10^{-1}$ m/s). Four flow regimes were observed after the sudden contraction, namely, huge-wave (HW), two-dimensional disturbance wave (DW_1) which is formed perpendicularly to a pipe axis having a distinct coherence, three-dimensional disturbance wave (DW_2) which is formed obliquely to the pipe axis, and bow wave (DW_3) which is transitional wave to "Misty wave" from DW_2. The disturbance wave occurrence frequencies were studied under various conditions of flow rate combination of air and water. DW_2 occurs over wide range of the superficial gas and liquid velocities. As the superficial gas velocity and the superficial liquid velocity increase, two-dimensional disturbance waves change into three-dimensional ones. The wave occurrence frequencies in the pipe with the sudden contraction ratio $A_C = 0.36$ were mostly higher than those in the pipe with $A_C = 0.64$.

Keywords: sudden contraction, disturbance wave, flow regime, wave occurrence frequency.

1 INTRODUCTION

Gas–liquid two-phase flows through orifices which are used for controlling pressure or flow rate, are important in multiphase flow measurements [1]. For instance, they are commonly found in pipelines in nuclear power, petroleum, geothermal and food processing industries. In order to predict cavitation erosion rate or improve the measurement accuracy and efficiency, many studies regarding the measurements of two-phase flows have been conducted numerically and experimentally [2]–[5]. A fundamental understanding of the gas–liquid two-phase flow is also essential for designing optimal energy-related apparatus in thermal energy plants, nuclear power plants, chemical plants as well as cooling and heating equipment. Flow pattern changes mainly depend on a number of parameters, such as flow direction, fluid property, flow conditions and channel configurations. Because it is, therefore, necessary to take the flow pattern and regime into account for predicting two-phase pressure drops in different pipe configurations and suitable operating conditions, a number of investigations have been made with regard to the classification of flow patterns in different channel configurations. Chang et al. [6] investigated air–water flows through horizontal and vertical U-bend rectangular channels. They analyzed the two-phase flow patterns by visualizing the interfacial flow transitions through the U-bend in order to understand the pressure drop and the heat transfer process. In addition, the effects of channel height-to-width ratio on the behaviour of air–water flows were demonstrated. They presented that the slug-annular flow in flow pattern maps shifted towards the higher superficial gas velocities as the height-to-width ratio decreased. Air–water flows through horizontal to vertical through a

WIT Transactions on Engineering Sciences, Vol 123, © 2019 WIT Press
www.witpress.com, ISSN 1743-3533 (on-line)
doi:10.2495/MPF190061

90-degree bend were analyzed by Saidj et al. [7]. They identified the flow patterns both in the horizontal pipe and the vertical pipe upstream, using the Probability Density Function (PDF) of the time series of the cross sectional averaged void fraction. Hsu et al. [8] examined two-phase flow patterns in 90-degree bends with small diameter tubes. They studied flow patterns in 90-degree bends installed upward horizontally, and downward. They observed a liquid flow reversal and a swirled motion for the upward flow arrangement. Whalley [9] studied air–water flows in a helically coiled tube. They showed the differences in the liquid film thickness and liquid film flow rate around the tube periphery between presented flows and vertical annular flows. Azzopardi and Whalley [10] investigated the effects of flow patterns in a T junction. They found that the entering flow direction of air–water flows was strongly dependent on the flow patterns upstream the junction. Saisorn and Wongwises [11] compared air–water flows through a horizontal circular micro-channel with two-phase flow through ordinarily size channels. They classified the observed flow patterns into three flow regimes and showed comparisons of the observed flow patterns with the flow regime maps proposed by some previous researchers. Few reports, however, have focused on the film mechanism of an annular mist flow in a horizontal pipe with a sudden contraction. Schmidt and Friedel [12] conducted extensive pressure drop measurements across a sudden contraction and provided a new pressure drop model. Chen et al. [13] studied the pressure drops and flow pattern changes across sudden contractions in small rectangular channels having different area ratios. However, a clear understanding of flow patterns is still expected and it is necessary to predict flow-pattern transitions in a horizontal pipe with a sudden contraction due to their physical and practical significance, as well as their feature complexity.

The authors have reported the flow regime transitions in a horizontal pipe with a sudden contraction and measured the water-drop size in the annular mist flow using immersion liquid method [14], [15]. In this paper, the effects of the contraction area ratio on film behaviour were investigated by flow observations and measurements of disturbance wave occurrence frequency in order to contribute the fundamental data concerning to film characteristics in air–water horizontal annular two-phase flow with a sudden contraction.

2 EXPERIMENTAL SET UP

A schematic diagram of the experimental apparatus is shown in Fig. 1. Compressed air was supplied from a compressor to the mixing device where it combines with water from a storage tank. The mixing device consisted of an annular section surrounding a porous sinter wall. The test section was made of transparent acrylic resin in order to allow flow behaviour observations and consisted of an inner diameter pipe d_1 = 50 mm, 4320 mm long upstream and d_2 = 30mm or 40mm downstream the horizontal system. The contraction area ratios $A_C = (d_2 / d_1)^2$ were obtained to be 0.36 and 0.64.

The experiments were carried out for various values of superficial gas and liquid velocities. The superficial gas velocity after mixing (j_{G1}) ranged from 18.02 to 24.41 m/s (the calculated velocity from the cross-sectional at the sudden contraction was j_{G2} = 27.98–67.39 m/s) and the superficial liquid velocity (j_{L1}) ranged from 8.62×10^{-2} to 1.23×10^{-1}m/s ($j_{L2} = 1.35\times10^{-1}$–$3.42\times10^{-1}$m/s) as summarised in Tables 1 and 2. The superficial gas and liquid velocities were calculated from mass flow rates (G_A and G_L), which were obtained from the changes (h_A and h_L) in pressure head upstream through a quadrant edged nozzle with the flow coefficients (α). The inlet temperature was measured using thermocouples to obtain the density and viscosity, from which the Reynolds numbers (Re_{G1} and Re_{L1}) were calculated as shown in Tables 1 and 2. A detailed flow observation using a high-speed video

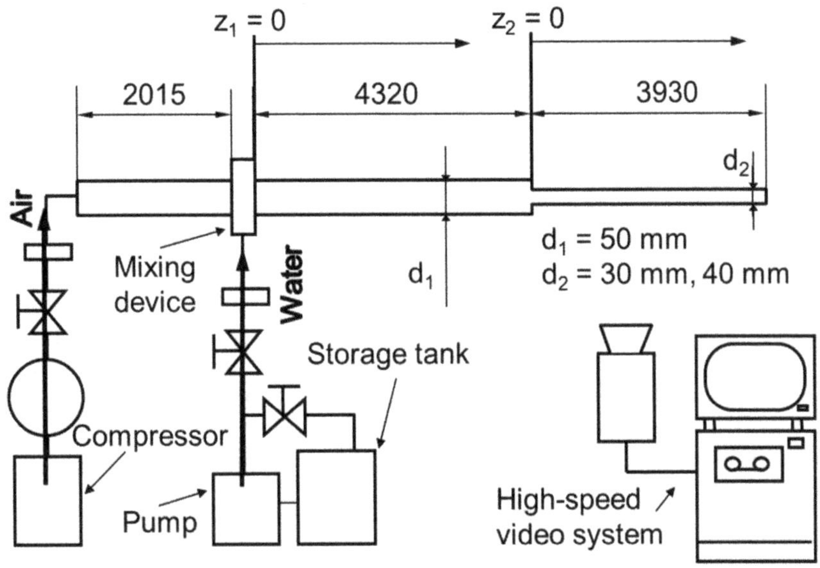

Figure 1: Experimental apparatus.

Table 1: Summary of the experimental conditions of gas phase.

h_A mmAq	G_A kg/s	α	j_{G1} m/s	j_{G2} m/s		Re_{G1}
				A_C=0.36	A_C=0.64	
250	4.23×10^{-2}	0.779	18.02	27.98	49.75	59237
350	5.02×10^{-2}	0.783	21.45	33.30	59.21	70328
450	5.72×10^{-2}	0.786	24.41	37.90	67.39	80085

Table 2: Summary of the experimental conditions of liquid phase.

h_L mmAq	G_L kg/s	j_{L1} m/s	j_{L2} m/s		Re_{L1}
			A_C=0.36	A_C=0.64	
50	1.67×10^{-1}	8.62×10^{-2}	1.35×10^{-1}	2.39×10^{-1}	4485
55	1.76×10^{-1}	9.00×10^{-2}	1.41×10^{-1}	2.50×10^{-1}	4683
60	1.84×10^{-1}	9.47×10^{-2}	1.48×10^{-1}	2.63×10^{-1}	4963
65	1.91×10^{-1}	9.80×10^{-2}	1.53×10^{-1}	2.72×10^{-1}	5136
70	1.99×10^{-1}	1.03×10^{-1}	1.60×10^{-1}	2.85×10^{-1}	5412
75	2.06×10^{-1}	1.06×10^{-1}	1.65×10^{-1}	2.93×10^{-1}	5544
80	2.13×10^{-1}	1.10×10^{-1}	1.72×10^{-1}	3.05×10^{-1}	5840
85	2.19×10^{-1}	1.13×10^{-1}	1.76×10^{-1}	3.13×10^{-1}	5984
90	2.23×10^{-1}	1.17×10^{-1}	1.82×10^{-1}	3.24×10^{-1}	6169
95	2.32×10^{-1}	1.19×10^{-1}	1.86×10^{-1}	3.31×10^{-1}	6302
100	2.38×10^{-1}	1.23×10^{-1}	1.92×10^{-1}	3.42×10^{-1}	6550

camera revealed the flow regime transitions at $z_2/d_2 = 10$, 40 and 80 denoting the normalised streamwise distance from the sudden contraction. Passing frequencies of disturbance waves were measured for ten minutes by visual observations at every measurement point under each condition. The average values of the measured frequencies which were taken for three minutes; one minute at the beginning, at the middle and at the end of the measurements, were used as disturbance wave occurrence frequencies, which will be explained in Section 3.2.

3 RESULTS AND DISCUSSIONS

3.1 Flow regime

The flow regime transitions were examined to elucidate the developing process of the two-phase flow. In a horizontal pipe a number of different flow patterns may be observed depending on the conditions of the flow rate combination of air and water [16]. Fig. 2(a) shows "Stratified flow" in which the two phases flow separately with a relatively smooth interface at low gas and liquid velocities. Fig. 2(b) shows "Wavy flow" in which the interface becomes disturbed by waves traveling in the direction of flow. At high gas flow rates the liquid flows in an annular film on the pipe wall with the gas moving rapidly in a central core. At low liquid flow rates the film is relatively smooth with small ripples as shown in Fig. 2(c). The liquid became entrained into the gas in droplet form, resulting Fig. 2(d). As the liquid rate is increased the surface of the film becomes distorted with the formation of "Huge waves" and "Disturbance waves".

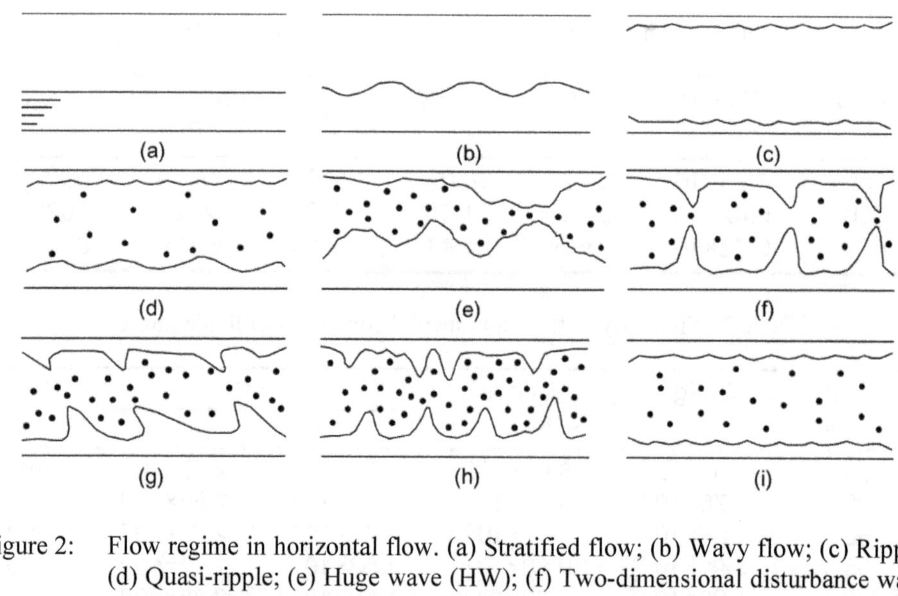

Figure 2: Flow regime in horizontal flow. (a) Stratified flow; (b) Wavy flow; (c) Ripple; (d) Quasi-ripple; (e) Huge wave (HW); (f) Two-dimensional disturbance wave (DW_1); (g) Three-dimensional disturbance wave (DW_2); (h) Quasi-bow wave (DW_3); and (i) Misty wave.

An annular mist flow, in which a liquid flows on the wall of a pipe as a film and a gas flows in the centre [17], was observed after the sudden contraction in the present experimental range, while "Wavy flow" was observed before the sudden contraction. The visual observation allowed a more detailed classification of the annular mist flows. Some photos of observed typical flow regimes are shown in Fig. 3. Four flow regimes were observed; huge-wave (HW), two-dimensional disturbance wave (DW_1), three-dimensional disturbance wave (DW_2) and quasi-bow wave (DW_3) which was a transitional wave to "Misty wave" from DW_2. DW_1 is formed perpendicularly to the pipe axis and coherent [18] around the pipe circumference, while DW_2 is distorted in the streamwise direction, then a part of the ringwise waves begins to be broken. These waves with no coherence around the pipe circumference three-dimensional disturbance waves were termed DW_2, as distinguished from DW_1. At high gas and liquid velocities, DW_3 which was transitional wave to "Misty wave" from DW_2 was observed. DW_3 is a comparable wave to "Bow wave" which has been observed in the literature [18] for two-phase flow through a large diameter vertical pipe.

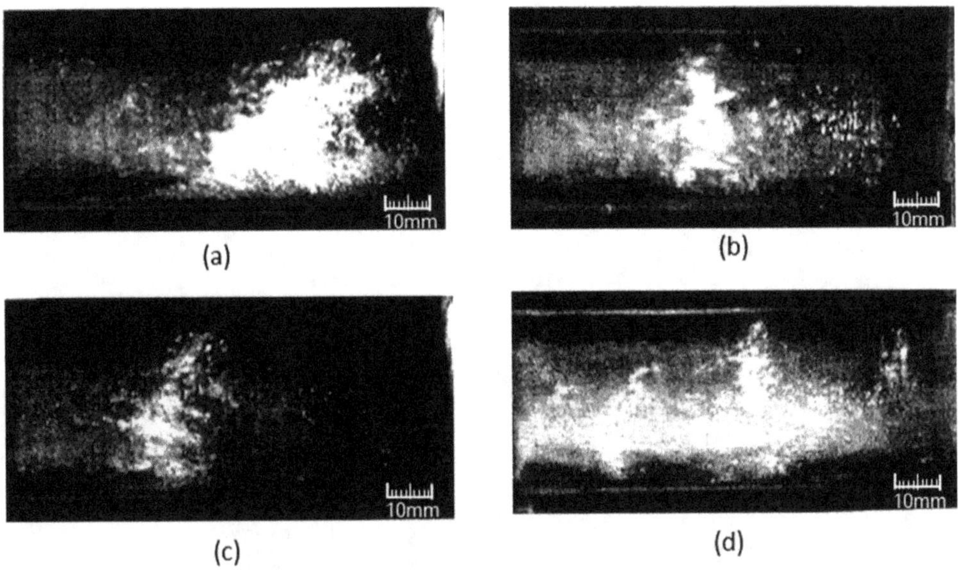

Figure 3: Photos of typical flow regime. (a) HW; (b) DW_1; (c) DW_2; and (d) DW_3.

Figs 4 and 5 show the observed flow regime maps at $z_2/d_2 = 10, 40, 80$ downstream the sudden contraction for the air–water flows in a horizontal pipe with a sudden contraction of $A_C = 0.36$ and 0.64, respectively. The flow pattern transition curves suggested by Mandhane et al. [19] and Taitel et al. [20] are illustrated by broken lines in Figs 4 and 5. The observed flows under the experimental conditions were categorized as annular regimes by them and classified more in detail into four flow regimes, as mentioned above. The flow regime DW_1+DW_2 is observed and subsequently DW_1+DW_2+DW_3, DW_2+DW_3 as the superficial gas and liquid velocities increase in Fig. 4. Moreover, as the normalized axial distance z_2/d_2 increases up to 80, the flow regime shifts from DW_1+DW_2 to DW_1+DW_2+DW_3 at the same j_{G1} and j_{L1}. DW_1 region narrows and alternately DW_3 region widens toward downstream after the sudden contraction. In Fig. 5(a) and (b), as the superficial gas and liquid velocities increase, the flow regime DW_1+DW_2 is observed and subsequently DW_1+DW_2+DW_3, which is the similar evolution occurred in Fig. 4, expect that no flow regime of DW_2+DW_3 is observed in Fig. 5. Fig. 5(c), in contrast, indicates the existence of HW. The flow regime HW+DW_1+ DW_2 is observed at low j_{G1} and high j_{L1}, $z_2/d_2 = 80$ in Fig. 5(c). HW occurrence at lower j_{G1} is observed at lower j_{L1}. When DW_3 occurs, HW disappears.

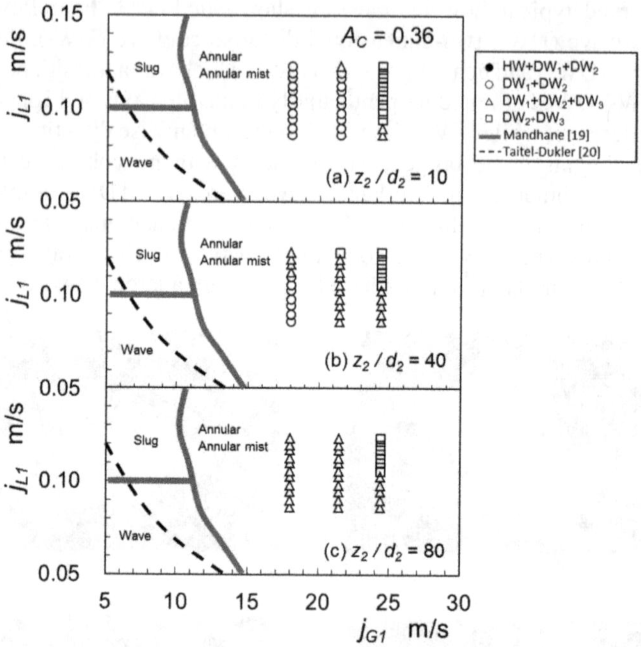

Figure 4: Flow regime map for A_C = 0.36.

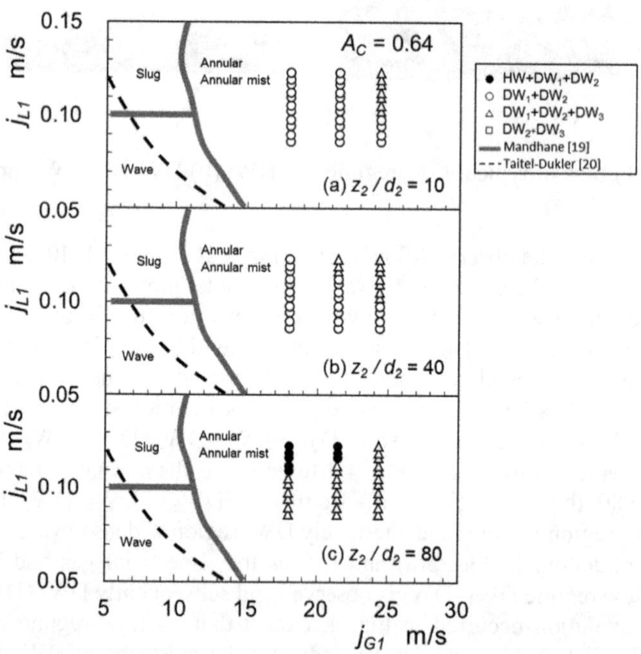

Figure 5: Flow regime map for A_C = 0.64.

3.2 Effect of the superficial gas velocity on mean frequencies for disturbance wave

Fig. 6(a) and (b) show the change in the mean frequency of disturbance wave occurrence with the superficial gas velocity j_{G1} at $j_{L1} = 1.10\times10^{-1}$m/s, $z_2/d_2 = 20$ and 80, respectively. The authors define the wave occurrence frequency as the mean number of waves per second passing an arbitrary point in the pipe. The frequencies of DW_1, DW_2, DW_3 and DW: the sum of these frequencies, are denoted by f_{D1}, f_{D2}, f_{D3} and f_D, respectively.

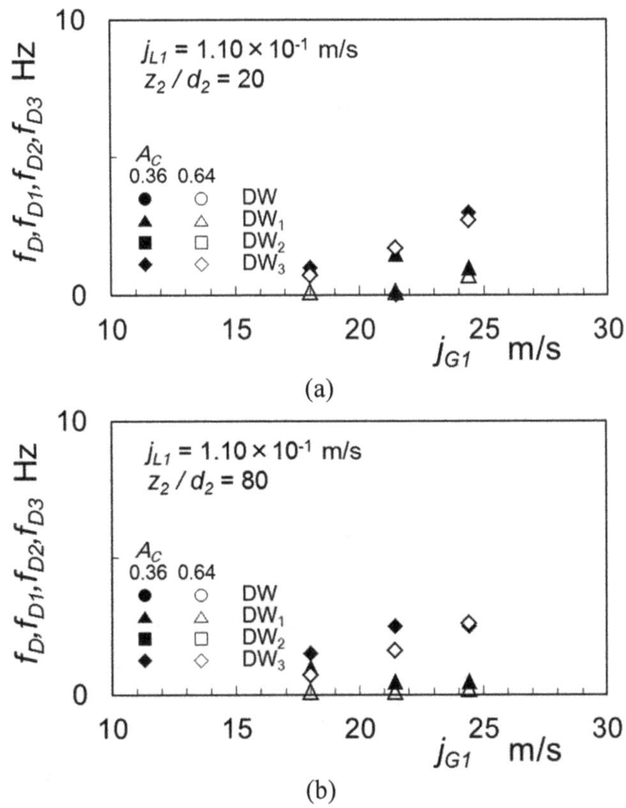

Figure 6: Effect of the superficial gas velocity on mean frequencies for disturbance waves ($j_{L1} = 1.10\times10^{-1}$ m/s). (a) $z_2/d_2 = 20$; and (b) $z_2/d_2 = 80$.

The mean frequency values obtained in flows through the sudden contraction of $A_C = 0.36$ are higher than those in the case of $A_C = 0.64$. DW_2 occurs over wide range of the superficial gas and liquid velocities in the both cases of $A_C = 0.36$ and 0.64. DW_2, DW_3 and DW increase with the superficial gas velocity. DW_1, in contrast, decreases with j_{G1}. DW_3 is observed at high velocity j_{G1}. At $z_2/d_2 = 80$ as shown in Fig. 6(b), the frequencies of DW_2 and DW for $A_C = 0.36$ are nearly constant up to the superficial gas velocity 21.45 m/s and decrease at $j_{G1} = 24.45$ m/s, while the frequency of DW_3 slightly increases. Thus, DW_2 decreases after the appearance of DW_3. On the other hand, the frequency of DW_2 and DW for $A_C = 0.64$ gradually increase. The increasing rate of DW at $z_2/d_2 = 80$ is smaller than that at $z_2/d_2 = 20$. The increasing rate of DW caused by the increasing j_{G1} is moderated with increasing z_2/d_2.

3.3 Effect of the superficial liquid velocity on mean frequencies for disturbance waves

Fig. 7(a) and (b) show the change in the mean frequency of disturbance wave with the superficial liquid velocity j_{L1} at j_{G1} = 21.45 m/s, z_2/d_2 = 20 and 80, respectively.

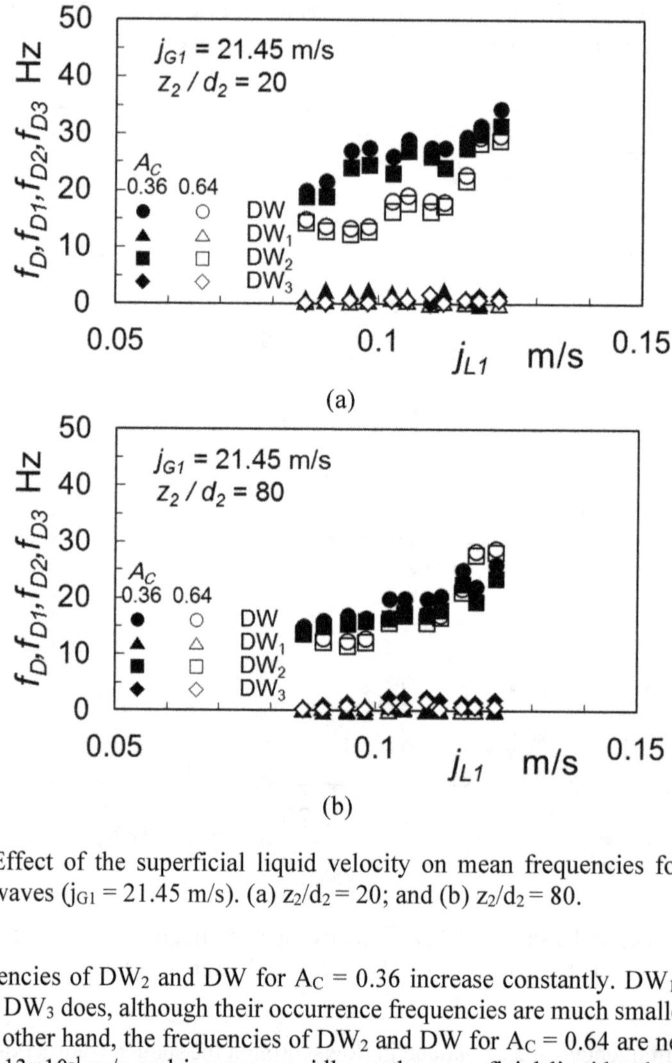

Figure 7: Effect of the superficial liquid velocity on mean frequencies for disturbance waves (j_{G1} = 21.45 m/s). (a) z_2/d_2 = 20; and (b) z_2/d_2 = 80.

The frequencies of DW_2 and DW for A_C = 0.36 increase constantly. DW_1 occurs more frequent than DW_3 does, although their occurrence frequencies are much smaller than that of DW_2. On the other hand, the frequencies of DW_2 and DW for A_C = 0.64 are nearly constant up to j_{L1} = 1.13×10^{-1} m/s and increase rapidly as the superficial liquid velocity increases. DW_1 and DW_3 occur over wide range of the superficial liquid velocity, although the occurrence frequencies of them are much smaller than that of DW_2. Fig. 7(a) clearly indicates effects of the contraction area ratio at low superficial liquid velocity.

Fig. 7(b) indicates that little influence of the sudden contraction area ratios on the mean frequencies of disturbance waves are observed. The frequencies of DW_2 in the both cases of A_C =0.36 and 0.64 increase with the superficial liquid velocity. It follows that the frequency of DW increases constantly. DW_3 for A_C = 0.36 is observed over wide range of the superficial liquid velocity at z_2/d_2 = 80, while DW_1 occurs at z_2/d_2 = 20. The mean frequency distributions become similar as z_2/d_2 increases.

3.4 Mean frequencies for disturbance waves in the normalized axial direction

Fig. 8 shows the mean frequencies for disturbance waves in the normalized axial direction. DW for $A_C = 0.36$ continually decreases, while that for $A_C = 0.64$ decreases up to $z_2/d_2 = 20$ and then stays quasi constant. DW_2, which mainly occurs in the normalized axial direction, decreases in a similar way as DW. DW_1 is much less than DW_2 and disappears at the downstream. DW_3 gradually increases in the downstream. The change tendency of disturbance waves with j_{G2} and j_{L2} are qualitatively the same in the experimental ranges.

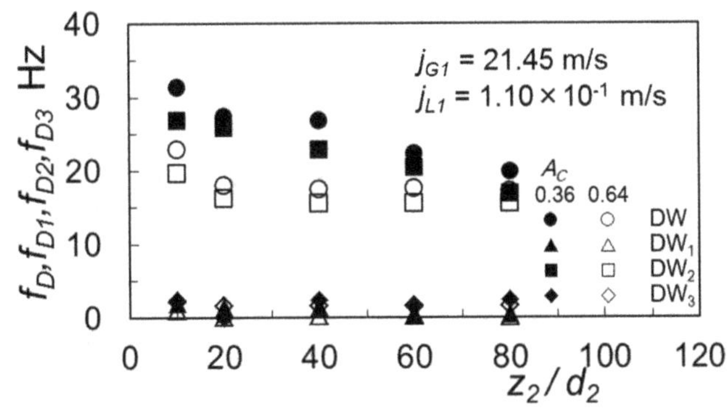

Figure 8: Mean frequencies for disturbance waves in the normalized axial direction.

4 CONCLUSIONS

The effects of the contraction area ratio on film behaviour of two-phase flow in a horizontal pipe were investigated. The following main results were obtained in the experimental conditions.

(1) Four flow regimes were observed; huge-wave, two-dimensional disturbance wave, three-dimensional disturbance wave and quasi-bow wave.
(2) The flow regime transits from $DW_1 + DW_2$ to $DW_1 + DW_2 + DW_3$ as the superficial gas and liquid velocities increase. In the case of $A_C = 0.36$, $DW_2 + DW_3$ is observed and the transition occurs at lower superficial gas and liquid velocities in contrast to the case of $A_C = 0.64$.
(3) The effects of the contraction area ratio become less as z_2/d_2 increases.
(4) DW_2 occurs mainly in the both cases of $A_C = 0.36$ and 0.64. The wave occurrence frequencies in the pipe with the sudden contraction ratio $A_C = 0.36$ were mostly higher than those in the pipe with $A_C = 0.64$.

REFERENCES

[1] Hewitt, G.F., *Measurement of Two Phase Flow Parameters*, Academic Press, pp. 73, 96, 97, 1978.
[2] Helbig, S. & Zarrouk, S.J., Measuring two-phase flow in geothermal pipelines using sharp edge orifice plates. *Geothermics*, **44**, pp. 52–64, 2012.
[3] Araoye, A.A., Badr, H.M. & Ahemd, W.H., Investigation of flow through multi-stage restricting orifices. *Annals of Nuclear Energy*, **104**, pp. 75–90, 2017.

[4] Roul, M.K. & Dash, S.K., Single-phase and two-phase flow through thin and thick orifices in horizontal pipes. *Journal of Fluid Engineering,* **134**(9), pp. 091301–091314, 2012.

[5] Yagi, Y., Murase, M., Sato, K. & Hattori, S., Mechanism of cavitation erosion at the exit of a long orifice. *Proceedings of ASME/JSME Joint Fluids Engineering*, FEDSM2003-45004, pp. 1359–1364, 2003.

[6] Chang, S.W., Chiang, K.F. & Lin, C.Y., Channel aspect ratio effect on thermal performance of air-water slug flow through U-bend channels. *International Journal of Thermal Sciences*,**76,** pp. 11–29, 2014.

[7] Saidj, F., Kibboua, R., Azzi, A., Ababou, N. & Azzopardi, B.J., Experimental investigation of air-water two-phase flow through vertical 90° bend. *Experimental Thermal and Fluid Science*, **57**, pp. 226–234, 2014.

[8] Hsu, L.C., Chen, I.Y., Chyu, C.M. & Wang, C.C., Two-phase pressure drops and flow pattern observations in 90° bends subject to upward, downward and horizontal arrangements. *Experiments Thermal and Fluid Science*, **68**, pp. 484−492, 2015.

[9] Whalley, P.B., Air-water two-phase flow in a helically coiled tube. *International Journal of Multiphase Flow*, **6**(4), pp. 345–356, 1980.

[10] Azzopardi, B.J. & Whalley, P.B., The effect of flow patterns on two-phase flow in a T junction. *International Journal of Multiphase Flow*, **8**(5), pp. 491–507, 1982.

[11] Saisorn, S. & Wongwises, S., An experimental investigation of two-phase air-water flow through a horizontal circular micro-channel. *Experimental Thermal and Fluid Science*, **33**, pp. 306–315, 2009.

[12] Schmidt, J. & Friedel, L., Two-phase pressure drop across sudden contractions in duct area. *International Journal of Multiphase Flow*, **23**(2), pp. 283−229, 1997.

[13] Chen, I.Y., Tseng, C.Y., Lin, Y.T. & Wang, C.C., Two-phase flow pressure change subject to sudden contraction in small rectangular channels. *International Journal of Multiphase Flow*, **35**, pp. 297−306, 2009.

[14] Fujimatsu, T., Kito, M. & Kondo, K., Film behavior of two-phase flow in a horizontal pipe with a sudden contraction. *Proceedings of the 5th International Conference on Jets, Wakes and Separated Flows (ICJWSF2015), Springer Proceedings in Physics*, **185**, pp. 493−500, 2016.

[15] Fujimatsu, T., Kito, M. & Kondo, K., Gas-liquid two phase flow in a horizontal pipe with a sudden contraction. *Abstract Book: International Conference on Advanced Technology of Experimental Mechanics (ATEM)*, p. 267, 2015.

[16] Govier, G.W. & Aziz, K., *The Flow of Complex Mixtures in Pipes*, Van Nostrand Reinhold Company: New York and London, pp. 503−589, 1972.

[17] Butterworth, D. & Hewitt, G.F. eds, *Two-phase Flow and Heat Transfer*, Oxford University Press, p. 20, 1977.

[18] Azzopardi, B.J. & Gibbons, D.S., Annular two phase flow in a large diameter tube. *The Chemical Engineer*, **398**, pp. 19−31, 1983.

[19] Mandhane, J.M., Gregory, G.A. & Aziz, K., A flow pattern map for gas-liquid flow in horizontal pipes. *International Journal of Multiphase Flow*, **1**(4), pp. 537–553, 1974.

[20] Taitel, Y., Lee, N. & Dukler, A.E., Transient gas-liquid flow in horizontal pipes; modelling the flow pattern transition. *AIChE J.*, **24**(5), pp. 921–934, 1978.

HIGH TEMPERATURE ALUMINIUM PLATE COOLING BY WATER IMPINGEMENT: NUMERICAL SIMULATION WITH ADAPTED CFD WATER COOLING MODELS AND EXPERIMENTAL VALIDATION

ROK KOPUN[1], STEPHAN UCSNIK[2], DONGSHENG ZHANG[3], PETER SAMPL[3], STEPHAN JÄGER[2] & REINHARD TATSCHL[3]
[1]AVL-AST d.o.o., Slovenia
[2]LKR Leichtmetallkompetenzzentrum Ranshofen GmbH, AIT Austrian Institute of Technology, Austria
[3]AVL List GmbH, Austria

ABSTRACT
This paper outlines the recently improved Computational Fluid Dynamics (CFD) model to simulate the jet impingement quenching cooling process, validated with experimental data. The main application area of the presented method is heat treatment of cast aluminium alloy parts, known as continuous horizontal chill casting process, mostly used in the automotive and aerospace industries, where accurate heat treatment prediction plays an important role in conceptual and thermal analysis. The comparison between measurement data and numerical results has been carried out to simulate the real time jet impingement cooling process of a high temperature aluminium plate using the commercial CFD code AVL FIRE™. Recently, simulation methodology has been applied to cast cooling processes, successfully demonstrating its ability to simulate cooling effects from 500°C down to room temperature in a realistic manner. The Eulerian multi-fluid modelling approach is used to handle the boiling flow and the heat transfer between the heated structure and the sub-cooled liquid. While for the fluid region governing equations are solved for each phase separately, only the energy equation is solved in the solid region. Heat transfer coefficients depend on the boiling regimes which are separated by the variable Leidenfrost temperature and the levitation force. Simulation results are compared with available measurement data on different positions along the aluminium plate. The temperature histories predicted by the presented model correlate very well with the provided measurement data at different monitoring locations.
Keywords: multiphase, CFD, continuous casting process, Leidenfrost temperature, levitation force.

1 INTRODUCTION

Optimization of heat transfer characteristics in automotive, aeronautic and process industries plays an important role in conceptual and thermal analysis. Hence, an efficient heat treatment technique, such as casting, are introduced in order to prevent component failure and to provide long-lasting products. Aluminium alloys are well established materials in those industry branches, therefore accurate heat treatment is essential.

Casting of Aluminium alloys for different semi-finishes for metal sheets is mostly conducted by means of the direct vertical semi-continuous or the continuous horizontal chill casting process. In this process the melt faces direct contact with the billet shaping casting mould. Primary cooling is responsible for the initial solidification of a thin outer shell of the semi-finished ingot. The subsequent energy deduction and ingot cooling happens through a secondary cooling with a time-dependent pulsed/continuous water impingement onto the solidifying outer shell. Material and mechanical properties of such produced alloys strongly depend on the casting process parameters and on the directly linked best suitable cooling conditions. These in turn have a strong influence on the micro-structure of the fast solidifying outer shell as well as of the internal ingot core.

www.witpress.com, ISSN 1743-3533 (on-line)
doi:10.2495/MPF190071

This paper presents the development, implementation and validation of enhanced computational fluid dynamics (CFD) multiphase models for numerical modelling and simulation of the cooling process of high temperature Aluminum plates by means of direct water impingement. The paper is split into three major sections. Section 2 presents the research and development work related to the CFD modelling of water jet impingement and heat transfer on high temperature Aluminum plates. Section 3 describes the experimental data record of the transient temperature distributions of water cooled high temperature Aluminum plates, and Section 4 shows the comparison of numerical with experimental results for validation of the newly developed CFD cast cooling model

2 CFD MODELING OF MULTIPHASE PLATE COOLING

The CFD code AVL FIRE™ solves the general conservation equations of mass, momentum and enthalpy plus additional transport equations for turbulence related quantities. The multiphase treatment is based on the Eulerian multi-fluid method that is the most general approach for simulating multiphase flows. The individual fluids (liquid/gaseous phases) are treated as continuous phases with conservation laws applied for each fluid. An ensemble averaging technique is adopted to remove the microscopic interfaces. This leads to macroscopic conservation equations which are analogous to their single-phase counterparts but differ in that a new variable volume fraction and additional exchange terms between the phases are introduced [1].

2.1 General multiphase CFD framework AVL FIRE™

The solution method adopted in AVL FIRE™ is based on a fully conservative finite volume approach. In order to offer full flexibility in terms of the structure and topology of the employed computational meshes, AVL FIRE™ allows for each computational cell to consist of an arbitrary number of cell faces.

All dependent variables for momentum, pressure, density, turbulent kinetic energy, dissipation rate, and scalar quantities are evaluated at the centres of the computational cells. A second-order midpoint rule is used for integral approximation and a second order linear approximation for calculation of any value at the cell-faces. A diffusion term is incorporated into the surface integral source after employment of the special interpolation practice. The rate of change is discretized by using implicit schemes, namely an Euler implicit scheme and a three-time level implicit scheme of second order accuracy.

Various initial and boundary conditions are available to match the setup of any simulation to the real flow problem. To meet the requirements of different application fields, AVL FIRE™ simulations may be set up in steady or transient (time stepping) mode. Due to the wide range of applicability of the multi-fluid method, the phase-exchange terms are modelled according to the different types of simulation problems, such as e.g. quenching, cavitation, flash boiling, etc.

2.2 Multiphase water impingement flow and heat transfer modeling

The Eulerian multi-fluid model considers each phase as interpenetrating continuum coexisting in the flow domain, with inter-phase transfer terms accounting for phase interactions. From the theoretical work of Drew and Passman [1], the ensemble averaged continuity, momentum and energy equations are presented as:

2.2.1 Continuity equation

$$\frac{\partial(\alpha\rho)_k}{\partial t} + \nabla \cdot \left(\alpha\rho\vec{V}\right)_k = \Gamma_{l\rightarrow k}, \tag{1}$$

subject to the compatibility condition

$$\alpha_l + \alpha_v = 1, \tag{2}$$

where α, ρ and $\vec{V}$ stand for phase volume fraction, density and velocity vector, respectively. The phase mass change rate (in this particular case, boiling) is Γ_l and subscript k is the phase index ($k = l$ or $k = v$). Subscripts v and l denote the vapour and liquid phases in the current work.

2.2.2 Momentum equation

$$\frac{\partial(\alpha\rho\vec{V})_k}{\partial t} + \nabla \cdot \left(\alpha\rho\vec{V}\vec{V}\right)_k = -\alpha_k \nabla p + \nabla \cdot (\alpha\tau)_k + (\alpha\rho)_k\vec{g} + M_{lk} + \vec{V}_i\Gamma_{l\rightarrow k}, \tag{3}$$

with p, $\vec{g}$, τ and $\vec{V}_i$ representing pressure, gravity vector, stress tensor and interfacial velocity, respectively. The interfacial momentum exchange ($\vec{M}_{lk}$) term is given by

$$\vec{M}_{lk} = \vec{F}_D + \vec{F}_{LEVI}, \tag{4}$$

where F_D indicates the drag force and $\vec{F}_{LEVI}$ is the levitation force.

2.2.3 Drag force
The drag force exerted on the continuous phase is given as

$$\vec{F}_D = \frac{3}{4}\alpha_d\rho_c\frac{C_D}{d_b}\left|\vec{V}_d - \vec{V}_c\right|\left(\vec{V}_d - \vec{V}_c\right), \tag{5}$$

where C_D is the drag coefficient depending on the droplet Reynolds number, and d_p stands for the water droplet diameter.

2.2.4 Levitation force
At the quench front of the water jet impingement cooling process considered in the present work, there is a narrow zone of vigorous boiling with massive production of finer droplets. These finer droplets are ejected almost perpendicular from the hot surface at the quench front, causing the liquid to break up into large irregular droplets. To accurately simulate free-surface liquid jet impingement quenching, it is of great importance to account for the repulsion of water droplets and/or film away from the hot surface and consequently, a force $\vec{F}_{LEVI}$, named as levitation force, is exerted on the liquid phase along the quench front. The model proposed by Kim [2] is adopted in this work, with

$$\vec{F}_{LEVI} = -0.39C_{LEVI}\frac{2\mu_v}{\chi\rho_v}\dot{m}\vec{n}_w, \tag{6}$$

where C_{LEVI} stands for the levitation coefficient, μ_v is the vapour phase viscosity, $\dot{m}$ is the mass change rate due to boiling and χ is the vapour film thickness, which typically in the range of 10–100 μm. It should be noticed that this kind of levitation force, eqn (6), is also found in the work of Erickson and Williams [3]. The vapour layer causes lifetime extension of Leidenfrost drops, and according to Bianche et al. [4], it is given by

$$\chi = \begin{cases} \left(0.75\frac{k_v\mu_v\Delta T}{gL\rho_v\rho_l Ca}\right)^{\frac{1}{4}}\sqrt{d_p} & d_p > Ca, \\ \left(\frac{k_v\mu_v\Delta T}{L\rho_v\sigma^2}g\rho_l d_p\right)^{\frac{1}{3}} d_p & d_p \le Ca. \end{cases} \quad (7)$$

Here Ca is the capillary length, given as

$$Ca = \sqrt{\frac{\sigma}{g\rho_l}}. \quad (8)$$

Fig. 1 shows the graphical implementation of the $\vec{F}_{LEVI}$, where left side of the figure represent the case without the force and where the right side shows the impact of the force (for general and direct casting scenario).

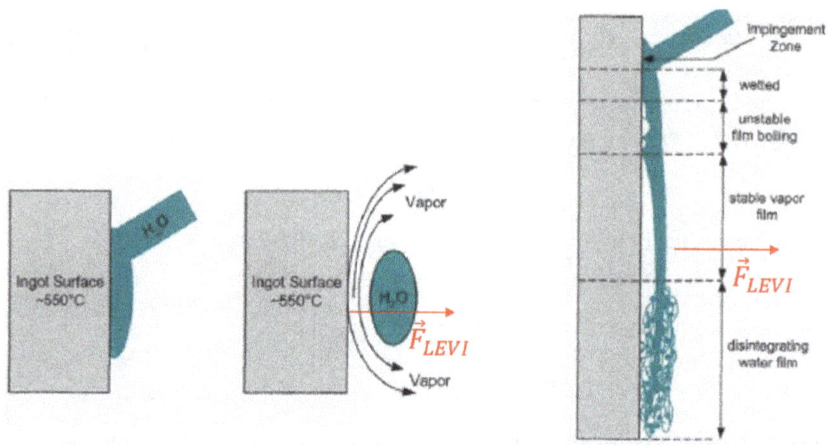

Figure 1: Implementation of $\vec{F}_{LEVI}$ force on general case and direct casting process [5].

2.2.5 Energy equation

In previous work [6], [7], vapour bubble condensation in the fluid domain is assumed to be negligible, the phasic static enthalpy equation is then described by

$$\frac{\partial(\alpha\rho h)_k}{\partial t} + \nabla\cdot\left(\alpha\rho\vec{V}h\right)_k = \nabla\cdot\left(\frac{\alpha\lambda_T}{C_p}\nabla h\right)_k + \frac{3}{4}\alpha_v\rho_l\frac{C_H}{d_b}(T_l - T_k) + \alpha_k\frac{\partial p}{\partial t} + S_h. \quad (9)$$

Thermal conductivity is presented as λ_T, C_H is the inter-phase heat transfer coefficient between phase k and phase l and other terms are lumped into the source term S_h

$$S_h = \nabla\cdot\left[\frac{\alpha\lambda_T}{C_p}(T - T_{ref})\nabla C_p\right]_k + \left(\alpha\vec{V}\right)_k\cdot\nabla p + \left(\alpha\tau:\nabla\vec{V}\right)_k. \quad (10)$$

2.2.6 Boiling model

Based on the assumption that the boiling heat transfer rate is proportional to the phase change rate, the mass transfer predominantly controls the heat transfer. Thus, the phase change rate due to boiling can be written as

$$\Gamma_c = \frac{C_m C_{boil} h_{boil}(T_w - T_{sat})}{H_{fg}}\frac{6\alpha_v}{d_b}, \quad (11)$$

where C_m, C_{boil}, h_{boil} and H_{fg} are the closure coefficient, the boiling correction coefficient, the boiling heat transfer coefficient and the latent heat of vaporization, respectively. The following sub-sections describe the models applied for heat transfer coefficient h_{boil} at the different boiling regimes.

2.2.7 Film boiling model

Bromley's model [8], originally applied for a horizontal tube, is employed to predict the film boiling heat transfer coefficient h_{boil}

$$h_{boil} = 0.62\left\{\frac{\lambda_v^3\rho_v(\rho_l-\rho_v)g[H_{fg}+0.4C_{p,v}(T_w-T_{sat})]}{\mu_v(T_w-T_{sat})d_b}\right\}^{0.25}, \tag{12}$$

where d_b is the length scale (here the vapour bubble diameter), λ_v is the vapour thermal conductivity and $C_{p,v}$ stands for the vapour specific heat.

2.2.8 Transition boiling mode

The heat transfer coefficient, h_{boil} in the transition boiling regime is given by

$$h_{boil} = \frac{q_{TB}}{T_w-T_{sat}}, \tag{13}$$

where the heat flux in transition boiling regime qTB is computed as follows

$$q_{TB} = \phi q_{CHF} + (1-\phi)\, q_{MHF}. \tag{14}$$

In eqn (14), q_{CHF} denotes the Critical Heat Flux, and q_{MHF} the Minimum Heat Flux. In a relation to the Zuber's correlation [9], the critical heat flux is computed as

$$q_{CHF} = 0.131\rho_v H_{fg}\left[\frac{\sigma g(\rho_l-\rho_v)g}{\rho_v^2}\right]^{0.25}, \tag{15}$$

where σ is the surface tension of the liquid phase. The Minimum Heat Flux correlation for the quenching system in eqn (14) is

$$q_{MHF} = 0.09K_{MHF}\rho_v H_{fg}\left[\frac{g(\rho_l-\rho_v)}{\rho_l+\rho_v}\right]^{0.5}\left[\frac{\sigma}{g(\rho_l-\rho_v)}\right]^{0.25}. \tag{16}$$

K_{MHF} serves as a model parameter and its typical value varies from 1 to 15. Further, more detailed information about the applied quenching model can be obtained from AVL FIRE™ multi-fluid model solver theory guide [10] and the reference [11].

2.2.9 Quench front temperature

During free-surface liquid jet impingement quenching, the quench front propagates in the direction of flow along the surface. Heat transfer occurs by boiling (nucleate, transition or film boiling) and/or single-phase forced convection upstream of the quenching front, while conduction and radiation heat transfer take place at dry zones of the surface downstream of the front. A transition between the boiling and conduction-radiation zones takes place when the surface temperature first drops below the quench front temperature. Quench front temperature is defined as the highest solid surface temperature when film boiling is taking place on the heated surface. Here, the correlation presented by Brenson [12] is employed to evaluate the quench front temperature. It is given by

$$T_{QFT} = T_{MFB,B} + \left(T_{MFB,B} - T_l\right)\sqrt{\frac{\rho_l C_{p,l}\lambda_l}{\rho_s C_{p,s}\lambda_s}}. \tag{17}$$

Based on Taylor instability theory, the minimum film boiling temperature $T_{MFB,B}$ reads

$$T_{MFB,B} = T_{sat} + 0.127 \frac{\rho_v h_{fg}}{\lambda_v} \left(\frac{g(\rho_l - \rho_v)}{\rho_l + \rho_v} \right)^{2/3} \left(\frac{\sigma}{(\rho_l - \rho_v)} \right)^{1/2} \left(\frac{\mu_v}{g(\rho_l - \rho_v)} \right)^{1/3} . \tag{18}$$

3 PLATE COOLING SET-UP AND RESULTS

The generation of temperature data during water jet impingement cooling was conducted using a special sensor-equipped test setup [13]. This test setup allows for the horizontal heating of metal coupon plates, the quick transfer from the oven to the impingement cooling to subsequently run the water cooling tests at defined water cooling parameters. Fig. 2(a)–(c) show the setup of the experimental test device.

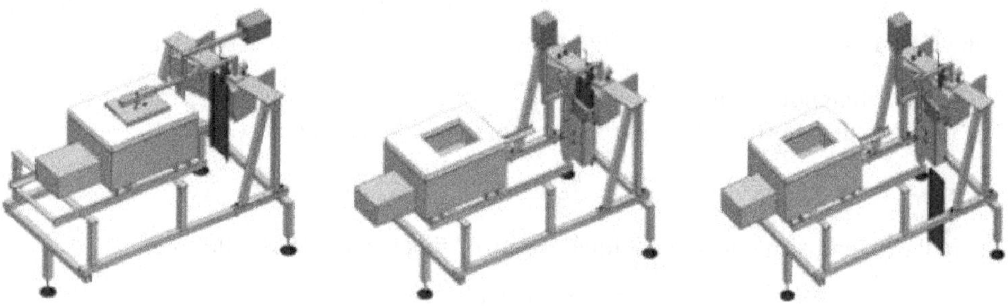

(a) (b) (c)

Figure 2: Experimental setup of the cooling device used for plate cooling tests. (a) Phase of coupon plate heating; (b) Phase of transfer to cooling position; and (c) Removal of shielding plate (brown) [13].

The setup is capable of varying the plate temperature in the oven as well as changing water cooling parameters such as mass flow, flow shape (pulsed, continuous) etc. at the cooling side. For the present investigations the cooling setup was further modified so to also vary the impingement angle and record its influence on the temperature graphs.

For the reference investigation, AA7075 Aluminium billets were vertical chill casted. Out of these billets, five rectangular coupon plates were manufactured (Fig. 3(a)). Dimensions of the coupon plates were 20 x 100 x 200 mm. The plates were equipped with 10 thermo-couples along the centre line, 3.0 mm under the impinged surface of the water-cooled plate (Fig. 3(b)). Thermo-couples T1–T2 were above the impingement zone IZ. T3 was at the IZ and remaining T4–T10 were positioned below the IZ. The coupon plates were heated up to 475°C and then transferred to the water cooling. For water cooling, room temperature water was used, the mass flow was continuous between 6–14 l/min and the splashing angle was between 11–23° down from the vertical. During the cooling the specimen faced no movement or transfer. The coupon plate remains vertically arranged in position. Only the cooling water rinses down the cooled surface.

The main function of the whole equipment is to guarantee the reproducible heating and cooling condition for the tests. This will secure the data (temperature) record which will be used for later comparison with numerical simulation results.

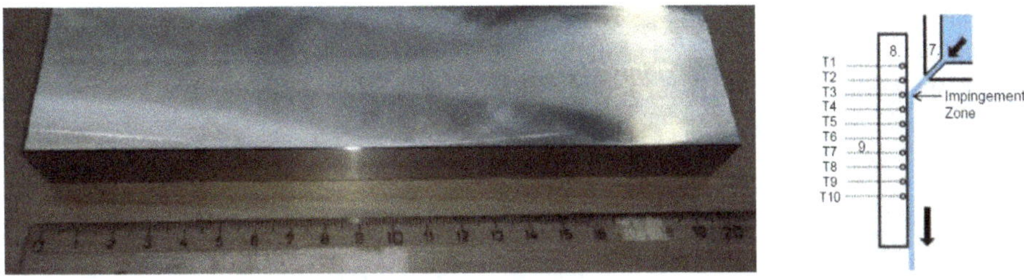

(a) (b)

Figure 3: (a) AA7075 coupon plate as cast with milled surfaces; and (b) Water cooling schema [12].

3.1 Experimental results

Fig. 4 shows the temperature results of test scenario A01 for thermo-couples T1–T10 over time. The graphs show an initial maximum temperature plateau at 475°C which is followed by the temperature decrease due to water impingement cooling. Thermo-couples T1 and T2 show a very smooth graph with the lowest temperature gradients for cooling as these points of the plate do not face direct contact with the cooling water. T3 (green) is directly at the impingement zone and faces the highest temperature gradient during water impingement in the initial face and goes down to 250°C within 5 seconds.

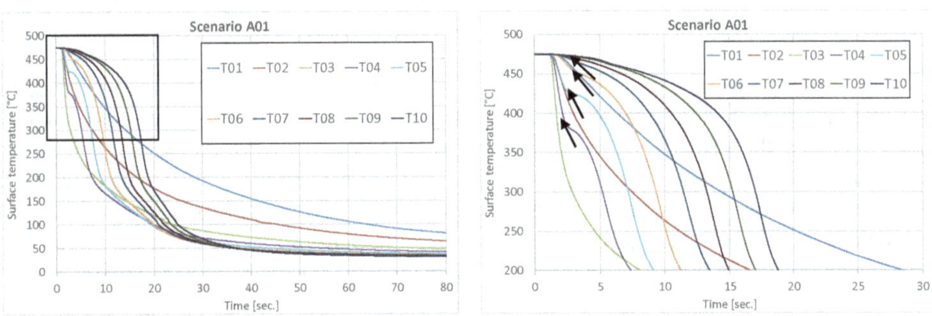

(a) (b)

Figure 4: (a) Temperature results of thermo-couples T1–T10 over time for cooling test scenario A01; and (b) Detail of T-t graphs for time interval 0–30 seconds with indicators for rewarming due to Leidenfrost isolation effect.

The initial cooling gradients of the remaining thermo-couples T5–T10 are smaller compared to T4. In addition, it is visible that these thermo-couples record a temporary delay of the cooling. These are indicated in Fig. 4(b). The delay of the cooling is due to the Leidenfrost effect. This builds a temporary isolation layer between cooling water and the hot coupon plate which rehashes the surface due to internal billet core energy conduction.

4 NUMERICAL MODELLING OF IMPINGEMENT WATER COOLING PROCESS

Fig. 5(a) displays the physical domain and the boundary conditions applied in the AVL FIRE™ CFD simulation. In the current study, one initial temperature and coolant mass flow rate for jet impingement quenching is investigated. For model simplification and faster simulation time, wall symmetry conditions are adopted. The total number of cells for the cooling plate piece (Fig. 5(b)) is about 350,000, with the solid domain consisting of approximately 25,000 cells.

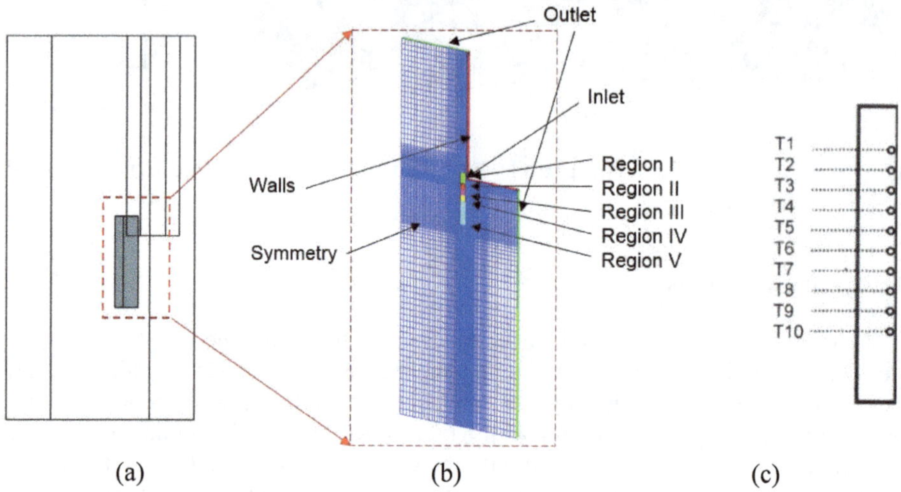

Figure 5: Physical domain. (a) Boundary conditions; (b) Temperature monitoring points; and (c) For investigated water impingement plate cooling configuration.

With reference to the liquid domain, prescribed atmospheric pressure is exerted at the outlet boundary on the top (green surface), and the inlet section is the given velocity boundary condition from the drilled hole at the side (black surface). In the current simulations, a constant impingement velocity of 14.14 m/s is used until the temperature of the quench piece reaches 23°C. The domain walls at the back and front are treated as adiabatic, whereas symmetry is applied on the sides. Temperature measurements along the presented piece were performed at different positions, and they are referred as T1, T2, T3, T4, T5, T6, T7, T8, T9, and T10, as shown in Fig. 5(c)).

The impingement area is divided into 5 regions according to the different cooling regimes, as proposed by Kim [2] (Fig. 5(b)). Region I is the region above the impingement zone. Region II and Region III are the regions where liquid hits the heated surface, whereas the Region IV and Region V are covering the parts bellow. Each region has its own quenching cooling parameters such as Leidenfrost Limit, Leidenfrost Variable Range, Transient Minimum Heat Flux Factor, Dry Phase Packing Limit and Transient Critical Heat Flux Factor. Details regarding water cooling parameters can be extracted from the AVL FIRE™ solver theory guide [10].

4.1 Numerical results

The liquid volume fraction and temperature distribution within the structure at different time instants are displayed in the Fig. 6. It can be seen that due to upcoming inlet velocity and newly introduced levitation force, coolant is pushed away from the heated plate, see Fig. 6(a), (b).

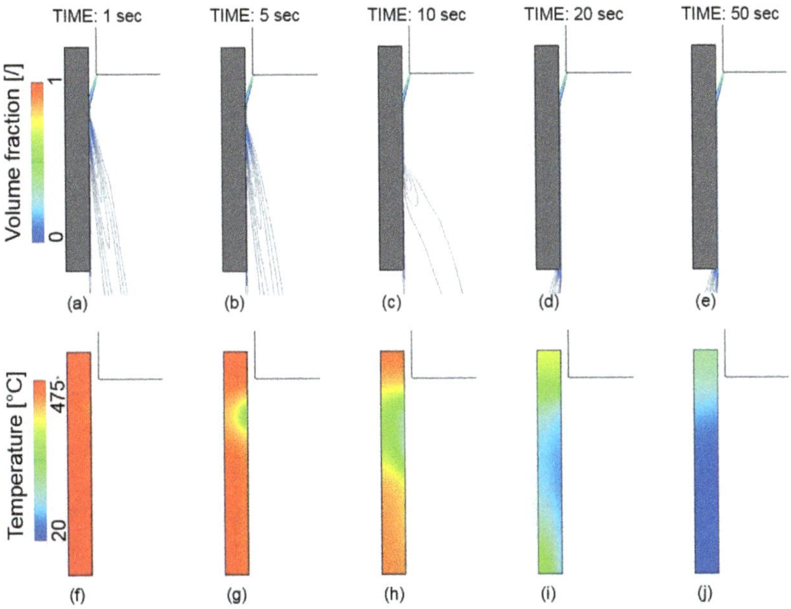

Figure 6: Water jet cooling of 475°C plate at different time steps; water volume fraction, resulting temperature distribution in the plate.

The solid plate starts to cool down short after the coolant hits the plate. At the time of 5 s, the surface temperature at the impingement zone is already around 360°C, whereas the other part of the solid plate still shares the temperature close to initial value, Fig. 6(g). As soon as the surface temperature drops below the Leidenfrost temperature, transition boiling and later nucleate boiling take place. This effect is visible in Fig. 6(d), (e), where due to lower solid temperature, levitation force is no longer active and that the coolant is no longer pushed away from the heated surface. This results in a faster cooling process. At the time of 20 s, it is observed in Fig. 6(i), that the impingement zone is already cooled down to ambient temperature, whereas the upper and lower parts of the solid still share higher temperature. A uniform cooling temperature is reached at the time of 50 s (see Fig. 6(j)). The newly implemented Levitation force is well visible on the Fig. 6(a)–(c), where some amount of water stays on the surface and cool down the solid plate, whereas the rest of the water is re-bounced from the heated surface, due to newly implementation. When the temperature falls below the Leidenfrost temperature, levitation force no longer visible, and all water which is entering the system stayed on the cooling plate (no re-bouncing effect is visible anymore – see Fig. 6(d), (e)).

4.2 Cooling model validation

The model capability to predict quench rates is validated through comparison of measured and numerically predicted temperature histories (see Fig. 7). From all monitoring points the simulated film and transition boiling regime are well described with the adopted jet impingement quenching model, where very good agreement with the available measurement data is achieved.

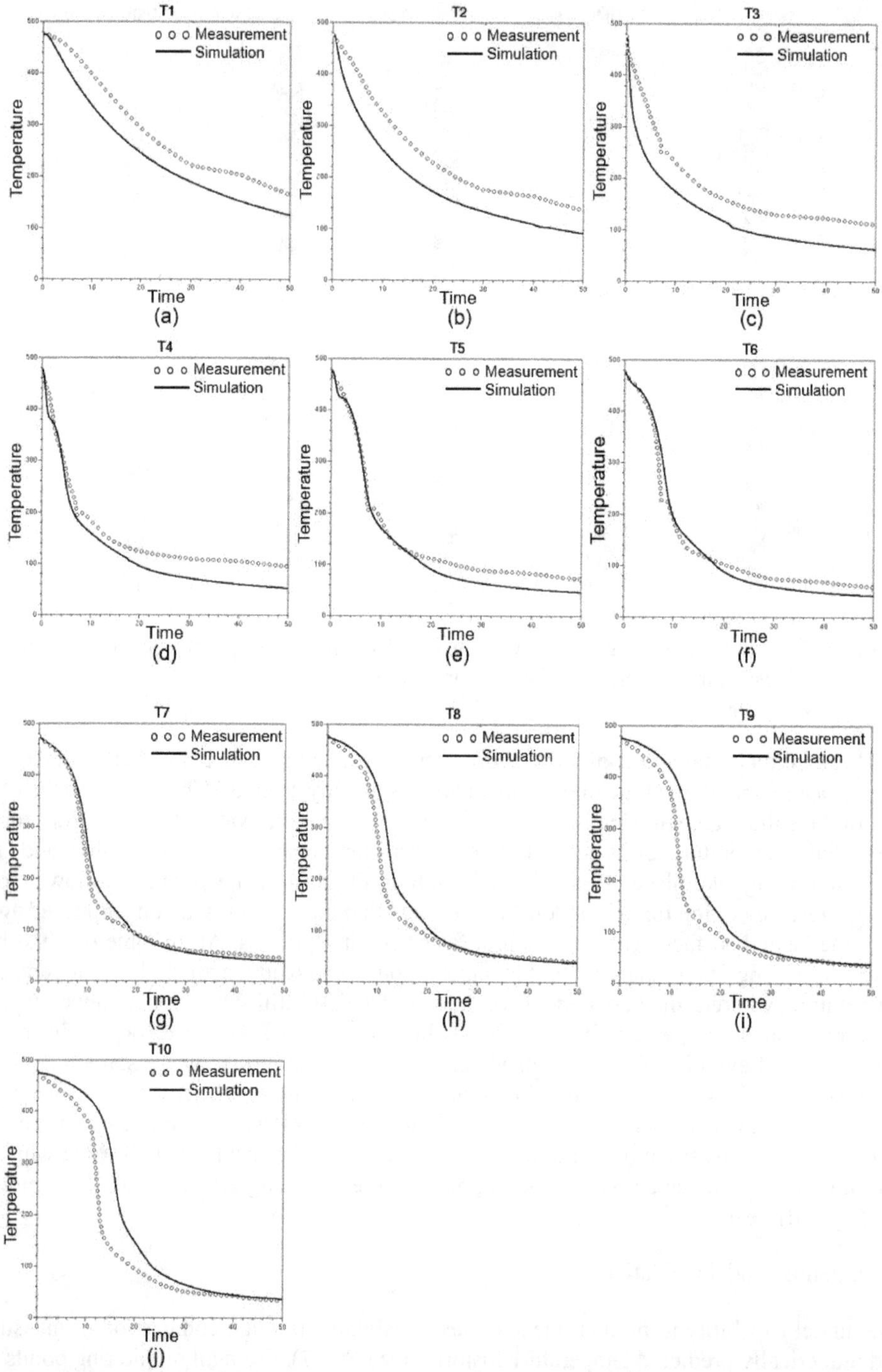

Figure 7: Comparison of numerically predicted and measured solid part temperature for jet impingement quenching model with water temperature of 20°C at monitoring points T1, T2, T3, T4, T5, T6, T7, T8, T9 and T10

5 SUMMARY

For improvement of the prediction of transient temperature distribution evolution during the vertical chill casting process of Aluminium alloys in the initial starting phase, it is essential to adopt highly accurate models for describing the heat flux away from cast billets, especially in the top hot solidifying outer billet shell at the phase of water impingement.

This paper presented the development, implementation and validation of an enhanced CFD multiphase heat transfer model in AVL FIRE™ for numerical modelling and simulation of the cooling process of high temperature Aluminium plates by means of direct water impingement. For this purpose, the AVL FIRE™ multiphase model was adapted and extended in order to capture the specific phenomena characteristic for high temperature Aluminum plates hit by cold water (e.g. liquid phase levitation, appearance of different boiling regimes, Leidenfrost effect, etc.). The resulting model was applied to a real test case scenario of water cooled coupon plates made of AA7075. The comparison of the numerical simulation results with recorded temperature data at selected thermo-couple positions shows good agreement between the numerical model and the test data especially for the points at and below the impingement zone. With the current implementation user has to define the impingement region out of existing model, but in future we would like to establish automatic approach, where the solver would recognize those regions, and automatically divide this into five regions and set all the needed data for simulation.

The newly developed model in AVL FIRE™ sets the basis for future research work so to exchange the generated transient heat flux with numerical finite element models of the continuous vertical chill casting process and to improve the temperature distribution especially in the starting phase of the process.

ACKNOWLEDGEMENTS

The authors would like to thank the Austrian Research Promotion Agency (FFG), the Federal Ministry for Transport, Innovation and Technology (BMVIT) and the State of Upper Austria for sponsoring this research work in the framework of COMET project Amoree. The authors would also like to acknowledge the support of the AMAG AG and of the Heat Transfer and Fluid Flow Laboratory, Brno University of Technology for conducting the experimental plate cooling tests.

REFERENCES

[1] Drew, D.A. & Passman, S.L., *Theory of Multi-Component Fluids*, Springer: New York, 1998.

[2] Kim, H., Floating phenomenon of a water drop on the surface of liquid nitrogen. *J. Korean Physical Society*, **49**, pp. 1335–1338, 2006.

[3] Erickson, D.W. & Williams, G.A., Experimental studies of the cryogenic Leidenfrost effect. http://reu.physics.ucla.edu/common/papers/2005/david_erickson.pdf. Accessed on: 14 Apr. 2019.

[4] Bianche, A.-L., Clanet, C. & Quéré, D., Leidenfrost drops. *Phys. Fluids*, **15**(6), pp. 1632–1637, 2003.

[5] Meenken, T., MSc thesis, Université du Québec à Chicoutimi, Canada, 2002.

[6] Kopun, R., Škerget, L., Hriberšek, M., Zhang, D.S., Stauder, B. & Greif, D., Numerical simulation of immersion quenching processes for cast aluminum part at different pool temperatures. *Appl. Therm. Eng.*, **65**, pp. 74–84, 2014.

[7] Srinivasan, V., Moon, K., Greif, D., Wang, D.M, & Kim, M., Numerical simulation of immersion quench cooling process using an Eulerian multi-fluid approach. *Appl. Therm. Eng.*, **30**, pp. 499–509, 2010.

[8] Bromley, L.A., Heat transfer in stable film boiling. *Chem. Eng. Progr.*, **46**, pp. 221–227, 1950.
[9] Zuber, N., On the stability of boiling heat transfer. *Trans. ASME*, **80**, 1958.
[10] AVL LIST GmbH, FIRE CFD Solver, Eulerian multi-fluid model. Solver Theory Guide, Graz, Austria, 2018.
[11] Kopun, R. & Škerget, L., Experimental and numerical investigation of boiling flow in a vertical pipe with phase change using multi-fluid modelling approach. *9th International Conference on Advances in Fluid Mechanics*, 2012.
[12] Brenson, P., Film boiling heat transfer from a horizontal surface. *ASME J. of Heat Transfer*, **83**, pp. 351–358, 1961.
[13] Guzej, M., Bartuli, E. & Kristof, O., Water jet cooling of Aluminum alloy. *Proceedings of Metal,* Brno, Czech Republic, 2015.

SECTION 2
NUMERICAL METHODS

MASS TRANSFER THROUGH FREE SURFACE BOUNDARY LAYERS USING A STATISTICAL APPROACH

FRANCISCO ANTONIO LOYOLA LAVIN[1] & HARRY EDMAR SCHULZ[2,3]
[1]Department of Mechanical Engineering University of São Paulo, Brazil
[2]Department of Hydraulics and Sanitary Engineering, University of São Paulo, Brazil
[3]Department of Hydraulics and Environmental Engineering, Federal University of Ceará, Brazil

ABSTRACT

Mass exchange through gas–liquid interfaces, whose liquid side has a turbulent nature, are still difficult to quantify due to the unclosed set of turbulence equations, which are also nonlinear. This paper describes an efficient method to overcome this difficulty, by substituting the statistical variables of the original equations by statistical relationships furnished by the Random Square Waves (RSW) tool. Oscillatory records are simplified using random square waves (ideal and binary), which allow a theoretical statistical treatment of the signals. This tool was applied to the concentration boundary layer at the gas–liquid interface. Normalized mass fluxes and mean concentration profiles were obtained using Taylor-series-based solutions, which allow for consideration of transient situations through the successive calculation of the higher order coefficients (derivatives). Comparisons with experimental data available in open literature are presented as a first evaluation of the Taylor series, showing promising results. This method is a viable tool, and this study shows novel conclusions that reproduce general tendencies observed in one-dimensional mass transfer phenomena in boundary layers.
Keywords: turbulent mass flux, concentration profile, gas–liquid interfaces, RSW, turbulence statistics.

1 INTRODUCTION

Physical phenomena that depend on gas–liquid free interfaces, such as the dissolution of substances into water bodies, still await a definitive quantification and understanding regarding the boundary layer details, mainly due the turbulent nature of the boundary layer.

In this study, the focus is the liquid side of the interface. It is known that, in addition to molecular diffusion, it is necessary to consider the mixing of the liquid promoted by the turbulent eddies created by the flow field, usually named as a turbulent diffusion [1].

Many studies correlate the so-called mass transfer coefficient and ad-hoc parameters, such as the boundary layer thickness, the mean time of exposure, or the renewal rate of the surface (see [2] and [3] for a broad review of conceptual models for interfacial transfer). A still limited number of measurements near the surface are available in the open literature [4–8], [12–14], [16], [17] conducted to confirm conclusions derived from conceptual and semi-empirical models, and to shed light on the details of the turbulent movements close to mobile interfaces. However, accurate results are still difficult to obtain due the small thickness (10–100 μm) of the concentration boundary layer [8], together with optical complexities linked to the measurement technique, such as noise, surface localization, and needed corrections due to laser attenuation and to optical blurring [10], [11].

Testing possibilities of numerical tools, DNS studies were done for example by [19], [20] for isotropic turbulence and a wide range of Schmidt and Reynolds numbers. The numerical results suggest that small eddies increase their effect on the mass flux as the Reynolds and Schmidt numbers increase. It is also suggested that numerical results agree with the surface divergence model proposed by [21] and given by $K_L \sim \sqrt{\varphi D}$, being K_L the mass transfer coefficient, D the molecular diffusivity coefficient, and $\varphi = -\left(\frac{\partial U}{\partial x} + \frac{\partial V}{\partial y}\right)$ the surface divergence. U, V are orthogonal superficial velocity components. It is assumed by several authors [18], [23] that φ has a key role in transfer processes. Because divergence considers the turbulent movement of the superficial velocity components, it points to the importance of elucidating how the structure of turbulence combines with the molecular diffusion to control the mass exchange in the adjacent region of the gas–liquid interface.

In a study of long maturation, Schulz and co-workers [9], [15], [16], [25–27] progressively presented the tool "Random Square Waves", which was applied to an interface boundary layer. The starting equation is the Reynolds averaged mass conservation equation, where the Reynolds decomposition of variables, given by a mean parcel and a fluctuating parcel, is the main assumption. For the scalar field the decomposition leads to $F = \bar{F} + f$, and for the velocity field it leads to $\boldsymbol{U} = (U, V, W) = (\bar{U} + u, \bar{V} + v, \bar{W} + w)$. Eqn (1) expresses the mass conservation equation in its usual form:

$$\frac{\partial F}{\partial t} + \boldsymbol{U} \cdot \nabla F = D \cdot \nabla^2 F. \tag{1}$$

F, $\bar{F}$, and f are the instantaneous, mean, and fluctuating scalar variable; U, $\bar{U}$, u; V, $\bar{V}$, v; W, $\bar{W}$, and w are the instantaneous, mean and fluctuating x, y, and z components of the velocity, respectively. When generating conservation equations for the statistical moments of the scalar fluctuations (moments given by θ=1, 2, 3…) eqn 2 is obtained (see [15]):

$$\overline{f^\theta}\frac{\partial \bar{F}}{\partial t} + \frac{1}{\theta+1}\frac{\partial \overline{f^{\theta+1}}}{\partial t} + \overline{f^\theta}[(\bar{U},\bar{V},\bar{W}) \cdot \nabla \bar{F}] + \frac{1}{\theta+1}[(\bar{U},\bar{V},\bar{W}) \cdot \nabla \overline{f^{\theta+1}}]$$
$$+(\overline{f^\theta u}, \overline{f^\theta v}, \overline{f^\theta w}) \cdot \nabla \bar{F} + \frac{1}{\theta+1}\nabla \cdot \overline{[(u,v,w)f^{\theta+1}]} = D\left[\overline{f^\theta}\nabla^2\bar{F} + \overline{f^\theta \nabla^2 f}\right]. \tag{2}$$

Most of the studies in the literature assume unidirectional mass flux trough a flat-interface, and no dominant velocity, so that eqn (2) can be simplified to eqn (3), taken here as the governing equation for the gas–liquid system:

$$\overline{f^\theta}\frac{\partial \bar{F}}{\partial t} + \frac{1}{\theta+1}\frac{\partial \overline{f^{\theta+1}}}{\partial t} + \overline{f^\theta w}\frac{\partial \bar{F}}{\partial z} + \frac{1}{\theta+1}\frac{\partial \overline{w f^{\theta+1}}}{\partial z} = D\left[\overline{f^\theta}\frac{\partial^2 \bar{F}}{\partial z^2} + \overline{f^\theta \frac{\partial^2 f}{\partial z^2}}\right]. \tag{3}$$

According to the authors' knowledge, studies conducted until now with eqn (3) assumed the condition of uniform attenuation/reduction of the wave amplitude (scalar fluctuations) along the boundary layer. Table 1 summarizes the main studies on RSW, all for a constant value of the reduction function α_f that quantifies attenuation. Although theoretical results for concentration profiles are in good agreement with measured data, the experiments of [16] already showed that α_f is not constant. In order to obtain a better understanding of the interfacial mass exchange process, this study presents the theoretical description of the RSW method including the variable reduction function α_f, and furnishes first results for transient problems with variable α_f.

2 GENERAL ASPECTS OF THE RSW METHOD

The Random Square Waves tool approximates fluctuating records by a bimodal random function as shown in Fig. 1, with two extreme values: F_p and F_n. Here F_p is the saturation concentration of a soluble gas in the liquid (taken at the surface of the liquid) and F_n ($< F_p$) is taken as the homogeneous gas concentration in the bulk liquid.

Table 1: Chronological development of RSW method.

Author	Regime-study type	Main results
[24]	Transient-theoretical	Introduced RSW tool by idealizing fluctuations and defining a 1st statistical function (partition n).
[9] [16]	Transient-theoretical and experimental	Definition of 2nd statistical function (reduction α_f). Measuring α_f profiles in oscillating grid tank using LIF.
[25]	Transient-theoretical	Definition of 3rd statistical function (superposition β). Concentration profile obtained numerically.
[15]	Transient-theoretical	Concentration profile obtained for a wide range of α_f values.
[26]	Stationary-theoretical	Theoretical solutions obtained for stationary regime.
[27]	Stationary-theoretical	Relevant functions in mass transport theoretically estimated: turbulent and diffusive flux, concentration central moments, velocity fluctuation RMS, among others.

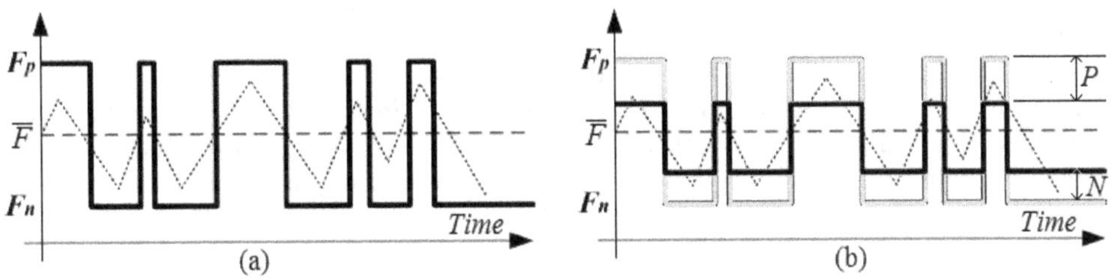

Figure 1: (a) RSW tool. Extreme values F_p and F_n (solid line) may not be attained by the real fluctuation (dotted line); (b) Mean amplitude reduction denoted by P and N.

The partition function n (see Table 1), ranging between 0 and 1, is the portion of time that the square wave is greater than $\bar{F}$ (or equals F_p). This definition implies that n coincides with the normalized concentration $n = (\bar{F} - F_n)/(F_p - F_n)$.

The reduction function α_f (see Table 1), ranging between 0 and 1, quantifies the decay of the amplitude using the variables P and N of Fig. 1(b), calculated as $N=\alpha_f n(F_p-F_n)$ and $P=\alpha_f (1-n)(F_p-F_n)$. $\alpha_f=0$ means no attenuation, and $\alpha_f=1$ means fully attenuation.

The superposition function β (see Table 1), ranging between 0 and 1, is the portion of time in which velocity and scalar fluctuations (w and f) are simultaneously positive.

Using n, α_f and β in the Reynolds averaged equations, [25] presented eqn (2) for any moment θ. Here eqns (4), (5), and (6) are used for θ=1, 2 and 3, respectively.

$$1 - n = S\frac{d^2 n}{dz^{*2}} - \frac{d\Phi}{dz^*}, \tag{4}$$

$$-n(1-n)A^2 + \Phi\frac{dn}{dz^*} + \frac{1}{2}\frac{d}{dz^*}[\Phi A(1-2n)] = Sn(1-n)A\frac{d^2[(1-2n)A]}{dz^{*2}}, \tag{5}$$

$$n(1-n)A^2[-(1-2n)A+1-n]+\Phi(1-2n)A\frac{dn}{dz^*}+\frac{d}{dz^*}\frac{\Phi A^2(1-3n+3n^2)}{3}$$
$$=Sn(1-n)A^2\left[\begin{array}{c}\{1-A(1-2n)\}\frac{d^2n}{dz^{*2}}+\{1-2n+2n^2\}\frac{d^2A}{dz^{*2}}+\\ 2\{2n-1\}\frac{dn}{dz^*}\frac{dA}{dz^*}\end{array}\right]. \tag{6}$$

Φ is the dimensionless turbulent flux given by eqn (7) ($\Phi=IJ^*$ in the original study of [15]). $A=1-\alpha_f$ and $S=D/K_LE$ is a dimensionless diffusivity, being E the length in z^* that implies in a proper variation of F. E is used to define the dimensionless distance $z^*=z/E$.

$$\Phi=\frac{n(n-1)(1-\alpha_f)\frac{\sqrt{\omega^2}}{EK_L}}{\sqrt{n(n-1)+\frac{\beta(1-\beta)}{(2\beta-1)^2}}}. \tag{7}$$

$\sqrt{\omega^2}$ is the root mean square of the velocity fluctuations. It can be seen that the 1D RSW problem reduces to the three unknown functions n, A, and Φ, thus needing only three equations to obtain a solution. In this sense, the RSW method leads to closed sets of equations for 1D turbulence problems.

The boundary conditions for the function n are: 1) first type: $n(z^*=0.0)=1.0$, $n(z^*\rightarrow\infty)=0.0$; and 2) second type: $dn/dz^*(z^*\rightarrow\infty)=0.0$. In boundary layer problems the conditions imposed at $z^*\rightarrow\infty$ are generally not useful, so that boundary layer thicknesses are commonly defined conditionally for $n=0.01$. In the present study it was adequate to apply boundary conditions as proposed in Fig. 2.

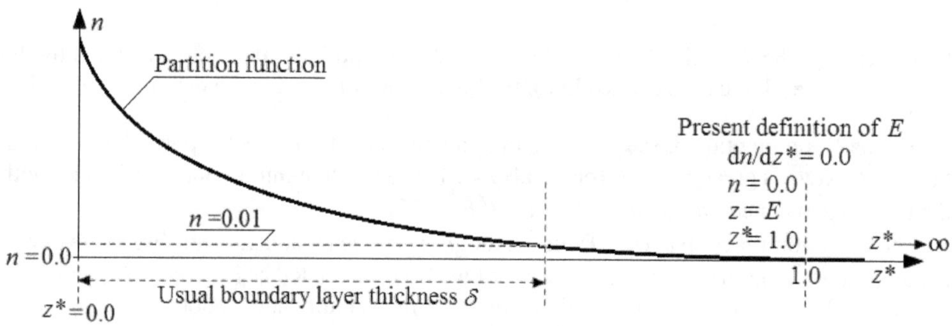

Figure 2: Boundary conditions adopted for defining E in the present study, conditioned by the evolution of the partition function n.

The condition $dn/dz^*(z^*=1.0)=0.0$ is also shown in Fig. 2. By evaluating eqns (5) and (6) at $z^*=1.0$ using the shown boundary conditions, the coupled eqns (8a) and (8b) are obtained

$$A_{z^*=1}\frac{d\Phi}{dz^*}_{z^*=1}+\Phi_{z^*=1}\frac{dA}{dz^*}_{z^*=1}=0, \tag{8a}$$

$$A_{z^*=1}\left[A_{z^*=1}\frac{d\Phi}{dz^*}_{z^*=1} + 2\Phi_{z^*=1}\frac{dA}{dz^*}_{z^*=1}\right] = 0. \tag{8b}$$

By assuming $A_{z^*=1}\neq 0.0$ and $\Phi_{z^*=1}\neq 0.0$, eqns (8a) and (8b) have the single solution $d\Phi/dz^*(z^*=1.0)=dA/dz^*(z^*=1.0)=0.0$. This led to take $\Phi_{z^*=1}$ and $A_{z^*=1}$ as parameters, with the simplified notation Φ_1 and A_1. The solution $d\Phi/dz^*(z^*=1.0)=0.0$ applied into eqn (4) produces $d^2n/dz^{*2}(z^*=1.0)=1/S$. By firstly derivating eqns (4), (5), and (6), then evaluating them at $z^*=1.0$ through the know results of d^2n/dz^{*2}, $d\Phi/dz^*$, and dA/dz^*, we obtain higher order derivatives at this position. Table 2 summarizes the derivatives of first, second, and third order.

Table 2: First, second and third order derivatives of the functions n, A and Φ at $z^*=1.0$.

Function	Order		
	1st	2nd	3rd
n	0	$\frac{1}{S}$	$-\left(\frac{\Phi_1}{A_1}\right)\left(\frac{1}{S}\right)^2(1-A_1)$
A	0	$-\frac{(1-A_1)}{S}$	$-\left(\frac{\Phi_1}{A_1}\right)\left(\frac{1}{S}\right)^2(1-A_1)^2-\left(\frac{A_1}{\Phi_1}\right)\left(\frac{1}{S}\right)(1-3A_1)$
Φ	0	$-\left(\frac{\Phi_1}{A_1}\right)\frac{(1-A_1)}{S}$	$\left(\frac{\Phi_1}{A_1}\right)^2\left(\frac{1}{S}\right)^2(1-A_1)^2-\left(\frac{1}{S}\right)(1+3A_1)$

This inductive process can be repeated until the desired accuracy (error) of the functions n, A, and Φ at a desired position is attained, related to the number of terms in Taylor series of these functions. The position for the control of the quality of the obtained solution is chosen as $z^*=0.0$, where it is known that $n=1.0$. The power series can be expressed for n, A and Φ as given in eqns (9a), (9b) and (9c), respectively:

$$n = n_{z^*=1} + (z^*-1)\frac{dn}{dz^*}_{z^*=1} + (z^*-1)^2\frac{d^2n}{dz^{*2}}_{z^*=1} + (z^*-1)^3\frac{d^3n}{dz^{*3}}_{z^*=1} + \cdots, \tag{9a}$$

$$A = A_{z^*=1} + (z^*-1)\frac{dA}{dz^*}_{z^*=1} + (z^*-1)^2\frac{d^2A}{dz^{*2}}_{z^*=1} + (z^*-1)^3\frac{d^3A}{dz^{*3}}_{z^*=1} + \cdots, \tag{9b}$$

$$\Phi = \Phi_{z^*=1} + (z^*-1)\frac{d\Phi}{dz^*}_{z^*=1} + (z^*-1)^2\frac{d^2\Phi}{dz^{*2}}_{z^*=1} + (z^*-1)^3\frac{d^3\Phi}{dz^{*3}}_{z^*=1} + \cdots. \tag{9c}$$

The first lower order derivatives of Table 2 convert eqns (9a), (9b), and (9c) into eqns (10a), (10b), and (10 c).

$$n = \frac{(z^*-1)^2}{S} - \left(\frac{\Phi_1}{A_1}\right)\frac{(1-A_1)(z^*-1)^3}{S^2} + \cdots, \tag{10a}$$

$$A = A_1 - \frac{1-A_1}{S}(z^*-1)^2 - \left[\left(\frac{\Phi_1}{A_1}\right)\left(\frac{1-A_1}{S}\right)^2 - A_1\left(\frac{1-3A_1}{S\Phi_1}\right)\right](z^*-1)^3 + \cdots, \tag{10b}$$

$$\Phi = \Phi_1 - \frac{\Phi_1(1-A_1)(z^*-1)^2}{SA_1} + (z^*-1)^3\left\{\left[\frac{\Phi_1(1-A_1)}{SA_1}\right]^2 - \frac{1+3A_1}{S}\right\} + \cdots. \quad (10c)$$

To obtain higher order coefficients a code was developed in the MATLAB® symbolic environment furnishing the vector $[n^{(p)}, A^{(p)}, \Phi^{(p)}]$, where (p) indicates de p^{th} order derivative. Eqns (10a), (10b), and (10c) also show that higher values of S imply in a lower number of terms in the three power series. This procedure (higher value of S) is used here to check the quality of the profiles obtained and the needed convergence.

Further, the experimental trend obtained by [16] for α_f was used as a boundary condition for A=1- α_f in the form $A(z^*=0)=0$, while the conclusions of the experimental and theoretical analysis of [11] were used as boundary conditions for Φ in the form $\Phi(z^*=0)=0$ and $d\Phi/dz^*(z^*=0)=0$. This implied adding a 19th parcel to the Taylor series for n, A, and Φ, and also a 20th parcel to the Taylor series of Φ.

3 OBTAINED PROFILES AND QUALITATIVE COMPARISON WITH EXPERIMENTAL DATA

Fig. 3 depicts the theoretical profiles of n, A, and Φ for $0.0\leq z/\delta\leq 1.0$ using 18 calculated parcels of the Taylor series, and the mentioned parcels for boundary conditions at z^*=0. The values of the control parameters A_1 and Φ_1 are shown in Fig. 3, while S was taken as S=1000. Experimental data of [9] are also shown for qualitative comparisons. To allow this comparison, the z axis is normalized with the usual boundary layer thickness δ at n=0.01. The positions z^* at which δ is attained are shown in Table 3. It must be commented that the data of [9] produce S~1.0, while the calculations were performed with S=1000 to facilitate the convergence of the Taylor series (reducing the number of parcels of the series). The present comparison is, thus, qualitative. It is firstly observed that the calculated n profile follows the general trend of experimental observations of the literature, i.e., decreases from its maximum value (n=1.0 at the interface) to the value n=0.01 at z/δ=1.0. It is further observed that, for $\Phi_1<10^3$, the different curves converge to the condition represented by the thin gray lines. Table 3 also shows that δ^* converges to ~0.2165 for the lower values of Φ_1.

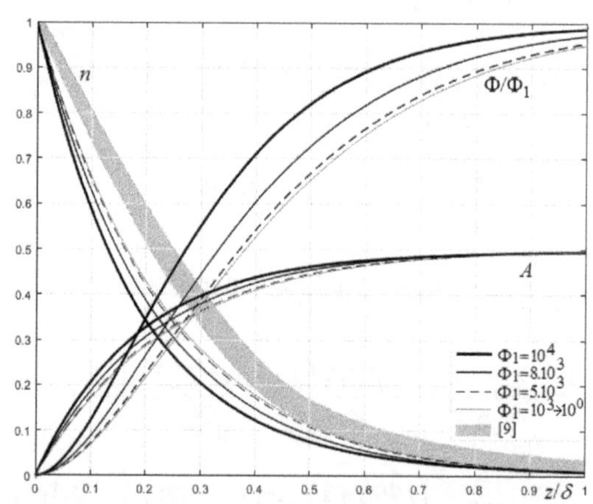

Figure 3: Profiles of variables related to mass transfer in the boundary layer: n, A and Φ/Φ_1. Parameter values: S = 1000, A_1 = 0.5 and Φ_1 shown in the figure. The gray cloud represents the data of [9].

Table 3: Control parameters A_1, Φ_1 and characteristic lengths δ, $\delta_{c'}$ of the boundary layer.

Φ_1	A_1=0.45		A_1=0.50		A_1=0.75	
	$\delta^*=\delta/E$ (n=0.01)	$\delta^*_{c'}=\delta_{c'}/E$ (n=0.01)	$\delta^*=\delta/E$ (n=0.01)	$\delta^*_{c'}=\delta_{c'}/E$ (n=0.01)	$\delta^*=\delta/E$ (n=0.01)	$\delta^*_{c'}=\delta_{c'}/E$ (n=0.01)
$1.0\ 10^4$	0.4010	0.1386	0.3145	0.08314	0.2190	0.07390
$8.0\ 10^3$	0.3058	0.08314	0.2513	0.07390	0.2181	0.07390
$5.0\ 10^3$	0.2284	0.07390	0.2229	0.07390	0.2173	0.07390
$1.0\ 10^3$	0.2169	0.07390	0.2169	0.07390	0.2166	0.07390
$1.0\ 10^2$	0.2170	0.07390	0.2165	0.07390	0.2165	0.07390
$1.0\ 10^1$	0.2165	0.07390	0.2165	0.07390	0.2165	0.07390
$1.0\ 10^0$	0.2165	0.07390	0.2165	0.07390	0.2165	0.07390

In Fig. 3, Φ/Φ_1 follows the expected behaviour of increasing turbulent mass flux with z/δ in the boundary layer. In real situations for fixed liquid volumes, the profile of Φ/Φ_1 firstly increases near the interface (in the boundary layer), and further decreases tending to zero in the bulk liquid (outside the boundary layer), as already discussed by [11].

From Fig. 3, A also increases with z/δ in the boundary layer, following experimental data obtained by [9] and shown in [16]. This result extends the discussion of [27], in which the focus was the stationary solution of RSW for constant A, showing now a wider range of application. A=0.0 is characteristic for damping (attenuation) of fluctuations due to molecular diffusivity near the interface. Because molecular diffusion is important in the so called "outer diffusive layer" (see the BOKVS interfacial structure in [22] and [27], for example), low values of A also occur in this layer.

In Table 3, $\delta^*_{c'}$ is the position of the peak at z^* of the function that expresses the RMS fluctuation of concentration, and used in the subsequent discussion.

Fig. 4 shows the same curves of Fig. 3, but for A_1=0.75. It can be seen that n, A, and Φ collapse to single curves. Because A is a measure of attenuation due to diffusivity, this result suggests that lower attenuations generate profiles less sensitive (or no sensitive) to variations of Φ_1, the turbulent mass flux taken as control parameter, for $1.0 \leq \Phi_1 \leq 10^4$ and S=1000.

Fig. 5(a) shows the comparison between measured and calculated concentration fluctuation intensities, given as the root mean square c' of the concentration fluctuations c, $c' = \sqrt{c^2}$, and the RSW equation $c' = A\sqrt{n(1-n)}$ (see [16]). The experimental data were obtained by [9] and discussed by [16] and [27]. The z axis is normalized with the position of the peak of c' in order to facilitate comparisons with measurements. As can be seen, the calculated profile of A follows the cloud of measured profiles.

Having the position of the peak of c' of Fig. 5(a), it was also possible to compare experimental and theoretical profiles of A against $z/\delta_{c'}$, as shown in Fig. 5(b). The growing of A with $z/\delta_{c'}$ for $z/\delta_c \to 0.0$ is observed in both the measured data and the theoretical curve, and they show similar slopes for the interval $0.0 < z/\delta_{c'} < \sim 1.3$. As differences between the two results it is observed that the cloud of data decreases for $z/\delta_{c'} > \sim 2.5$, which is not followed by the calculated curve. This last theoretical result is more closely related to the boundary conditions applied at z^*=1, and to the imposed value S=1000 for the momentary convergence analysis, which not necessarily reproduce the data of [9].

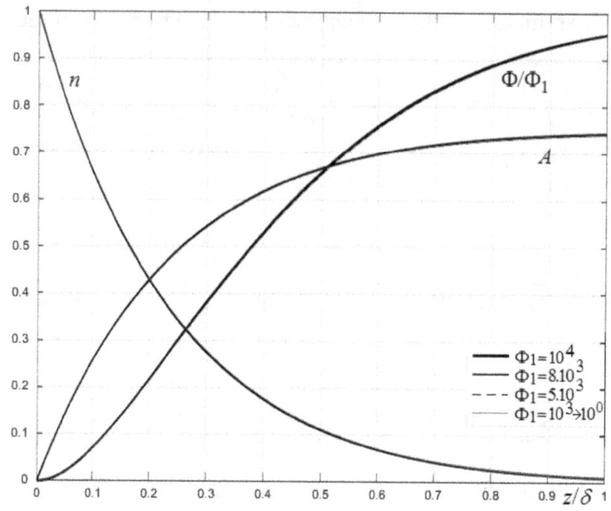

Figure 4: Profiles of variables related to mass transfer in the boundary layer: n, A and Φ/Φ_1. Parameter values: S = 1000, A_1 = 0.75 and Φ_1 shown in the figure. The evolution of the different variables collapse to single curves.

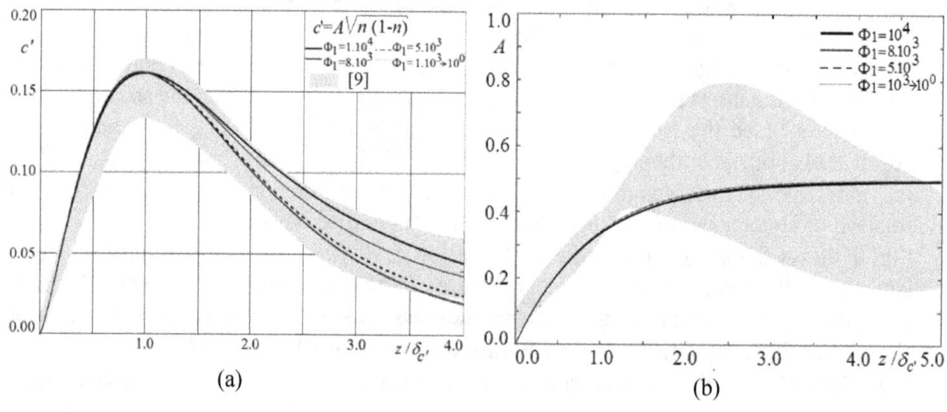

Figure 5: The comparison between measured and calculated concentration fluctuation intensities: (a) Profiles of the experimental $c' = \sqrt{\overline{c^2}}$ and the theoretical RSW equation $c' = A\sqrt{n(1-n)}$ for A_1=0.5; (b) Profiles of experimental and theoretical A. The solid line is the calculated curve for A_1=0.5. The gray cloud represents data of [9].

In the present study Φ/Φ_1 increases with the second power of z^* for $z^*\rightarrow 0.0$, one of the possibilities discussed by [11]. This result shows that the proposed Taylor series of Φ converges, and that the condition $d\Phi/dz^*$=0.0 at z^*=0.0 is adequate to obtain realistic trends. The proportionality between Φ/Φ_1 and z^{*2} for free slip surfaces may thus be incorporated by the present formulation. This result is shown in Fig. 6, where the theoretical results of Φ/Φ_1 obtained for A_1=0.5 are plotted against z^* in logarithm scale.

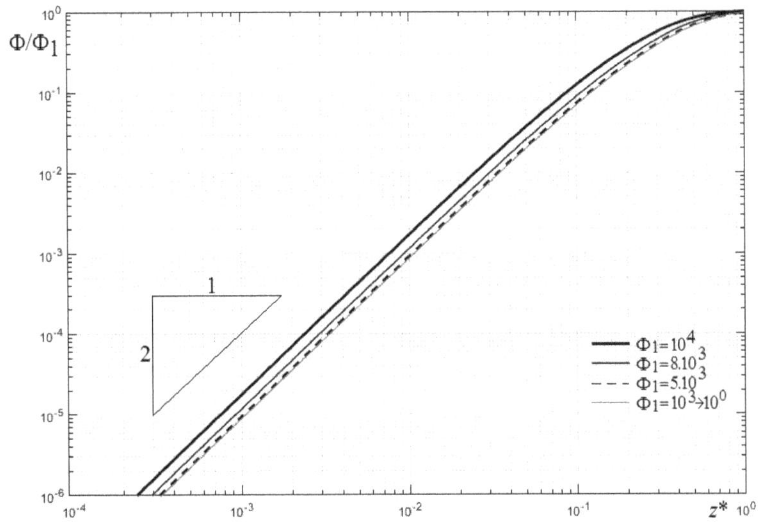

Figure 6: Calculated profile of Φ/Φ_1. The trend to the slope 2.0 close to the interface shows the adequacy of the result, considering discussions and experimental data of the literature (see [11]), and the convergence of the method. A_1=0.5.

The results presented in Figs 3–6 express the evolution of relevant variables in concentration boundary layers. It is shown that the RWS tool is adequate to quantify scalar boundary layers, being dependent of a number of control parameters (A_1, Φ_1, and S). These parameters allow quantifying the whole set of coefficients of the Taylor series that expresses n, (partition function) A (complementary of the reduction function), and Φ (nondimensional turbulent mass flux). In the present study 18 coefficients were calculated for each series. By reducing the value of S, more parcels should be added to the solution.

4 CONCLUSIONS

The mathematical tool of Random Square Waves (RSW) for the study of gas exchanges across a turbulent gas–liquid interface was described. It was shown that the originally open set of statistical equations for turbulence can be closed by applying this tool. The condition of transient one dimensional mass transfer considering a variable reduction function α_f was adopted for the calculations. It was shown that the RSW tool for 1 D transfer depends on three functions: the partition function n, the reduction function a_f (or its complementary A=1-a_f), and the turbulent flux Φ. As a consequence, only three equations are needed to calculate the turbulent parameters in the boundary layer.

The calculations show the first results of a novel form of quantifying the turbulent parameters n, A, and Φ in the concentration boundary layer, for which Taylor series are used.

To obtain the successive derivatives of the variables under study that are needed for the Taylor series, adequate boundary conditions for the normalized concentration profile n were imposed at the normalized position z^*=1.0, in the form: $n^*(z^*=1.0)=0.0$ and $dn/dz^*(z^*=1.0)=0.0$. Because the derivatives involve negative powers of the parameter S of the RSW tool, the present first calculations were performed using S=1000, so that only 18 parcels were used to quantify the coefficients of the Taylor series of n, A, and Φ. To consider boundary conditions at z^*=0, one parcel was added to the series of n and A and two parcels were added to the series of Φ. All derivatives depend on the values of A and Φ at z^*=1.0, expressed as A_1 and Φ_1, respectively. In the present study three values of A_1 were tested, together with seven values of Φ_1, as shown in Table 3.

It was observed that the Taylor series applied to the RSW tool furnished good results:

1. The calculated profile of the normalized concentration, n (or the partition function), showed the general behaviour observed in experimental results of the literature, decreasing from the value n=1.0 at the interface to the value n=0.01 at the end of the boundary layer.
2. The calculated profile of the intensity of the concentration fluctuations c' superposed very well measured profiles.
3. The profiles of the functions A increased with the distance to the interface, following experimental results found in the literature. Close to the interface (z^*=0) the calculated evolution of A superposed well measured evolutions of the literature. Differences observed between theoretical results and experimental data for great distances to the interface may be related to the boundary conditions adopted for z^*=1.0. This situation indicates possible future steps in this research line.
4. The calculated profiles of the functions Φ are adequate. In the present calculations the turbulent mass flux Φ increased with the second power of the distance to the interface, a possibility that is discussed in the literature.
5. The RSW tool showed to be adequate for boundary layer studies.

Having obtained the above described positive results of the first application of Taylor series to the RSW tool, the reproduction of real conditions will be now pursued related to the value of S and alternative boundary conditions for z^*=1.0. It implies in lowering the value of S, which results in using a larger number of terms in the Taylor series. Both measures are necessary and can now be conducted in the light of the present positive results.

ACKNOWLEDGEMENTS

The authors gratefully acknowledge CNPq (National Council for Scientific and Technological Development, Brazil) for the financial support. The first author thanks CAPES for the fellowship PPG-AEM-EESC-USP. The second author mentions the process CNPq 307105/2015-6 that assures the continuity of the research line in statistical turbulence and RSW.

REFERENCES

[1] Gulliver, J.S., (eds), *Introduction to Chemical Transport in the Environment*, Cambridge University Press: New York, 2007.

[2] Jorgensen, S.E. & Gromiec, M.J., Chapter 3: Reaeration. *Developments in Environmental Modelling*, Elsevier: Amsterdam, vol. **14**, pp. 33–64, 1989.

[3] Bennett, J.P. & Rathbun, R.E., *Reaeration in Open-Channel Flow*, Washington, DC: US Government Printing Office, 1972.

[4] Janzen, J.G., Jirka, G.H. & Schulz, H.E., Details of air–water gas transfer: experiments with precision – in Portuguese. *Proceedings of the 16th Brazilian Conference on Water Resources* (*XVI Simpósio Brasileiro de Recursos Hídricos*), pp. 1–17, 2005.

[5] Janzen, J.G., Schulz, H.E. & Jirka, G.H., Details of air–water gas transfer – in Portuguese. *Brazilian Journal of Water Resources*, **11**(4), pp. 153–161, 2006.

[6] Chu, C.R. & Jirka, G.H., Turbulent gas flux measurements below the air–water interface of a grid-stirred tank. *International Journal of Heat and Mass Transfer*, **35**(8), pp. 1957–1968, 1992.

[7] Herlina, *Gas Transfer at the Air–Water Interface in a Turbulent Flow Environment*, Dissertation series, Karlsruhe: Institute for Hydromechanics of the University of Karlsruhe, 2005.

[8] Herlina & Jirka, G.H., Application of LIF to investigate gas transfer near the air–water interface in a grid-stirred tank. *Experiments in Fluids*, **37**(3), 341–349, 2004.

[9] Janzen, J.G., *Gas Transfer Near the Air–Water Interface in Oscillating-Grid Tanks and Proprieties of Isotropic Turbulent Flows* (in Portuguese), Dissertation, São Paulo: São Carlos School of Engineering at the University of São Paulo, 2006.

[10] Janzen, J.G., Herlina, H., Jirka, G.H., Schulz, H.E. & Gulliver, J.S., Estimation of mass transfer velocity based on measured turbulence parameters. *American Institute of Chemical Engineers Journal*, **56**(8), pp. 2005–2017, 2010.

[11] Janzen, J.G., Schulz, H.E. & Jirka, G.H., Turbulent gas flux measurements near the air–water interface in an oscillating-grid tank. *Proceedings of the 2010 Gas Transfer at Water Surfaces Conference*, pp. 65–77, 2011.

[12] Herzog, A.G., *Imaging of Water-Sided Gas-Concentration Fields at a Wind-Driven, Wavy Air–Water Interface*, Dissertation, Heidelberg: Faculties for the Natural Sciences and for Mathematics Institute of the Ruperto-Carola University of Heidelberg, 2010.

[13] Friedl, F., *Investigating the Transfer of Oxygen at the Wavy Air–Water Interface Under Wind-Induced Turbulence*, Dissertation, Heidelberg: Faculties for the Natural Sciences and for Mathematics Institute of the Ruperto-Carola University of Heidelberg, 2013.

[14] Krah, N., *Visualization of Air and Water-Sided Concentration Profiles in Laboratory Gas Exchange Experiments*, Dissertation, Heidelberg: Faculties for the Natural Sciences and for Mathematics Institute of the Ruperto-Carola University of Heidelberg, 2014.

[15] Schulz, H.E., Lopes Jr., G.B., Simões, A.L.A. & Lobosco, R.J., (eds), Chapter 1: One-dimensional turbulent transfer using random square waves – scalar/velocity and velocity/velocity interactions. *Hydrodynamics: Advanced Topics*, Rijeka: InTech, pp. 3–34, 2011.

[16] Schulz, H.E. & Janzen, J.G., Concentration fields near air–water interfaces during interfacial mass transport: oxygen transport and random square wave analysis. *Brazilian Journal of Chemical Engineering*, **26**(3), pp. 527–536, 2009.

[17] Lacassagne, T., El-Hajem, M., Morge, F., Simoens, S. & Champagne, J.Y., Study of gas liquid mass transfer in a grid stirred tank. *Oil & Gas Science and Technology–Revue d'IFP Energies nouvelles*, **72**(1), p. 7, 2017.

[18] Tamburrino, A. & Martínez, N., Wave and wind effects on the oxygen transfer across an air-water interface: An experimental study. *The Canadian Journal of Chemical Engineering*, **95**(8), pp. 1594–1604, 2017.

[19] Herlina, H. & Wissink, J.G., Direct numerical simulation of turbulent scalar transport across a flat surface. *Journal of Fluid Mechanics*, **744**, pp. 217–249, 2014.

[20] Herlina, H. & Wissink, J.G., Simulation of air–water interfacial mass transfer driven by high-intensity isotropic turbulence. *Journal of Fluid Mechanics*, **860**, pp. 419–440, 2019.

[21] McCready, M.J., Vassiliadou, E. & Hanratty, T.J., Computer simulation of turbulent mass transfer at a mobile interface. *AIChE Journal*, **32**(7), pp. 1108–1115, 1986.

[22] Brumley, B.H. & Jirka, G.H., Near-surface turbulence in a grid-stirred tank. *Journal of Fluid Mechanics*, **183**, 235–263, 1987.

[23] Sirkar, K.K. & Hanratty, T.J., Relation of turbulent mass transfer to a wall at high Schmidt numbers to the velocity field. *Journal of Fluid Mechanics*, **44**(3), pp. 589–603, 1970.

[24] Schulz, H.E. & Schulz, S.A.G., Modelling below-surface characteristics in water reaeration. *Proceedings: Water Pollution: Modelling, Measuring and Prediction*, pp. 441–454, 1991.
[25] Schulz, H.E., Simões, A.L. & Janzen, J.G., Statistical approximations in gas–liquid mass transfer. *Proceedings of the 2010 Gas Transfer at Water Surfaces Conference*, pp. 208–221, 2011.
[26] Schulz, H.E. & Gonçalves, B.B., Solutions of scalar mean profiles close to gas–liquid interfaces under turbulent free slip motion. *Proceedings of the Eight Conference on Computational Methods in Multiphase Flow*, pp. 33–44, 2015.
[27] Schulz, H.E., Lavín, F.A.L. & Gonçalves, B.B., Turbulence aspects of mass transfer in the thin interfacial region of the concentration boundary layer in gas–liquid systems. *International Journal of Computational Methods and Experimental Measurements*, pp. **6**(1), pp. 186–197, 2018.

ANALYTICAL MODEL AND NUMERICAL STABILITY ANALYSIS FOR FALLING LIQUID FILM REGIMES IN VERTICAL PIPES

JOSÉ LUIS MUÑOZ-COBO[1], SUSANA M. IGLESIAS[1,2], DANY S. DOMINGUEZ[1,2], ALBERTO ESCRIVÁ[1] & CÉSAR BERNA[1]
[1]Institute for Energy Engineering, Universitat Politècnica de València, Spain
[2]Exact and Technological Sciences Department, Universidade Estadual de Santa Cruz, Brazil

ABSTRACT

Gravity-driven falling film flows in vertical pipes are relevant in many industrial processes like evaporators, chemical reactors, and condensers. The wave formation and propagation processes, which continuously take place in the film, have a strong influence on the flow hydrodynamics and the heat and mass transfer phenomena. Several researchers have been studying the wave behaviour in these multiphase regimes through experimental works and computational fluid dynamics simulation techniques. In some simplified cases, like high viscosity fluids and infinity inclined plates, analytical solutions have been obtained. In this work, we propose an analytical model for falling film flow regimes in vertical pipes. Starting from the incompressible axisymmetric Navier–Stokes equations in cylindrical coordinates, we consider the force balance in the fluid, an asymptotic long-wave approximation and the first-order perturbation approximation for axial velocity. From this balance, we obtain a partial differential equation that describes the interface behaviour through the film thickness. The resulting equation can be solved using a numerical approach. The main resulting equation represents a stiff problem, thus, we perform a stability analysis using the fluid viscosity as a parameter. Finally, we set the model validity conditions and suggest some actions to improve the numerical strategy in order to better describe low viscosity fluids.

Keywords: falling liquid film, analytical model, vertical pipes, Navier–Stokes equations.

1 INTRODUCTION

The dynamics of core-annular flows are relevant in many industrial and biological applications, such as oil recovery, nuclear power plants and human body's airways. In these regimes, one fluid phase fills the core, and the other, more viscous, occupies the annulus surrounding the pipe wall. When core-phase is air, motionless or with small velocity, the volume of annular-phase is small if compared with the pipe dimensions and only the gravity force acts on the fluid, then we have a gravity-driven falling film. In addition, films flowing down on vertical pipes are very unstable and develop nonlinear waves in the interface surface. These waves have a strong influence in the flow hydrodynamics, enhancing the heat and mass transfer phenomena.

Several researchers have been studying the wave behaviour in these multiphase regimes using experimental, analytical and computational fluid dynamics (CFD) approaches. For example, Mascarenhas and Mudawar [1] using experimental and CFD techniques to study the falling heated water film in turbulent regimes, analyzing the influence of interfacial waves on mass, momentum and heat transfer. Camassa et al. [2], conducted experiments and constructed an analytical model to study the mechanism by which forced steady airflow transports annular layers of high-viscosity fluids (low Reynolds), conditions corresponding to lung airways. The model developed shows the different mass transport regimes in the wave propagation. Also, Camassa et al. [3] use a numerical model to predict the speed and the wavelength of the most unstable mode of these viscous film flows on small pipes in gravity-driven conditions. The model results are validated using experimental measurements. The

WIT Transactions on Engineering Sciences, Vol 123, © 2019 WIT Press
www.witpress.com, ISSN 1743-3533 (on-line)
doi:10.2495/MPF190091

direct numerical simulation (DNS) of the full Navier–Stokes equations, was used in [4], to study the interfacial instabilities in thin-film flows. This approach is difficult to perform due to the complexity of free surface and the needed of extremely finer meshes in space and time. Muñoz-Cobo et al. [5], applies the perturbation theory, the long-wave approximation and the velocity expansion in harmonic functions to generate an analytical model for gravity-driven film flow on inclined plates. The present work extends this work, proposing an analytical model for falling films flows regimes in vertical pipes. To construct the model, we consider the force balance in the fluid, an asymptotic long-wave approximation and the first-order perturbation approximation for axial velocity. From this balance, we obtain a partial differential equation that describes the interface behavior through the film.

At this point, we discretized the problem using a uniform spatial grid. Then, we integrate the interface equation in one arbitrary spatial cell, obtaining an ordinary differential equations system (ODES), for the film thickness at each grid cell. The resulting ODES is time integrated using an Adams–Moulton type semi-implicit predictor-corrector method to generate the numerical results. The resulting model presents a high level of stiffness leading to numerical instabilities in some conditions. At the end of this work, we perform numerical experiments using fluid viscosity as a parameter to set the model validity range.

This paper is arranged as follows. In the next section, we offer the mathematical and physical foundations of the proposed analytical model. Section 3 describes the numerical procedure and the algorithm for solving the model. In Section 4, we present the numerical experiments results and finally, in Section 5 we offer the research conclusions and suggestions for future developments.

2 MATHEMATICAL MODEL FORMULATION

In order to develop the analytical model for falling films in vertical pipes, we consider an axisymmetric cylindrical domain with a down annular flow regime as shown in Fig. 1(a).

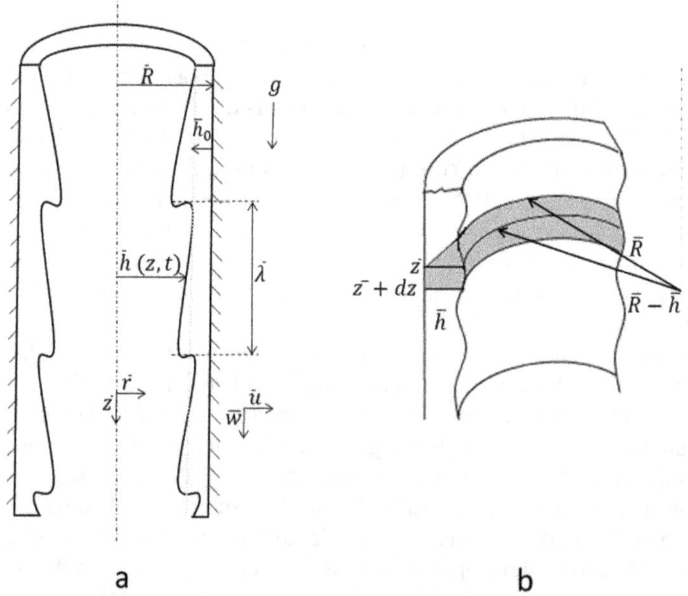

Figure 1: (a) Problem domain representation with the main variables; and (b) Detail of a differential film volume.

As a starting point, we consider the incompressible axisymmetric Navier–Stokes equations in cylindrical coordinates in the form

$$\bar{r}^{-1}\partial_{\bar{r}}(\bar{r}\bar{u}) + \partial_{\bar{z}}\bar{w} = 0, \tag{1}$$

$$\partial_{\bar{t}}\bar{u} + \bar{u}\partial_{\bar{r}}\bar{u} + \bar{w}\partial_{\bar{z}}\bar{u} = -\frac{1}{\rho}\partial_{\bar{r}}\bar{P} + \nu\left(\frac{1}{\bar{r}}\partial_{\bar{r}}(\bar{r}\partial_{\bar{r}}\bar{u}) + \partial_{\bar{z}\bar{z}}\bar{u} - \frac{\bar{u}}{\bar{r}^2}\right), \tag{2}$$

$$\partial_{\bar{t}}\bar{w} + \bar{u}\partial_{\bar{r}}\bar{w} + \bar{w}\partial_{\bar{z}}\bar{w} = -\frac{1}{\rho}\partial_{\bar{z}}\bar{P} + \nu\left(\frac{1}{\bar{r}}\partial_{\bar{r}}(\bar{r}\partial_{\bar{r}}\bar{w}) + \partial_{\bar{z}\bar{z}}\bar{w}\right) + g, \tag{3}$$

where $\bar{r}$ and $\bar{z}$ are the radial and axial coordinates, respectively; $\bar{u}$ is the radial velocity component; $\bar{w}$ is the downward direction velocity component; $\bar{P}$ is the pressure; finally, ρ, ν, and g indicates the fluid density, fluid kinematic viscosity and gravity acceleration, respectively. Besides, in Fig. 1(a) $\bar{h}(z,t)$ indicates the film thickness, and $\bar{h}_0$ is the Nusselt thickness of the flat film [6]. In our notation, the overbars denote dimensional quantities. At the pipe wall, we consider no-slip boundary conditions

$$\bar{w}|_{\bar{R}} = \bar{u}|_{\bar{R}} = 0. \tag{4}$$

In order to obtain the free surface boundary conditions, we regard some physical conditions at the interface. For the normal stress component, we consider that the sum of forces at the interface is zero and we obtain

$$\bar{P}|_{\bar{R}-\bar{h}} - \bar{P}_c + \sigma\left\{\frac{(\partial_{\bar{z}\bar{z}}\bar{h})}{\left(1+(\partial_{\bar{z}}\bar{h})^2\right)^{\frac{3}{2}}} + \frac{1}{\bar{R}-\bar{h}}\right\}_{\bar{R}-\bar{h}} - \frac{2\mu}{1+(\partial_{\bar{z}}\bar{h})^2}\left\{\partial_{\bar{r}}\bar{u} + (\partial_{\bar{z}}\bar{h})(\partial_{\bar{z}}\bar{u} + \partial_{\bar{r}}\bar{w}) + (\partial_{\bar{z}}\bar{h})^2\partial_{\bar{z}}\bar{w}\right\}_{\bar{R}-\bar{h}} = 0, \tag{5}$$

here $\bar{P}_c$ is the annular-phase pressure, σ and μ indicate the fluid surface tension and the fluid dynamic viscosity respectively. The eqn (5) is similar to Young–Laplace equation [7]. Then, for the tangential stress component, we use the stress continuity condition [3] at the interface to get

$$\left[1-(\partial_{\bar{z}}\bar{h})^2\right]\{\partial_{\bar{z}}\bar{u} + \partial_{\bar{r}}\bar{w}\}_{\bar{R}-\bar{h}} + 2\partial_{\bar{z}}\bar{h}\{\partial_{\bar{z}}\bar{w} - \partial_{\bar{r}}\bar{u}\}_{\bar{R}-\bar{h}} = 0. \tag{6}$$

We can obtain an equation to govern the interface using a film mass balance in the interval $[z,\ z+dz]$. The geometry of the mass balance is displayed in Fig. 1(b). Considering that the mass flow variation over time in the film differential volume is equal to the difference between the flow entering by the top surface and the flow leaving by the bottom surface, then we have

$$\partial_{\bar{t}}\left[2\pi\rho\bar{h}\left(\bar{R}-\frac{\bar{h}}{2}\right)\right] = -\partial_{\bar{z}}\left[\rho\int_{\bar{R}-\bar{h}}^{\bar{R}} 2\pi\bar{r}\bar{w}d\bar{r}\right]. \tag{7}$$

Solving the integral in eqn (7) using the Leibnitz rule and considering the mass conservation eqn (1), we obtain

$$\partial_{\bar{t}}\bar{h}(\bar{z},\bar{t}) = -\left[\bar{w}|_{\bar{R}-\bar{h}}\,\partial\bar{z}\,\bar{h} + \bar{u}|_{\bar{R}-\bar{h}}\right]. \tag{8}$$

The eqn (8) is a partial differential equation that describes the fluid interface as a function of the time and z-coordinate, some authors call this equation, a kinematic boundary condition [3]. Is possible to obtain the integral formulation for the interface equation. For that, we define $\bar{q}$ as the mass flow rate through a radian of the circumference in the form

$$\bar{q} = \int_{\bar{R}-\bar{h}}^{\bar{R}} \bar{r}\bar{w}(\bar{r},\bar{z},\bar{t})d\bar{r}, \tag{9}$$

then, we calculate the time derivative on the eqn (7) left side. Considering eqn (9) for the right side and grouping, the interface integral formulation is written as

$$(\bar{R}-\bar{h})\partial_t \bar{h} + \partial_{\bar{z}}\bar{q} = 0. \tag{10}$$

At this stage, we introduce the dimensionless parameters to transform the equations system into a non-dimensional form. This dimensionless scheme does not depend on mass flow quantities and was proposed by Ruyer-Quil and Manneville [8]. The dimensionless parameters are

$$L_c = \left(\frac{\nu^2}{g}\right)^{1/3}, T_c = \left(\frac{\nu}{g^2}\right)^{1/3}, \; U_r = (\nu g)^{1/3}, P_r = \rho(\nu g)^{2/3}, K_a = \frac{\sigma}{\rho \nu^{4/3} g^{1/3}}, \tag{11}$$

where L_c and T_c are the characteristic longitude and time, U_r and P_r are the reference velocity and pressure and K_a is the Kapitza number. The mass conservation and momentum eqns (1)–(3) in the dimensionless form are written as

$$r^{-1}\partial_r(ru) + \partial_z w = 0, \tag{12}$$

$$\partial_t u + u\partial_r u + w\partial_z u = -\partial_r P + r^{-1}\partial_r(r\partial_r u) + \partial_{zz}u - \frac{u}{r^2}, \tag{13}$$

$$\partial_t w + u\partial_r w + w\partial_z w = -\partial_z P + r^{-1}\partial_r(r\partial_r w) + \partial_{zz}w + 1. \tag{14}$$

The wall boundary condition (eqn (4)), appear as

$$w|_R = u|_R = 0, \tag{15}$$

and the boundary conditions at the fluid interface (eqns (5) and (6)) become

$$P|_{R-h} - P_c + Ka\left\{\frac{\partial_{zz}h}{(1+(\partial_z h)^2)^{\frac{3}{2}}} + \frac{1}{R-h}\right\}_{R-h}$$
$$-\frac{2}{1+(\partial_z h)^2}\{\partial_r u + (\partial_z h)(\partial_z u + \partial_r w) + (\partial_z h)^2 \partial_z w\}_{R-h} = 0, \tag{16}$$

$$[1-(\partial_z h)^2]\{\partial_z u + \partial_r w\}_{R-h} + 2\partial_z h\{\partial_z w - \partial_r u\}_{R-h} = 0. \tag{17}$$

The PDE interface equation (eqn (8)) and the corresponding integral form (eqn (10)) are transformed into

$$\partial_t h(z,t) = -[w_{R-h}\, \partial z\, h + u|_{R-h}], \tag{18}$$

$$(R-h)\partial_t h + \partial_z q = 0, \tag{19}$$

respectively. In eqn (19), the parameter q adopts the non-dimensional form of eqn (9)

$$q = \int_{R-h}^{R} rw(r,z,t)dr. \tag{20}$$

The interface integral eqn (19) relates the film thickness h, with the axial velocity w. Then, if we obtain one approximation for the axial velocity, we can evaluate the integral term in eqn (20) and solve the resultant PDE to get the film thickness.

Now, we regard the long-wave approach [3], [5], [6], this is, when the film surface perturbations wavelength (λ), is greater than the film thickness. Thus, the perturbation parameter is defined by

$$\varepsilon = \frac{h}{\lambda} \ll 1. \tag{21}$$

Rewriting the constitutive eqns (12)–(14), the corresponding boundary conditions eqns (15)–(17) and the interface eqn (19) and considering a long-wave approximation, we generate the general perturbative equations. Here, the terms involving temporal and z-spatial derivatives, are first order in ε.

From these considerations, the resulting balance equations are

$$r^{-1}\partial_r(ru) + \varepsilon\partial_z w = 0, \tag{22}$$

$$\varepsilon\partial_t u + u\partial_r u + \varepsilon w\partial_z u = -\partial_r P + r^{-1}\partial_r(r\partial_r u) + \varepsilon^2\partial_{zz}u - \frac{u}{r^2}, \tag{23}$$

$$\varepsilon\partial_t u + u\partial_r u + \varepsilon w\partial_z u = -\varepsilon\partial_z P + r^{-1}\partial_r(r\partial_r w) + \varepsilon^2\partial_{zz}w + 1. \tag{24}$$

The perturbative equations for the normal (eqn (16)) and tangential (eqn (17)) stress at the film surface are written as

$$P|_{R-h} - P_c + Ka\left\{\frac{\varepsilon^2\partial_{zz}h}{[1+\varepsilon^2(\partial_z h)^2]^{\frac{3}{2}}} + \frac{1}{R-h}\right\}_{R-h}$$

$$-\frac{2}{1+\varepsilon^2(\partial_z h)^2}\{\partial_r u + \varepsilon(\partial_z h)(\varepsilon\partial_z u + \partial_r w) + \varepsilon^3(\partial_z h)^2\partial_z w\}_{R-h} = 0, \tag{25}$$

$$[1-\varepsilon^2(\partial_z h)^2]\{\varepsilon\partial_z u + \partial_r w\}_{R-h} + 2\varepsilon\partial_z h\{\varepsilon\partial_z w - \partial_r u\}_{R-h} = 0. \tag{26}$$

For the wall boundary condition in eqn (15) and the interface eqn (19), their perturbative form is equal to the corresponding dimensionless equations. At this point, we have all the equations in the perturbative mode. It is possible to find the problem's solutions, this is, velocities and pressure, using the first-order expansion expresses as

$$u = u_0 + \varepsilon u_1, \tag{27}$$

$$w = w_0 + \varepsilon w_1, \tag{28}$$

$$p = p_0 + \varepsilon p_1. \tag{29}$$

Now, we write the zero-order perturbative equations. To perform this step, we substitute the expansion eqns (27)–(29) in the perturbative eqns (22)–(26) and (15) and take the limit $\varepsilon \to 0$.

For the continuity eqn (22) we obtain

$$r^{-1}\partial_r(ru_0) = 0, \tag{30}$$

here, considering the wall boundary conditions, we have that $u_0(r) = 0$. This result is consistent with the problem phenomenology where the changes in the radial direction are lower in order than the changes in the axial direction. For the momentum eqns (23) and (24) and considering the previous result for u_0, we have

$$\partial_r P_0 = 0, \tag{31}$$

$$\bar{r}^{-1}\partial_r(r\partial_r w_0) = -1, \tag{32}$$

for the normal stress eqn (25) we obtain

$$P_0|_{R-h} = P_c - Ka\left\{\partial_{zz}h + \frac{1}{R-h}\right\}_{R-h}, \tag{33}$$

here, we use a serial expansion to retain the axial curvature term according with [9]. For the tangential stress eqn (26) the results is

$$\partial_r w_0|_{R-h} = 0, \tag{34}$$

and the non-slip wall boundary conditions appear as

$$w_0|_R = u_0|_R = 0. \tag{35}$$

We integrate the eqn (32) to obtain the zero-order moment for radial velocity w_0,

$$w_0(r,z) = \frac{(R-h)^2}{2} log\left(\frac{r}{R}\right) + \frac{(R^2-r^2)}{4}, \tag{36}$$

finally, using eqn (31) and the respective boundary condition (33), we have P_0 expressed as

$$P_0(z,t) = P_c - Ka\left\{\partial_{zz}h + \frac{1}{R-h}\right\}. \tag{37}$$

At this point we have the zero-order equations and their corresponding solutions. Now, we proceed with the first-order formulation.

We substitute the first-order expansions (27)–(29) in eqns (22)–(26) and (15) and consider the first order terms, to obtain the first order constitutive equations. So, the first order continuity equation appears as

$$r^{-1}\partial_r(ru_1) + \partial_z w_0 = 0, \tag{38}$$

in (38), we use eqn (36) for w_0 and integrate to obtain u_1 in the form

$$u_1(r,z,t) = \frac{\partial_z h}{r}(R-h)\left\{\frac{r^2}{2} log\left(\frac{r}{R}\right) + \frac{R^2-r^2}{4}\right\}. \tag{39}$$

The corresponding first-order radial momentum equation appear as

$$\partial_r P_1 = \bar{r}^{-1}\partial_r(r\partial_r u_1) - \frac{u_1}{r^2}, \tag{40}$$

the first-order axial momentum equation is written in the form

$$\bar{r}^{-1}\partial_r(r\partial_r w_1) = \partial_z P_0 + \partial_t w_0 + u_1\partial_r w_0 + w_0\partial_z w_0, \tag{41}$$

at the interface the first-order tangential stress equation is written as

$$\partial_r w_1|_{R-h} = 0, \tag{42}$$

and the boundary condition at wall

$$w_1|_R = 0. \tag{43}$$

Using the solutions for P_0, (eqn (37)), w_0 (eqn (36)) and u_1 (eqn (39)), we integrate the eqn (41) to obtain the first-order component of radial velocity (w_1). After some algebraic work, the result appears as

$$\begin{aligned} w_1(r,z,t) = \; & Ka\partial_{zzz}h\left\{\frac{R^2}{4}-\frac{r^2}{4}+\frac{(R-h)^2}{2}log\left(\frac{r}{R}\right)\right\}+ \\ & Ka\partial_z h\left\{\frac{1}{2}log\left(\frac{r}{R}\right)+\frac{R^2}{4(R-h)^2}-\frac{r^2}{4(R-h)^2}\right\}+ \\ & \frac{\partial_z h}{128}(R-h)\left\{r^4+11R^4-12r^2R^2+8r^2R^2log\left(\frac{r}{R}\right)\right\}+ \\ & \frac{\partial_z h}{32}(R-h)^3\left\{5R^2-5r^2+(6r^2+4R^2)log\left(\frac{r}{R}\right)-\right. \\ & \qquad (r^2-R^2)\left[8log\left(\frac{R-h}{R}\right)\right]-4r^2log^2\left(\frac{r}{R}\right)+ \\ & \qquad \left. log\left(\frac{r}{R}\right)log\left(\frac{R-h}{R}\right)(8r^2-4R^2)\right\} \\ & \frac{\partial_z h}{16}(R-h)^3\left\{R^2[log^2(r)-log^2(R)]-2R^2log\left(\frac{r}{R}\right)log(R-h)\right\}+ \\ & \quad \frac{\partial_z h}{8}(R-h)^5\left\{log\left(\frac{r}{R}\right)\left[log\left(\frac{R-h}{R}\right)-2log^2\left(\frac{R-h}{R}\right)+\frac{3}{4}\right]\right\}. \end{aligned} \tag{44}$$

Eqns (36) and (44) represent the first-order approximation for radial velocity. Using these approximations in the interface integral eqn (19) we can solve it, through a numerical procedure obtaining the liquid interface behavior. In the next section, we describe the numerical approach.

3 NUMERICAL PROCEDURE

To analyze the film hydrodynamics, specifically the wave evolution under different geometrical and physical conditions, we have developed a numerical method to integrate the eqn (19). We set the calculation domain to a test pipe with $0 \leq z \leq L$ of length. Firstly, we discretize this domain using a uniform grid of N spatial cells, with a length of $\Delta z = L/N$. The spatial cell boundaries are given by $z_i = (i-1)\Delta z$, with $i = 1{:}\,N+1$. The coordinates of each spatial cell were taken at the spatial cell centers and denoted by overdot, i.e. $\dot{z}_i = (i-1/2)\Delta z$ with $i = 1{:}\,N$. So, an overdot in a magnitude with sub-index i means that magnitude is being calculated at the center of the i-th spatial cell, located between the boundaries z_i and z_{i+1}. In the other hand, a magnitude with sub-index i without overdot is related to the i-th cell boundary.

In the second step, we spatially discretize the PDE (eqn (19)). Then, we apply the averaging operator over a spatial cell

$$\dot{f_i} = \frac{1}{\Delta z}\int_{z_i}^{z_{i+1}} f(z)dz, \tag{45}$$

in eqn (19). The result is written as

$$d_t\dot{h}_i(t) = -\frac{q_{i+1}(t) - q_i(t)}{[R - \dot{h}_i(t)]\Delta z}, \quad i = 1 \dots N. \tag{46}$$

here the mass flow by a radian values $q_i(t)$ are obtained integrating eqn (20) over the film and evaluating for the corresponding z_i. To integrate eqn (20), we use the approximation for radial velocity defined by eqns (36) and (44). The first and third z-derivatives that appears in $q_i(t)$ are numerically evaluated at the boundaries of each spatial cell using an appropriate difference scheme [10].

The ordinary differential equation system represented by eqn (46) has one equation per spatial cell with the corresponding average film thickness $\dot{h}_i(t)$ as unknown. To express the film thickness at the cell boundaries $h_i(t)$ in terms of the values at the node centers, we consider a linear approximation in the form

$$h_i(t) = \left(\dot{h}_{i-1}(t) + \dot{h}_i(t)\right)/2, \quad i = 2 \dots N, \tag{47}$$

for the boundaries $h_1(t)$ and $h_{N+1}(t)$ values, we derive appropriate linear schemes using the left/right cell centers values. The linear approximation is a good approach for a fine uniform spatial grid [11].

The coefficients analysis and the preliminary calculations show that the ODE system (46) is a stiff problem [12]. A strategy to reduce the stiffness is to decrease the coefficients order, in consequence, we manipulate (46), extracting R^5 factor in the right side numerator, and introducing a variable change $\dot{h}_i(t) = \dot{h}'_i(t)/R$, the result is

$$d_t\dot{h}'_i(t) = -\frac{R^3\left[q'_{i+1}(t) - q'_i(t)\right]}{\left[1 - \dot{h}'_i(t)\right]\Delta z}. \tag{48}$$

To solve the system of eqn (48), we use a semi-implicit fourth order Adams–Moulton predictor-corrector method, with the prediction step based on the Adams–Bashforth algorithm [5], [12]. To initialize the Adams–Moulton multi-step method, we perform four iterations using a classical Runge–Kutta method [11]. As the initial condition for the problem, we use a flat Nusselt profile in all domain, adding at the first-quarter domain inlet, a small multiple modes Fourier perturbation [2] expressed as

$$h(z,0) = h_0 - \textstyle\sum_{k=1}^{K} b\cos(2\pi kz + \alpha_k), \quad 0 \le z \le L/4, \tag{49}$$

where K is the Fourier modes number, typically $K = 6$, the amplitude b satisfy $(b < 0.05h_0)$, and α_k is a random phase shift for each mode.

4 NUMERICAL RESULTS

To evaluate the analytical model and numerical procedure stability, we select a one-wave problem, this is, we study the initial perturbation over the pipe without introducing new inlet perturbations. We perform simulations for six different fluids and observe the solution stability. The selected fluids and their properties are listed in Table 1. Preliminary simulations showed that the stability is directly related to the fluid viscosity, for that reason we choose fluids with viscosities varying from 1E-2 to 1E-7.

Table 1: Fluid types and physical properties used in numerical experiments.

Fluid	T (°C)	ν (m²/s)	ρ (kg/m³)	σ (N/m)
Water	25	8.930E-07	9.970E+02	7.200E-02
Fuel oil light	20	1.650E-05	9.100E+02	2.300E-02
Castor oil	30	5.800E-04	9.550E+02	3.900E-02
Glycerine	20	1.183E-03	1.261E+03	6.340E-02
Fuel oil heavy	20	8.000E-03	9.900E+02	2.300E-02
Silicone oil [3]	-	1.330E-02	9.700E+02	2.150E-02

The selected simulation domain is a pipe with 2.1E-2m radius and 2.5m length. The initial Nusselt thickness corresponds to 1E-3m and the initial perturbation amplitude was $b = 0.03$. With this domain, we perform a mesh optimization study. Thus, empirically determine the maximum space and temporal steps (Δz and Δt), with which we can solve the problem without significant result differences for thinner meshes. For the spatial mesh the optimum step for all the fluids was $\Delta z = 1E - 3m$. The temporal step for silicone oil, fuel oil heavy, glycerin, and castor oil was $\Delta t = 5.0E - 5s$ while for fuel oil light and water was $\Delta t = 5.0E - 7s$.

Fig. 2. shows the wave temporal evolution for castor oil fluid. At the time $t = 0$, the initial condition, a multiple cosinusoidal perturbation is observed in the domain first quarter, being its maximum amplitude about $1.15h_0$. In the final three quarters of this profile, we have a flat behavior. Observing the time evolution we can notice that after the first seconds, the wave begin to change its shape, compressing and increasing its height, until it reach the maximum

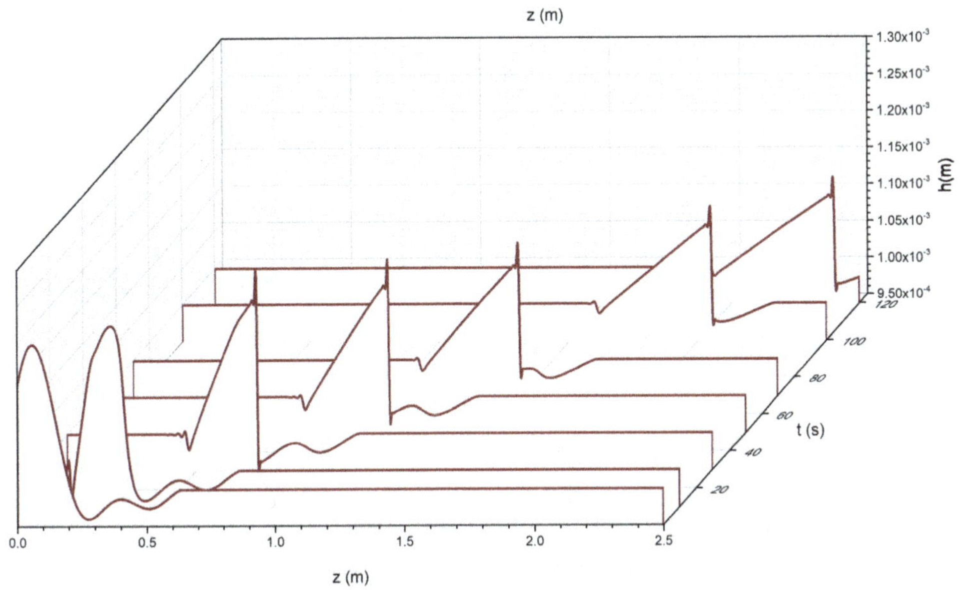

Figure 2: Temporal wave evolution for the castor oil fluid.

($1.25h_0$ approximately at $t_m = 25$s). From the maximum amplitude point, the wave holds its shape, and the amplitude slowly decrease until $t_s = 115$s and $z_s = 2.3m$ position. Next, these values stay stationary until the wave leaves the domain. Thus, the wave evolution reached the steady state.

The wave qualitative behavior of the others evaluated fluids is similar, varying the domain position and time where the wave becomes steady. In Table 2 we show the amplitude, position and time were the wave reaches the maximum (h_m, z_m, t_m). In addition, in Table 3 we offer this values for the steady state (h_s, z_s, t_s) within the time when the wave leaves the domain (t_T). Finally, Fig. 3 shows the fluid interfaces in the steady state beginning and data as film thickness and domain positions are offered.

Table 2: Wave evolution features for study fluids at maximum amplitude.

Fluid	h_m (m)	z_m (m)	t_m (s)
Castor oil	1.236E-03	5.895E-01	2.250E+01
Glycerine	1.247E-03	5.710E-01	4.400E+01
Fuel oil heavy	1.237E-03	5.770E-01	2.980E+02
Silicone oil	1.238E-03	5.930E-01	5.520E+03

Table 3: Wave evolution features for study fluids at reach steady stage, and total time.

Fluid	h_s (m)	z_s (m)	t_s (s)	t_T (s)
Castor oil	1.122E-03	2.290E+02	1.150E+02	1.540E+02
Glycerine	1.147E-03	1.861E+00	1.830E+02	3.120E+02
Fuel oil heavy	1.179E-03	1.430E+00	7.550E+02	1.840E+03
Silicone oil	1.183E-03	1.293E+00	1.489E+03	3.251E+03

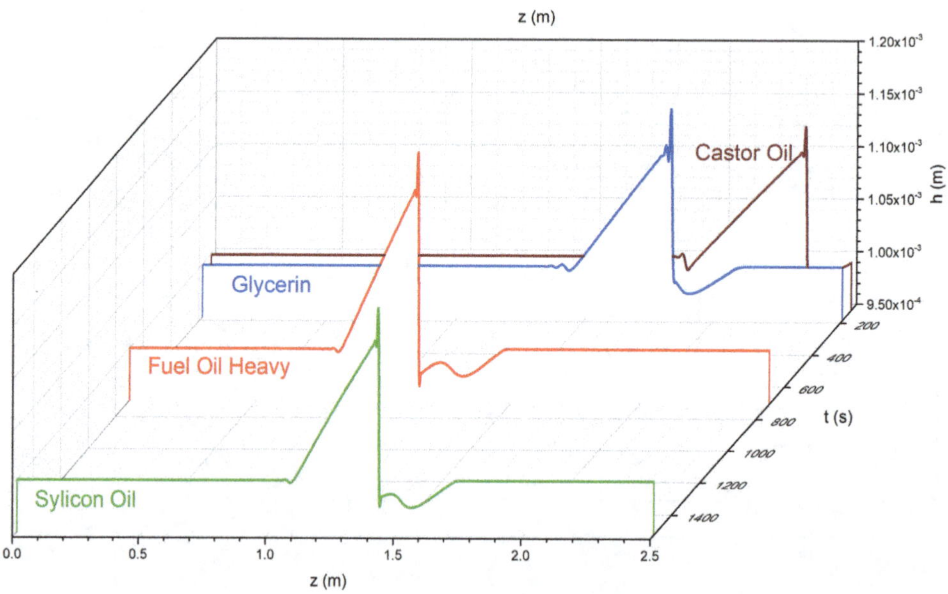

Figure 3: Fluid profiles when the wave evolution reach steady state.

The solutions for water and fuel oil light showed up unstable, for that reason, this data is not reported in Fig. 3 and Tables 2 and 3. For these low viscosity fluids, the algorithm results were not satisfactory.

Table 2 shows that the maximum wave height and its position do not depend on the fluid, being governed only by geometry and initial conditions. However, the time taken for the wave deformation, and the time needed to reach the maximum amplitude increase considerably with the fluid viscosity

Analyzing the Table 3 and Fig. 3, we can verify that when the fluid viscosity increases, the distance travelled by the wave to become steady decreases, while increasing the maximum wave height in steady state. Moreover, the time to reach this state and for the wave leaves the pipe, are significantly higher with the viscosity increment. The wave evolution in each fluid and the behavior dependence on kinetic viscosity, is consistent with the problem physic.

5 CONCLUSIONS AND FUTURE WORKS

In this work, we present an analytical model to simulate the wave evolution in gravity driven falling films in vertical pipes. The model is represented by an integral-differential equation that describes the film thickness as a function of the axial velocity.

The proposed numerical strategy discretizes the problem and solves the resultant ODE system using a semi-implicit fourth order predictor-corrector method. The analytical model and numerical procedure were evaluated solving a one-wave problem for fluids with different viscosities.

The model and numerical approach offered satisfactory results for medium and high viscosity fluids using fine temporal and spatial meshes. Nevertheless, for low viscosity fluids was not possible to reproduce the wave evolution, being the results unstable. The ODE system associated with this model is stiff, feature that is highlighted for low viscosity fluids. In this situation, the proposed numerical procedure is not appropriate.

In the short-term, this research future steps involve more robust numerical strategies, recommended for high stiffness problems, and techniques to evaluate the ODE's numerical stability. Therefore, we are expecting to get a numerical procedure capable to solve the model for low viscosity fluids with a good performance using thick spatial and temporal steps.

As a long-term objective, we need to evaluate the proposed model in more complex regimes involving multiple waves. In addition, we will use available experimental results to validate the model.

ACKNOWLEDGEMENTS

The authors are indebted to the plan of I+D support of the EXMOTRANSIN project ENE2016-79489-C2-1-P. Authors, Susana M. Iglesias and Dany S. Dominguez, acknowledge the Universidade Estadual de Santa Cruz for supporting this research.

REFERENCES

[1] Mascarenhas, N. & Mudawar, I., Study of the influence of interfacial waves on heat transfer in turbulent falling films. *International Journal Heat Mass Transfer*, **67**, pp. 1106–1121, 2013.

[2] Camassa, R., Forest, M.G., Lee, L., Ogrosky, H.R. & Olander, J., Ring waves as a mass transport mechanism in air-driven core-annular flows. *Physical Review E*, **86**(6), p. 66305, 2012.

[3] Camassa, R., Ogrosky, H.R. & Olander, J., Viscous film flow coating the interior of a vertical tube. Part 1: Gravity-driven flow. *Journal of Fluid Mechanics*, **745**, pp. 682–715, 2014.

[4] Ramaswamy, B., Chippada, S. & Joo, S., A full-scale numerical study of interfacial instabilities in thin-film flows. *Journal of Fluid Mechanics*, **325**, pp. 163–194, 1996.

[5] Muñoz-Cobo, J.L., Miquel, A., Berna, C. & Escrivà, A., Spatial and time evolution of non linear waves in falling liquid films by the harmonic expansion method with predictor-corrector integration. *Proceedings of the 12th International Conference on Heat Transfer, Fluid Mechanics and Thermodynamics*, pp. 91–97, 2016.

[6] Kalliadasis, S., Ruyer-Quil, C., Scheid, B. & Velarde, M.G., *Falling Liquid Films*, Springer-Verlag: London, 2012.

[7] Chen, T., Chiu, M.-S. & Weng, C.-N., Derivation of the generalized Young–Laplace equation of curved interfaces in nanoscaled solids. *Journal of Applied Physics*, **100**(7), p. 74308, 2006.

[8] Ruyer-Quil, C. & Manneville, P., Modeling film flows down inclined planes. *European Physics Journal B*, **6**, pp. 277–292, 1998.

[9] Craster, R.V. & Matar, O.K., Dynamics and stability of thin liquid films. *Review of Modern Physics*, **81**, pp. 1131–1198, 2009.

[10] Minkowycz, W.J., Sparrow, E.M. & Murthy, J.Y., *Handbook of Numerical Heat Transfer*, John Wiley & Sons: Hoboken, NJ, 2000.

[11] Hoffman, J.D. & Frankel, S., *Numerical Methods for Engineers and Scientists,* 2nd ed., Marcel Dekker: New York, 2001.

[12] Atkinson, K., Han, W. & Stewart, D., *Numerical Solution of Ordinary Differential Equations*, John Wiley & Sons: Hoboken, NJ, 2009.

LUBRICATION OF GEARBOXES: CFD ANALYSIS OF A CYCLOIDAL GEAR SET

FRANCO CONCLI[1], LORENZO MACCIONI[1] & CARLO GORLA[2]
[1]Faculty of Science and Technology, Free University of Bolzano/Bozen, Italy
[2]Dipartimento di Meccanica, Politecnicodi Milano, Italy

ABSTRACT

In recent years, the increasing demand for more and more compact and efficient solutions has highlighted the need to have appropriate tools in order to optimize the internal design, to avoid thermal problems, ensuring proper lubrication and to increase the reliability of the systems. Typical high power density gearbox designs are based on planetary, harmonic and cycloidal architectures. Although many analytical and numerical models are already available for the prediction of the power losses related to gear meshing (sliding), bearings and seals, literature is lacking in terms of hydraulic power loss models (deep lubrication, churning, windage and squeezing). Some numerical multiphase CFD and experimental studies on parallel axis and planetary gear sets have been already performed by the authors in previous research. The aim of this paper is to extend the applicability of the previously developed numerical techniques to cycloidal architectures, taking into account the typical lubricants used for these type of drives. With respect to the load independent power losses (related to the interaction of the mechanical component and the surrounding lubricant), the cycloidal gear set has been numerically simulated with an especially developed CFD code implemented in the OpenFOAM® environment. A specific mesh handling technique allows us to manage the topological changes of the domain ensuring the numerical stability of the simulation and the correct calculation of the complex multiphase flows that take place in gearboxes. The results have been compared with those already available for other gear architectures with similar performances (dimensions, reduction ratios and loads).

Keywords: cycloidal drive, CFD, power losses, efficiency maps, multiphase.

1 INTRODUCTION

Nowadays, a big selection of high power density gearboxes are available on the market and, in a business-to-business context, designers could choose the ideal solution based on a series of attributes such as output speed or gear ratio (GR), torque, backlash, overall dimensions, axis orientation, working conditions, cost, efficiency, etc.

Typical competing solutions are: traditional planetary gearboxes, cycloidal architectures and harmonic drives. For high GRs, the most promising architectures seem to be the cycloid and harmonic ones (having the same number of stages) while for low GRs the harmonics are not competitive solutions. For GRs between 30 and 130, cycloidal and harmonic solutions have approximately the same power density [1]. Below a GR of 30, the kinematic of the harmonic drive is not working. In addition, the tolerances required to ensure the correct operation of such solution are of an order of magnitude higher than those needed in cycloidal and planetary gear sets [1]. This affects significantly costs and reliability. For this reason, for GRs between 10 and 30, only planetary and cycloidal architectures can be effectively adopted.

A cycloidal gear has the following main elements: an input shaft with an eccentric cam, a cycloidal disk, an external (fixed) housing with a series of rollers that engaged with the cycloid disk, an output shaft able to extract the rotation motion of the cycloid disk (usually through hole-pin). The input shaft moves the centre of the cycloid disk along a circular trajectory. The cycloidal disk, performing this eccentric motion, engages with the rollers positioned on the external housing. It has an opposite rotating direction with respect the input

www.witpress.com, ISSN 1743-3533 (on-line)
doi:10.2495/MPF190101

shaft. Holes in the cycloidal disk drag the pins, allowing the output shaft to extract this motion. Since this solution is dynamically unbalanced, additional components or modifications are often introduced, such as, for example, counterweights or an additional cycloidal out of phase (180°) disk.

The GR depend on the number of the cycloidal lobes Z_1 and the number of rollers Z_2. It is possible to note how the maximum GR can be achieved when the difference $Z_2 - Z_1 = 1$. Therefore, this is the solution usually considered when not differently explained.

$$GR = \frac{\omega_{input}}{\omega_{output}} = \frac{Z_1}{Z_2 - Z_1} \tag{1}$$

From a theoretical point of view, in a cycloidal gear set, each pin is in contact with the corresponding lobe, but only half of the rollers transfer forces [2]. However, the presence of backlash between rollers and cycloidal disk (for compensating manufacturing tolerances and to guarantee better lubrication and mounting conditions [3], causes that not all the engaged flanks transmit the theoretical load. A numerical and experimental analysis of a cycloidal reducer with a single contact lobe-roll (to simulate the worst use condition) was presented by Sensinger and Lipsey [1]. It is shown that the maximum stress occurs in the contact point between the lobe and the roller. Indeed, cycloidal profiles present only compression stresses, thus ensuring a better resistance to overloads (up to 500%). The investigations have also shown that just in the contact areas where the force has the maximum value, the equivalent curvature has its maximum value and this is exceptionally favourable from the point of view of the loading capacity of the cycloidal disk.

However, the efficiency of a cycloidal gear is generally not a competing factor compared with traditional planetary solutions.

2 POWER LOSSES OF GEARS

The power losses in geared transmissions can be classified, according to Niemann, into load dependent and load independent one (sub-index $|_0$). The load dependent power losses are proportional to the transmitted torque, while the load independent one are not directly related to the load. However, the level of torque affects the operating temperatures and the lubricant properties (density and mostly viscosity) having an influence also on the so-called load independent power losses. A further classification can be made according to the mechanical component that generates the losses:

$$P_L = P_{LG} + P_{LG0} + P_{LB} + P_{LB0} + P_{LS} + P_{LX}. \tag{2}$$

The sub-indexes $|_G$, $|_B$, $|_S$ and $|_X$ stand for gears, bearings, seals and other generic components respectively.

In the years, several authors have published experimentally derived or analytical results for most of the losses. The load dependant power losses of gears (P_{LG}) are due to sliding in the contact. The ISO 6336 standard [4] provides an accurate model to predict such component of the losses. Examples of particular designs with improved meshing efficiency have been shown [5]–[7].

Several research activities were carried out for what concerns the bearing losses. The SKF model [8] or the Schaeffler [9] one, for example, provide accurate results both for the load dependant (P_{LB}) and for the load independent (P_{LB0}) power losses of bearings. For the seal losses (P_{LS}), a general method is shown in [10]. What is still missing is a general method for the prediction of the load independent power losses of gears (P_{LG0}).

In the past, several authors proposed empirical models based on experimental evidences. Daily and Nece [11] and Mann and Marston [12] proposed models for the churning losses of smooth discs. These results were extended by Ohlendorf [13] who proposed the first churning model for a (single) gear. The next improvement was done by Richter [14]. Its model is the first that take into consideration the mutual effect of two meshing gears. Terekhov [15] performed systematic research in order to understand the influence of the rotational speed, the module, the immersion depth, the gear diameter, the tooth face width, the gear ratio and the oil viscosity on the churning and squeezing power losses for splash lubricated gears. The formulation proposed by Terekhov was improved by Walter and Langenbeck [16] and Walter [17]. One of the most complete empirical models was published in 1975 by Mauz [18]. More recent experimental studies were performed by Diab et al. [19] and Höhn et al. [20]. However, all these models are applicable only to spur gears and for a limited range of operating conditions. The limitations of the empirical models have been overcome with the introduction, in the recent years, of the computational fluid dynamic and the recent development in computer science. The first numerical works were presented by Diab et al. [19] and Marchesse et al. [21]. Further studies were performed by Gorla et al. [22], [23] and Concli et al. [24]. The main limitation of the numerical models is related to the computational effort required. Some research was conducted by the author in order to improve the computational efficiency of the numerical methods [25], [26]. SPH based methods were presented by Groenenboom et al. [27], Liu et al. [28], Ji et al. [29], Shi et al. [30], Rahmatjan and Geni [31], Imin and Geni [32], Zhigang et al. [33] and Liu et al. [34]. A review article showing advantages and limitation of all the available models was presented by Concli and Gorla [35].

This lack of literature, valid for gears in general, is even more noticeable for cycloidal architectures, where the sources of loss are due to friction of bearings (on the eccentric input shaft), friction between the rollers and the cycloidal disk, friction in the holes-pins contacts (or any solution adopted to extract the motion from the cycloidal disk) and hydraulic losses.

Although the first three points of the bullet list have been widely discussed in the literature, the hydraulic losses have not been studied at all.

3 FV APPROACH

In the past some authors have already studied the lubrication of planetary gearboxes with a finite volume (FV) approach [36]–[43]. In this study, the already presented global remeshing method (GRA) [44] was applied to a cycloidal gear-set.

The solver adopted is a solver for 2 incompressible fluids, which captures the interface using a VOF method with optional mesh motion. It relies on the numerical solution of a mass and a momentum conservation equation:

$$\frac{\partial \rho}{\partial t} + \nabla \cdot (\rho \boldsymbol{v}) = 0, \tag{3}$$

$$\frac{\partial (\rho \boldsymbol{v})}{\partial t} + \nabla \cdot (\rho \boldsymbol{v}\boldsymbol{v}) = -\nabla p + \nabla \cdot [\mu(\nabla \boldsymbol{v} + \nabla \boldsymbol{v}^T)] + \rho \boldsymbol{g} + \boldsymbol{F}, \tag{4}$$

where ρ is the density, $\boldsymbol{v}$ is the velocity vector, μ is the viscosity of the lubricant, $\boldsymbol{g}$ is the gravitational acceleration and $\boldsymbol{F}$ represents the external forces. These equations are applied to each cell of the computational grid. The solution of the system is performed with a PIMPLE (merged PISO-SIMPLE) algorithm. This conjugate the advantages in terms of computational efficiency of the SIMPLE scheme with the capability of the PISO one to be time-conservative.

An additional equation is used to take into account the presence of two phases (air and lubricant):

$$\frac{\partial}{\partial t}\alpha + \frac{\partial}{\partial x_i}(\alpha u_i) = 0, \tag{5}$$

where α is a scalar called volume fraction. The averaged properties (ϕ) of the mixture in each cell of the domain are calculated as an α-averaged value of the properties of air and lubricant:

$$\phi = \phi_{lub} \cdot \alpha + \phi_{air} \cdot (1 - \alpha). \tag{6}$$

4 MESH HANDLING

In CFD simulations of gearboxes, the computational domain intrinsically changes its topology during the calculation. This implies a severe mesh distortion and the need to update the grid after few time steps. In the past, the authors have used commercial software capable to handle automatically the grid update. Even if this approach is effective, it is not efficient. The user, in fact, loses the control of the mesh size and element shape. Small cells were inevitably produced and this causes a significant increment in the computational effort required, not so much for the increased number of elements, but for the need, in order to ensure the stability of the solution, to reduce drastically the time step (to keep the Co number equal to 1). The authors have shown how, with the so called global remeshing approach (GRA) [45], it is possible to have a much higher control of the grid parameters, ensuring huge time savings for the computation thanks to the bigger possible time steps.

The grid is regenerated completely after few time steps (Fig. 1) and the results of the previous computation mapped on the new mesh.

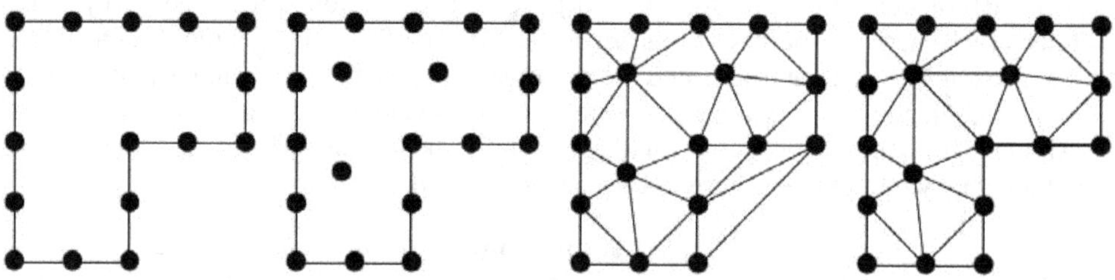

Figure 1: Steps performed by the Delaunay algorithm [45].

In OpenFOAM® two meshing tools are available. BlockMesh, is capable of creating parametric structured grids but is not suitable to handle complex geometries. The second tool, SnappyHexMesh, is capable to generate 3-dimensional grids starting from triangulated surface geometries. The grid approximately conforms to the surface by iteratively refining a starting structured grid and morphing the resulting split-hex mesh to the surface. This approach is robust but also very time consuming and not suitable for the purposes of this study.

For this reason, the grid (2-dimensional) was created following a top down strategy that starts by computing the corner points, and discretize the edges and the faces with an advancing front surface mesh generator. A fast Delaunay algorithm generates the mesh.

Eventually it fails for the last elements, back-tracking rule based algorithm takes over [39], [46]. The Delaunay algorithm subdivides the edges into segments according to the local seed prescribed element size. The faces are seeded with points. These are connected together during the so-called triangulation phase. If the results of the triangulation is a grid that exceed the boundaries, the protruded elements are removed.

Fig. 2 shows the flow chart of the GRA. A first mesh is generated and the simulation performed up to the point in which the quality pf the grid becomes lower than a threshold. At this point a new grid is generated starting from the topology of the domain of the previous time step. The results are mapped on the new grid. The simulation restarts. The procedure is repeated up to when the last time step is reached.

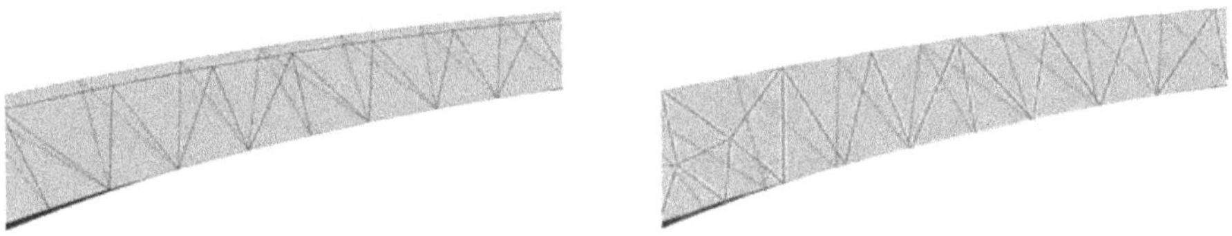

(a) (b)

Figure 2: Example of mesh update. (a) Grid before the deformation (blue) and after several time steps (red); and (b) Deformed mesh of the previous time step (red) and new mesh (green).

5 CYCLOIDAL GEAR SET

The profile of a cycloidal gear set, with a unitary difference between number of rollers and number of lobes, can be expressed by the coordinates C_x and C_y (with the origin in the centre of the cycloid) as function of the parameter ϕ (angle of input shaft) that can vary between 0 and 2π.

$$C_x = \mathrm{R} \cdot \cos(\phi) - R_r \cdot \cos(\phi + \psi) - \mathrm{e} \cdot \cos((Z_1 + 1) \cdot \phi), \tag{7}$$

$$C_y = -\,\mathrm{R} \cdot \sin(\phi) + R_r \cdot \sin(\phi + \psi) + \mathrm{e} \cdot \sin((Z_1 + 1) \cdot \phi), \tag{8}$$

$$\text{where } \psi = \tan^{-1}\left(\frac{\sin(Z_1 \cdot \phi)}{\cos(Z_1 \cdot \phi) - \frac{\mathrm{R}}{\mathrm{e} \cdot (Z_1 + 1)}}\right), \tag{9}$$

where R is the distance between the centre of each roller and the input shaft axis, R_r is the radius of rollers, e is the eccentricity of the input shaft and Z_1 is the number of lobes of the cycloid (one less of the number of rollers in this case) [47].

With the aim of comparing the numerical hydraulic losses of a traditional planetary gear-set (already studied in [36]) with a potentially competing cycloidal gear set, geometrical data were chosen in order to have the same gear ratio, comparable dimensions and same nominal torque.

$$R = 40 \text{ mm};\ R_r = 5 \text{ mm};\ e = 2 \text{ mm};\ Z_1 = 10 \tag{10}$$

A radial gap of 0.1 mm was selected. The pin-hole connection and any additional balancing cycloid were not considered.

6 RESULTS AND DISCUSSION

Simulations were performed with a 2D model. The lateral backlashes between the internal shaft and the external housing were not modelled. This is an acceptable assumption if the axial width of the gear-set is significant. The meshing between the two cycloids (internal and external ones) produces a significant pumping effect [48]–[50]. This can be observed in Figs 3–5. The volume between two lobes reduces suddenly causing an overpressure in the gap (Fig. 6). Part of the lubricant in squeezed out. The successive opening of the gap (separation of the surfaces belonging to the internal and the external cycloids) causes a successive pressure decrease (Fig. 7). Vortices take place. The circulation involves mainly the region filled with air while the lubricant seem to have a laminar flow.

The pressure is constant in each gap.

Similar squeezing/pocketing effects were observed for planetary gears, but only near the meshing region (Fig. 8). The much bigger open spaces in a traditional planetary configuration, promote a better circulation of the lubricant and the pumping effects results reduced with respect to the cycloidal gear set.

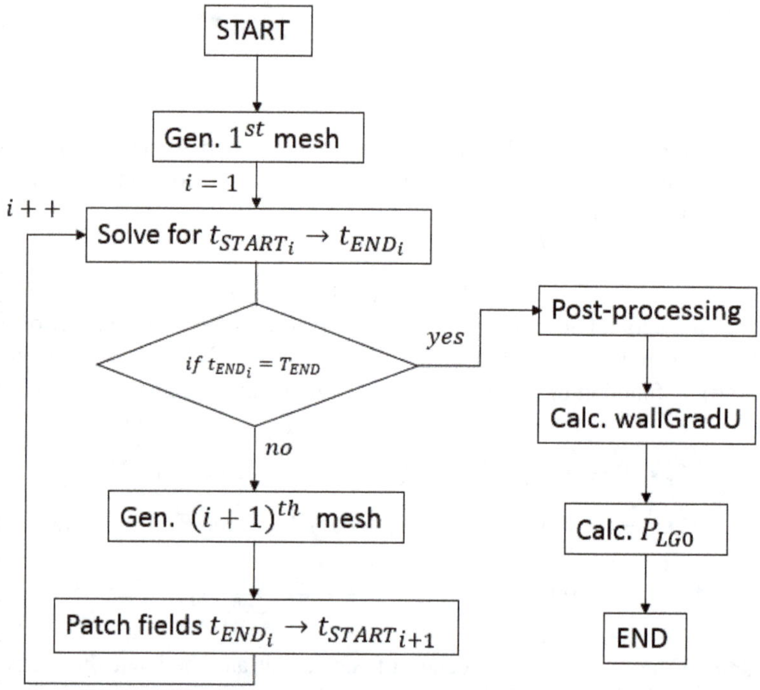

Figure 3: Flow chart of the GRA.

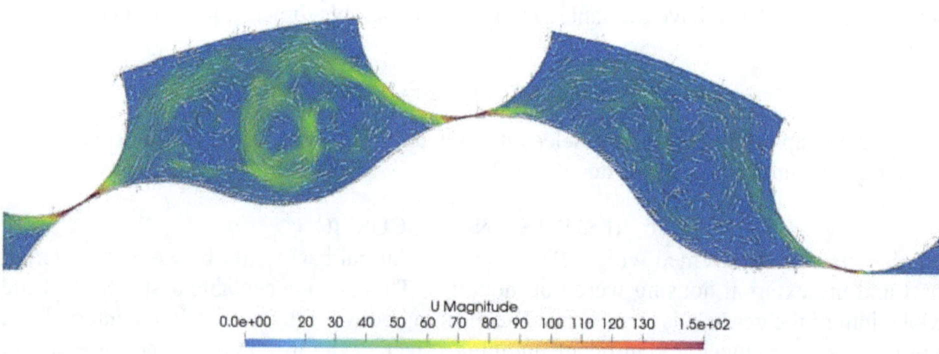

Figure 4: Velocity contour plot and oil distribution.

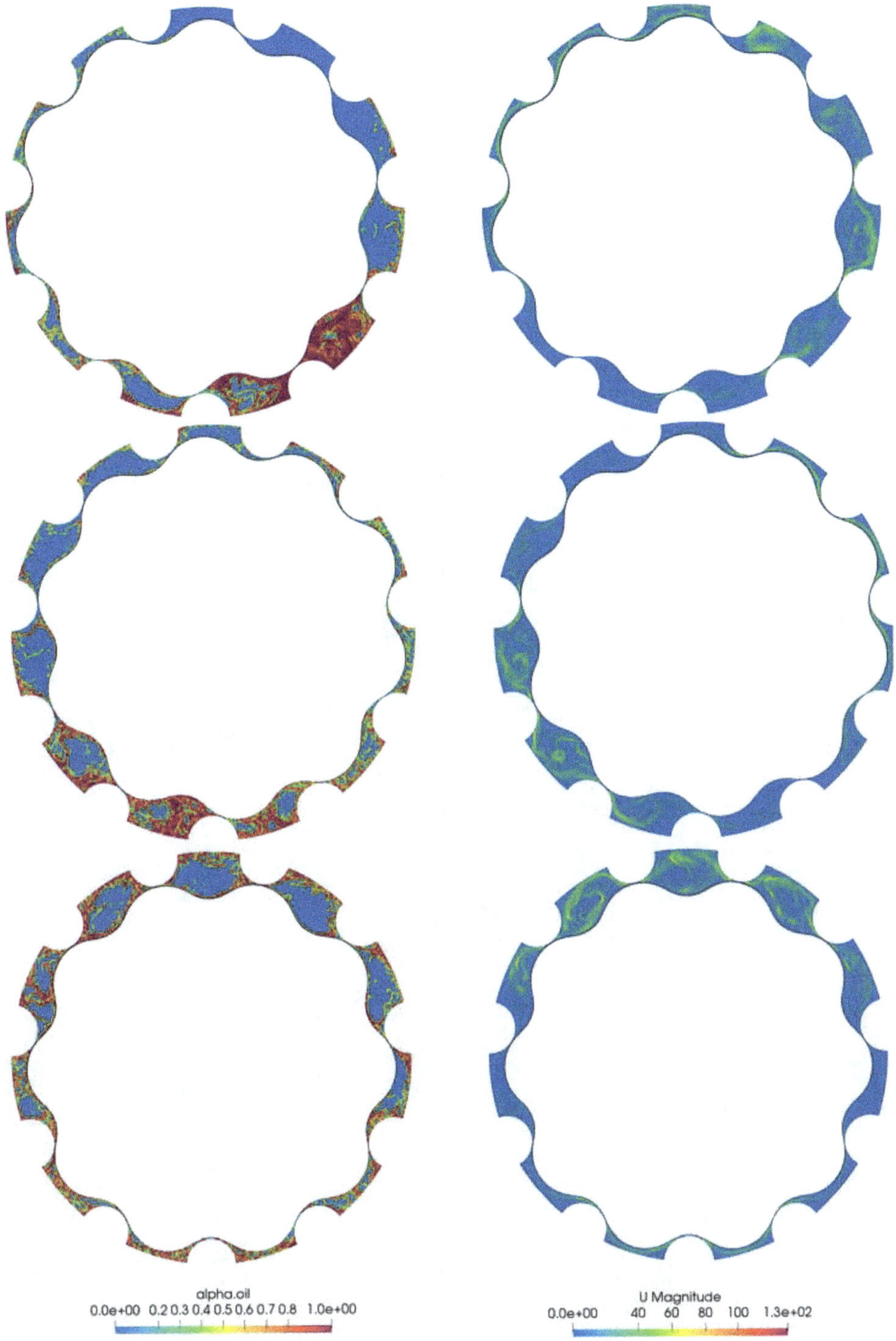

Figure 5: Details of the circulation effects: on the right a gap that is closing, on the left a gap that is opening. The squeezing effect can be clearly observed.

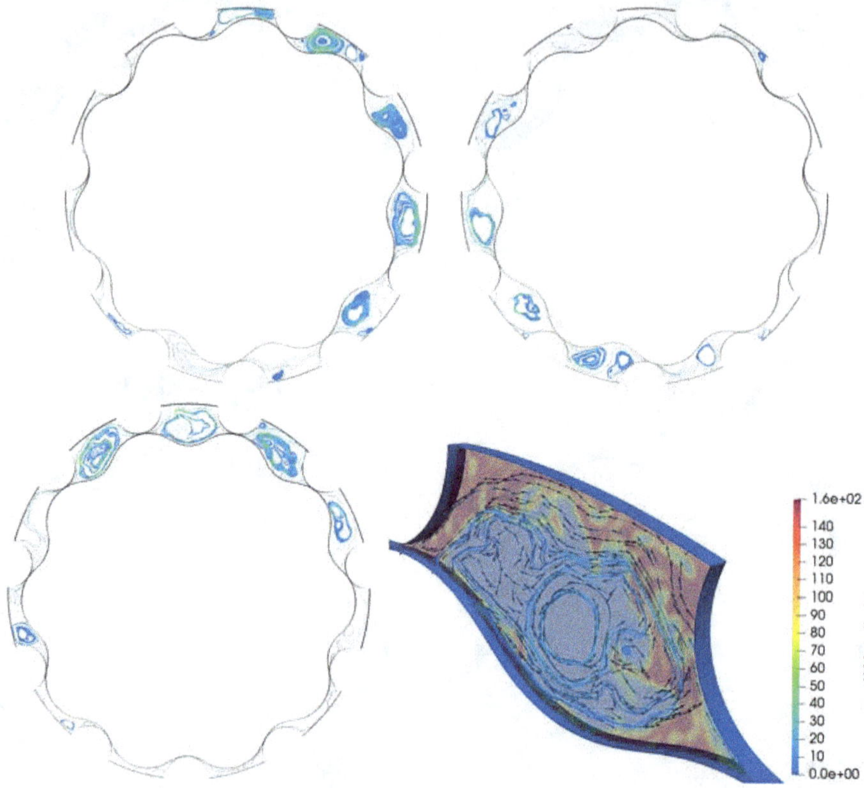

Figure 6: Streamlines and oil volume fraction contour.

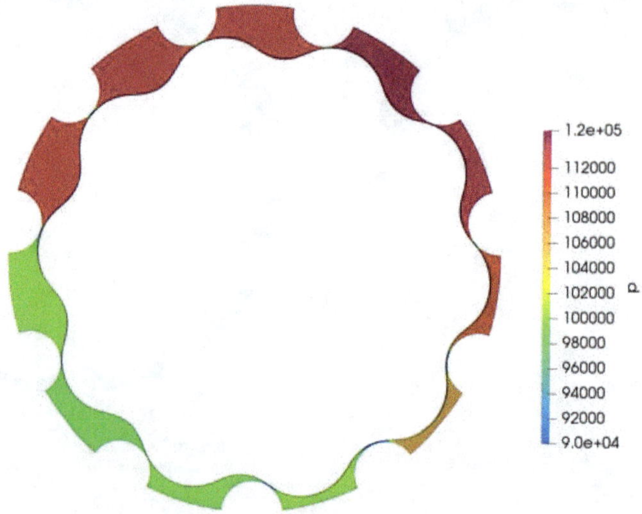

Figure 7: Pressure distribution in the different gaps.

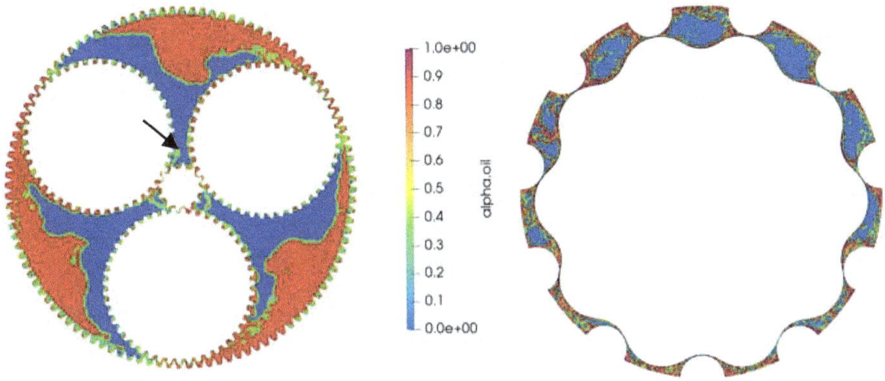

Figure 8: Comparison between a traditional planetary gear-set [6] and a cycloidal configuration (reduction ratio 10).

7 COMPUTATIONAL EFFORT

This preliminary simulation was performed on a single core 9.6GFLOP hardware. The simulation took approximately 3 days to perform a complete rotation of the internal cycloid.

8 CONCLUSIONS

The present paper reports some preliminary results concerning the lubrication of a cycloidal gear set. The adopted model is 2D. This implies that the axial flows were neglected. This assumption is acceptable if the axial with of the gears is significantly higher with respect to the lateral backlash. Beside the geometrical simplifications, the proposed approach has allowed the simulation of a complex kinematic in a reasonable amount of time, ensuring a very stable numerical convergence. Based on this promising results, the presented numerical approach will be extended and used for deeper studies of the lubrication of cycloidal gear sets.

The 2D results have shown that while in planetary gear sets, the main load independent power losses of gears are due to churning (of the planet carrier and planets), in cycloidal configurations, significant pumping/squeezing effect took place in the gaps near the meshing region due to sudden variations of the volumes and high pressure gradients. Vortices appears, especially in the opening gaps (where the lubricant/air mixture is squeezed from the previous gap that is reducing its volume). These vortices involve mainly the air while the lubricant flux seems to be laminar ($Re < 2E3$).

REFERENCES

[1] Sensinger, J.W. & Lipsey, J.H., Cycloid vs. harmonic drives for use in high ratio, single stage robotic transmissions. *2012 IEEE Int. Conf. Robot. Autom.*, pp. 4130–4135, 2012.

[2] Gorla, C., Davoli, P., Rosa, F., Longoni, C., Chiozzi, F. & Samarani, A., Theoretical and experimental analysis of a cycloidal speed reducer. *J. Mech. Des. Trans. ASME*, **130**(11), pp. 1126041–1126048, 2008.

[3] Blagojevic, A. et al., Numerical and experimental analysis of the cycloid disc stress state. *Tech. Gaz.*, **21**(2), pp. 337–382, 2014.

[4] ISO 6336, 2006.

[5] Xia, Y. et al., A new topology of CMG for high torque and low loss. *2016 Prog. Electromagn. Res. Symp. PIERS 2016 – Proc.*, **65**(11–12), pp. 78–82, 2016.
[6] Concli, F., Low-loss gears precision planetary gearboxes: Reduction of the load dependent power losses and efficiency estimation through a hybrid analytical–numerical optimization tool. *Forsch. im Ingenieurwesen/Engineering Res.*, 2017.
[7] Anderson, N.E. & Loewenthal, S.H., Design of spur gears for improved efficiency. *J. Mech. Des. Trans. ASME*, **104**(4), pp. 767–774, 1982.
[8] www.SKF.com.
[9] www.schaeffler.it.
[10] Niemann, G. & Winter H., *Maschinenelemente*, 2005.
[11] Daily, J.W. & Nece, R.E., Chamber dimension effects on induced flow and frictional resistance of enclosed rotating disks. *J. Fluids Eng. Trans. ASME*, **82**(1), pp. 217–230, 1960.
[12] Mann, R.W. & Marston, C.H., Friction drag on bladed disks in housings as a function of reynolds number, axial and radial clearance, and blade aspect ratio and solidity. *J. Fluids Eng. Trans. ASME*, **83**(4), pp. 719–723, 1961.
[13] Ohlendorf, H., *Verlustleistung und Erwärmung von Stirnrädern*, 1958.
[14] Richter, W., *Stirnradgetriebe, Zahnreibung, Verlustleistung und Erwärmung*, 1964.
[15] Terekhov, A.S., Hydraulic losses in gearboxes with oil immersion. *Russ. Eng. J.*, **55**(5), pp. 7–11, 1975.
[16] Walter, P. & Langenbeck, K., *Anwendungegrenzen für die Tauchschmierung von Zahnradgetrieben, Plansch- und Quetschverluste bei Tauchschmierung*, 1982.
[17] Walter, P., *Untersuchung Zur Tauchschmierung Von Stirnrädern Bei Umfangsgeschwindigkeiten Bis 60 M/S*, 1982.
[18] Mauz, W., Hydraulische Verluste von Strinradgetrieben bei Umfansgs-geschwindigkeiten bis 60 m/s. *Hydraul. Verluste von Stirnradgetrieben Bei Umfangsgeschwindigkeiten Bis 60 M/s*, 1987.
[19] Diab, Y., Ville, F., Velex, P. & Changenet, C., Windage losses in high speed gears-preliminary experimental and theoretical results. *J. Mech. Des. Trans. ASME*, **126**(5), pp. 903–908, 2004.
[20] Höhn, B.-R., Michaelis, K. & Otto, H.-P., Influence on no-load gear losses. *Ecotrib. 2011 Conf. Proc.*, **2**, pp. 639–644, 2011.
[21] Marchesse, Y., Changenet, C., Ville, F. & Velex, P., Investigations on CFD simulations for predicting windage power losses in spur gears. *J. Mech. Des. Trans. ASME*, **133**(2), 2011.
[22] Gorla, C. et al., Hydraulic losses of a gearbox: CFD analysis and experiments. *Tribol. Int.*, **66**, 2013.
[23] Gorla, C. et al., CFD simulations of splash losses of a gearbox. *Adv. Tribol.*, 2012.
[24] Concli, F. et al., Load independent power losses of ordinary gears: Numerical and experimental analysis. *5th World Tribology Congress, WTC 2013*, **2**, 2013.
[25] Concli, F., Gorla, C., Della Torre, A. & Montenegro, G., Windage power losses of ordinary gears: Different CFD approaches aimed to the reduction of the computational effort. *Lubricants*, **2**(4), pp. 162–176, 2014.
[26] Concli, F., Gorla, C., Della Torre, A. & Montenegro, G., Churning power losses of ordinary gears: A new approach based on the internal fluid dynamics simulations. *Lubr. Sci.*, **27**(5), 2015.
[27] Groenenboom, P.H.L., Mettichi, M.Z. & Gargouri, Y., Simulating oil flow for gearbox lubrication using smoothed particle hydrodynamics. *Proc. Int. Conf. Gears 2015*, 2015.

[28] Liu, Z., Shen, Y. & Rinderknecht, S., Theoretical and experimental investigation on power loss of vehicle transmission synchronizers with spray lubrication. *SAE Tech. Pap.*, **2019**(January), 2019.
[29] Ji, Z., Stanic, M., Hartono, E.A. & Chernoray, V., Numerical simulations of oil flow inside a gearbox by Smoothed Particle Hydrodynamics (SPH) method. *Tribol. Int.*, **127**, pp. 47–58, 2018.
[30] Shi, Y., Li, S., Chen, H., He, M. & Shao, S., Improved SPH simulation of spilled oil contained by flexible floating boom under wave–current coupling condition. *J. Fluids Struct.*, **76**, pp. 272–300, 2018.
[31] Rahmatjan, I. & Geni, M., SPH algorithm for proper meshing and coupling contact of gears. *Zhendong yu Chongji/Journal Vib. Shock*, **34**(12), pp. 65–69, 2015.
[32] Imin, R. & Geni, M., Stress analysis of gear meshing impact based on SPH method. *Math. Probl. Eng.*, **2014**, 2014.
[33] Zhigang, Y., Imin, R. & Geni, M., Study on the gear modeling in SPH analysis. *Adv. Mater. Res.*, **33–37**, pp. 773–778, 2008.
[34] Liu, H. et al., Numerical modelling of oil distribution and churning gear power losses of gearboxes by smoothed particle hydrodynamics. *Proc. Inst. Mech. Eng. Part J J. Eng. Tribol.*, **233**(1), pp. 74–86, 2019.
[35] Concli, F. & Gorla, C., Windage, churning and pocketing power losses of gears: Different modeling approaches for different goals (Wirkungsgrad und Verluste von Zahnradgetrieben: Verschiedene Methoden für verschiedene Anwendungen). *Forsch. im Ingenieurwesen/Engineering Res.*, **80**(3–4), pp. 85–99, 2016.
[36] Concli, F. & Gorla, C., Numerical modeling of the churning power losses in planetary gearboxes: An innovative partitioning-based meshing methodology for the application of a computational effort reduction strategy to complex gearbox configurations. *Lubr. Sci.*, 2017.
[37] Concli, F. & Gorla, C., CFD simulation of power losses and lubricant flows in gearboxes. *American Gear Manufacturers Association Fall Technical Meeting 2017*, 2017.
[38] Concli, F., Conrado, E. & Gorla, C., Analysis of power losses in an industrial planetary speed reducer: Measurements and computational fluid dynamics calculations. *Proc. Inst. Mech. Eng. Part J J. Eng. Tribol.*, **228**(1), 2014.
[39] Concli, F., Thermal and efficiency characterization of a low-backlash planetary gearbox: An integrated numerical–analytical prediction model and its experimental validation. *Proc. Inst. Mech. Eng. Part J J. Eng. Tribol.*, **230**(8), 2016.
[40] Drewniak, J., Kopec, J. & Zawislak, S., Kinematical and efficiency analysis of planetary gear trains by means of various graph-based approaches. *Mech. Mach. Sci.*, **34**, pp. 263–284, 2016.
[41] Yin, H.B., Li, S.L., Zhang, H., Zhao, X.Y. & Zhang, J., The power loss and efficiency analysis of a 3DOFs planetary gear box. *Adv. Mater. Res.*, **834–836**, pp. 1285–1289, 2013.
[42] Concli, F. & Gorla, C., Computational and experimental analysis of the churning power losses in an industrial planetary speed reducer. *WIT Transactions on Engineering Sciences*, vol. 74, WIT Press: Southampton and Boston, 2012.
[43] Gorla, C. & Concli, F., Influence of lubricant temperature, lubricant level and rotational speed on the churning power loss in an industrial planetary speed reducer: Computational and experimental study. *Int. J. Comput. Methods Exp. Meas.*, **1**(4), pp. 353–366, 2013.

[44] Concli, F. & Gorla, C., Numerical modeling of the power losses in geared transmissions: Windage, churning and cavitation simulations with a new integrated approach that drastically reduces the computational effort. *Tribol. Int.*, **103**, 2016.
[45] Concli, F., Della Torre, A., Gorla, C. & Montenegro, G., A new integrated approach for the prediction of the load independent power losses of gears: Development of a mesh-handling algorithm to reduce the CFD simulation time. *Adv. Tribol.*, **2016**, 2016.
[46] Schöberl, J., An advancing front 2D/3D-mesh generator based on abstract rules. *Comput. Vis. Sci.*, **1**(1), pp. 41–52, 1997.
[47] Sensinger, J.W., Unified approach to cycloid drive profile, stress, and efficiency optimization. *J. Mech. Des. Trans. ASME*, **132**(2), pp. 245031–245035, 2010.
[48] Concli, F. & Gorla, C., Oil squeezing power losses in gears: A CFD analysis. *WIT Transactions on Engineering Sciences*, vol. 74, WIT Press: Southampton and Boston, 2012.
[49] Concli, F. & Gorla, C., Analysis of the oil squeezing power losses of a spur gear pair by mean of CFD simulations. *ASME 2012 11th Biennial Conference on Engineering Systems Design and Analysis, ESDA 2012*, **2**, 2012.
[50] Gorla, C. & Concli, F., A CFD analysis of the oil squeezing power losses of a gear pair. *Int. J. Comput. Methods Exp. Meas.*, **2**(2), pp. 157–167, 2014.

COMPUTATIONAL INVESTIGATIONS OF POLYMER SHEET BREAKUP FOR OPTIMIZATION OF DEVOLATILIZATION PROCESSES IN STEAM CONTACTORS

BRAD SHINDLE[1] & ABHILASH J. CHANDY[2]
[1]Department of Mechanical Engineering, The University of Akron, USA
[2]Department of Mechanical Engineering, Indian Institute of Technology Bombay, Mumbai, India

ABSTRACT

Polymer devolatilization is a vital process in polymer manufacturing and is significantly impactful on the successful creation of high quality polymers, meeting both rigorous product specifications and regulatory requirements. One of the devolatilization methods that is the focus of this research is steam stripping, where superheated steam is used to remove any unwanted substances, such as volatiles and solvents, from the polymer mixture. This polymer mixture, referred to as "cement" and comprised of polymer and a volatile cyclohexane solvent, mixes with superheated steam and undergoes breakup into smaller droplets followed by phase change resulting in the removal of cyclohexane. The objective of the current study is embodied in two steps. The first step involves the development of a computational fluid dynamics (CFD)-based multiphase model that solves for the initial breakup of the liquid polymer mixture by steam. As part of this effort, detailed parametric studies are conducted to determine the effects of different contactor geometries on the initial sheet breakup, and the potential impact on the final polymer product quality. The second step then uses the resulting diameter distribution to model the multiphase heat and mass transfer of the polymer mixture including evaporation. Specifically, 3D CFD calculations are carried out using a Eulerian–Lagrangian approach, where the superheated steam is modelled as the continuous phase and tracked in a Eulerian frame, while the cement droplets are treated using a Lagrangian tracking method, thereby providing distributions of particle sizes, temperatures, and solvent content in the contactor. Results helped optimize the devolatilization process in terms of steam savings and volatile content in the final polymer.
Keywords: polymer devolatilization, CFD, multiphase.

1 INTRODUCTION

The quality of each polymer is paramount to the successful use of the material for its intended application, some of which range from agricultural and biomedical solutions to aerospace components and even to modern day clothing and accessories. High quality polymers, in particular, require significant refinement to meet product specifications and environmental regulations. Perhaps the most significant manufacturing process affecting quality is the process of polymer devolatilization. Essentially, polymer devolatilization is the removal of unwanted byproducts in the polymer mixture, such as monomers, solvents, and any other volatile products introduced when making the polymer [1].

The devolatilization process occurs inside a pressurized vessel called a steam contactor, to cause the polymer mixture to disintegrate into smaller liquid structures or droplets. These droplets continue to break apart as they are carried through the contactor by the superheated steam, which also serves as the heat source necessary to further evaporate the solvents. The resulting product mixture, comprised of a higher polymer content and a lower volatile content than at the start of the process, then undergoes additional processing to create granular polymer particles called "crumb".

The liquid polymer mixture in this research, hereon referred to as "cement", is comprised of polymer and cyclohexane, a hydrocarbon solvent. The superheated steam and cement are

WIT Transactions on Engineering Sciences, Vol 123, © 2019 WIT Press
www.witpress.com, ISSN 1743-3533 (on-line)
doi:10.2495/MPF190111

both injected into the annular space of the contactor via circumferential slits. The cement is injected downstream of the steam. The resulting interaction can be categorized into two different multiphase flow phenomenon: primary and secondary sheet breakup. The first occurs in the section of the contactor nearest the inlets as the cement comes into contact with the high pressure and temperature steam. Upon contact, the liquid column begins to shear apart into liquid ligaments and particles. The second phenomena occurs downstream as the droplets continue to experience heat transfer and aerodynamic forces from the steam, leading to further potential breakup and evaporation of the cyclohexane. The complexity and computational expense associated with modeling the multiphase flow interaction necessitates the development of a separate model for the secondary, multiphase flow regime.

Most polymers contain low-molecular weight components like unreacted monomers, solvents, and other by-products. These are collectively known as volatiles and are separated from the polymer for a variety of reasons, including regulatory requirements, polymer quality, and odor elimination [1]. This separation process is known as polymer devolatilization and can be accomplished utilizing either static or dynamic equipment, depending on the polymer mixture composition (e.g. volatile mass fraction) [2]. Static methods, as in this research, are particularly effective for the devolatilization of low-viscosity, high volatile mass fraction mixtures by causing the solvent to evaporate rapidly or flash [1], [3]. The flash occurs when the vapor pressure, p^*, of the volatile in the polymer mixture exceeds its own local partial pressure, p_v [4]. Because the vapor pressure is heavily dependent on temperature, the operating temperature of the polymer mixture must be high enough (without resulting in thermal degradation) to allow $p^* - p_v > 0$ upon injection into the static devolatilization equipment. The temperature increase is provided by a heated stripping agent like the superheated steam in this research. Despite being used for its simplicity and lower operating costs, the geometry and operating conditions of the static equipment must be optimized to prevent over- or under-exposure to the heated stripping agent and inconsistent final product quality [3].

Changes to the actual equipment and operating conditions in such manufacturing facilities would result in higher costs and loss of productivity, and hence, computational fluid dynamics (CFD) is a very cost-effective tool to evaluate a number of different modifications. Previous research by the same authors focused on developing a CFD model to simulate flow conditions in this problem [5, 6]. The model of this multiphysics process was used to optimize the consumption of costly superheated steam and evaluate the affects of geometry changes on droplet evaporation in the mixing and nanogap zones of the contactor [3], [7]. The objective of this paper is to further build on that research by assessing the impact of contactor design on the initial breakup of the polymer mixture near its inlet and on how much polymer mixture sticks to the walls of the vessel.

2 GEOMETRY

The steam contactor is a pressurized vessel where superheated steam and cement are mixed to accomplish the polymer devolatilization process. The dimensions of the steam contactor are shown in Fig. 1, which shows a cross-sectional view of the contactor, where the annular space is created by a plug (shown as hollow space in the figure) that fills the center of the contactor [3]. The total length of the contactor is 1320 mm with a maximum external diameter of 183 mm.

The process begins with the injection of superheated steam through a circumferential, annular slit at a 45° angle. Just downstream of the steam injection, cement is also injected into the annular space of the contactor through a circumferential slit. Injected normal to the contactor wall and into the incoming flow of steam, the cement is immediately affected by

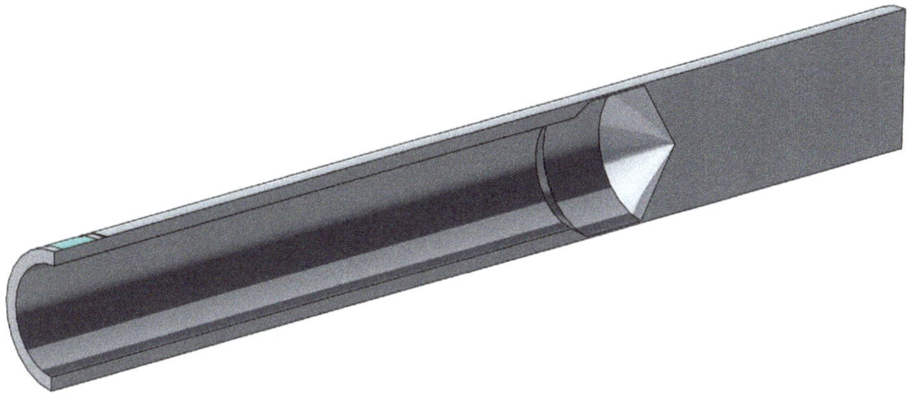

Figure 1: Cross-sectional view of entire G1 contactor.

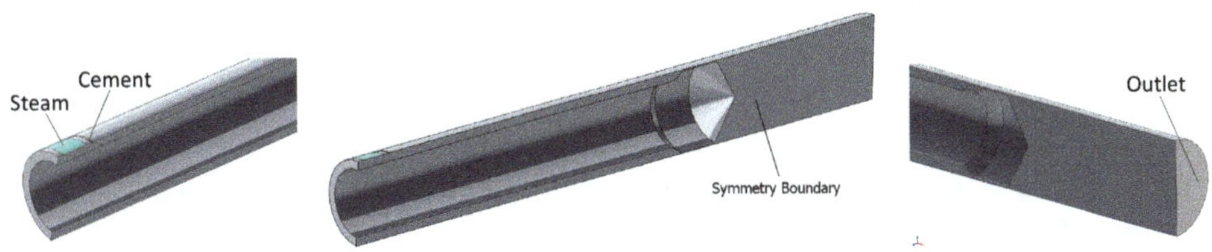

Figure 2: Geometry callouts for (left) steam and cement inlets, (center) symmetry boundary, and (right) outlet.

the momentum and temperature of the steam. Because the composition is 86% cyclohexane solvent and 14% polymer, the solvent begins to flash upon contact, starting the disintegration of the cement liquid column.

To balance computational efficiency and accuracy, primary breakup and the resulting multiphase flow after breakup are evaluated in two different models. Primary breakup is evaluated using a full three-dimensional model of the first 254 mm of the contactor. Secondary breakup is calculated via an axisymmetric 3D model of the entire contactor as shown in Fig. 2 [3]. The operating conditions (in bold) in Table 1 for the original G1 contactor based on current production settings. The boundary conditions were chosen based on prior research and results obtained through this study, which are described later in the thesis [3].

Table 1: Operating and boundary conditions of the contactor.

Parameter	Steam inlet	Cement inlet	Outlet
Boundary	Pressure inlet	Velocity inlet	Pressure outlet
Velocity	—	3.16 m/s	—
Pressure	2100 kPa	—	**140** kPa
Temperature	**599** K	**433** K	**366** K

The original contractor geometry is modified such that it includes an additional steam inlet slightly downstream of the current cement inlet. Depicted by Fig. 3, this inlet is intended to add additional heat to the flow and prevent the upstream steam inlet from *pressing* the cement onto the wall. Similar to the upstream steam inlet, the downstream inlet is a

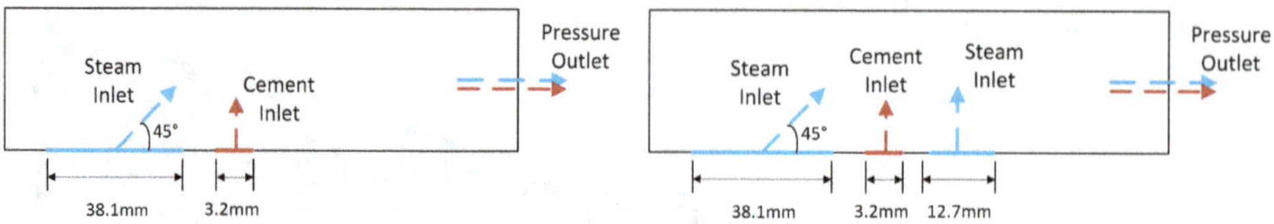

Figure 3: Plane schematic views of the bottom near the inlets for the original geometry (left) and the modified geometry (right).

circumferential, annular slit with the same operating and boundary conditions. Boundary conditions are altered in the modified geometry in order to maintain necessary production parameters, namely a cement mass flow rate of 3.9 kg/s and a steam-to-cement ratio of 0.6. These conditions were obtained as part of a separate optimisation effort related to cost savings with reducing the use of superheated steam [3].

This additional inlet is modeled in such a way that the steam is injected normal to the contactor walls.

3 FORMULATION

The governing equations in this study involve the Favre-averaged Navier–Stokes (FANS) equations given by

$$\frac{\partial \overline{\rho}}{\partial t} + \frac{\partial}{\partial x_i}(\overline{\rho}\tilde{u}_i) = 0, \tag{1}$$

$$\frac{\partial}{\partial t}(\overline{\rho}\tilde{u}_i) + \frac{\partial}{\partial x_i}(\overline{\rho}\tilde{u}_j\tilde{u}_i) = -\frac{\partial \overline{P}}{\partial x_i} + \frac{\partial}{\partial x_j}\left[\overline{t_{ji}} - \overline{\rho u_j'' u_i''}\right], \tag{2}$$

$$\frac{\partial}{\partial t}\left[\overline{\rho}\left(\overline{e} + \frac{\tilde{u}_i\tilde{u}_i}{2}\right) + \frac{\overline{\rho u_i'' u_i''}}{2}\right] + \frac{\partial}{\partial x_j}\left[\overline{\rho}\tilde{u}_j\left(\tilde{h} + \frac{\tilde{u}_i\tilde{u}_i}{2}\right) + \tilde{u}_j\frac{\overline{\rho u_i'' u_i''}}{2}\right]$$
$$= \frac{\partial}{\partial x_j}\left[-q_{L_j} - \overline{\rho u_j'' h''} + \overline{t_{ji}u_i''} - \overline{\rho u_j'' \tfrac{1}{2}u_i'' u_i''}\right] + \frac{\partial}{\partial x_j}\left[\tilde{u}_i\left(\overline{t_{ij}} - \overline{\rho u_i'' u_i''}\right)\right], \tag{3}$$

and also the ideal gas equation given by

$$P = \overline{\rho}R\tilde{T}. \tag{4}$$

In eqns (1), (2), and (3), $(\ldots)''$ represents the fluctuating quantity [8], [9]. ρ is the density, u is the velocity, P is the pressure, t_{ji} is the viscous stress tensor, e and h are the specific internal energy and specific enthalpy, respectively, and q_{L_j} is the laminar mean heat flux. Eqn (4) is the Favre-averaged ideal gas equation used to model the continuous steam phase. In this problem, the Ma attained inside of the contactor is above 1, which makes the flow compressible, and hence modeling the steam as an ideal gas is apropos. The viscous stress tensor is written as: Turbulence modeling is achieved ion the simulations here through the RNG $k-\epsilon$ model. This model has shown to demonstrate improved accuracy for capturing the effect of swirl on turbulence and hence is appropriate for this simulation [10]–[13]. In addition, a species transport model is used to solve for the steam and cyclohexane vapor species using a convection-diffusion equation [13].

3.1 Volume of fluid interface tracking method for primary breakup

To perform a simulation showing the primary breakup of the cement injection within the G1 contactor, a separate model is used. In this model, the continuous steam phase is already in motion flowing through the contactor's main body, and the cement is added in as a liquid. As the liquid enters the contactor, it is quickly sheared apart into droplets. Since the interface between the gas and cement phase is not dispersed, an optimized version of the Volume of Fluid (VOF) interface tracking method, called the Coupled Level Set VOF (CLSVOF) [14] is employed here. This method combines the Level-Set (LS) method, first introduced by Li et al. [15], and the VOF tracking method. The volume fraction equation tracks the interface between the gas and liquid phases. A solution is found by solving the continuity equation for the volume fraction of one of the phases. For the ith phase, the continuity equation to solve the volume fraction is given as

$$\frac{1}{\rho_i}\left[\frac{\partial}{\partial t}\left(\alpha_i\rho_i\right) + \nabla\cdot\left(\alpha_i\rho_i\vec{v}_u\right) = \sum_{p=1}^{n}\left(\dot{m}_{pi} - \dot{m}_{ip}\right)\right]. \tag{5}$$

Here, the term α_i represents volume fraction of the ith fluid within the cell, thus α_i is zero when the cell is empty of the ith, 1 when the cell is completely full of the ith fluid, and between 0 and 1 when the cell contains and interface between the ith fluid and the other fluid. Additionally, $\dot{m}_{ip}$ is the mass transfer from phase i to phase p and $\dot{m}_{pi}$ is the mass transfer from phase p to phase q. More details can be found in [16].

3.2 Discrete phase model for secondary phase change

To predict the trajectory of the discrete phase particles in the secondary phase change step of the model, a discrete phase model (DPM) formulation is used, which integrates the force balance on the particle in a Lagrangian reference frame. This force balance equates the particle inertia with the forces acting on the particle, and can be written as

$$\frac{du_p}{dt} = F_D(\vec{u} - \vec{u}_p) + \frac{\vec{g}(\rho_p - \rho)}{\rho_p}, \tag{6}$$

where $F_D(\vec{u} - \vec{u}_p)$ is the drag force per unit particle mass, given by

$$F_D = \frac{18\mu}{\rho_p d_p^2}\frac{C_D Re}{24}, \tag{7}$$

and $\vec{u}$ is the continuous phase velocity, $\vec{u}_p$ is the particle velocity, μ is the dynamic viscosity of the fluid, ρ is the density, ρ_p is the density of the particle, and d_p is the particle diameter. Also, a high-Mach number drag law is also used here to account for the high-speed flows. In the context of the DPM, a convection/diffusion controlled vaporization model is used. In addition, boiling rate and energy equations are solved as well. Using the velocity fluctuations caused by turbulence, the particle trajectories are tracked as the particles flow through the domain, through the use of the stochastic discrete random walk (DRW) model is used [17]. To account for the coupling between the gas and particle phase, momentum exchange, heat exchange, and mass exchange are all calculated within the simulation.

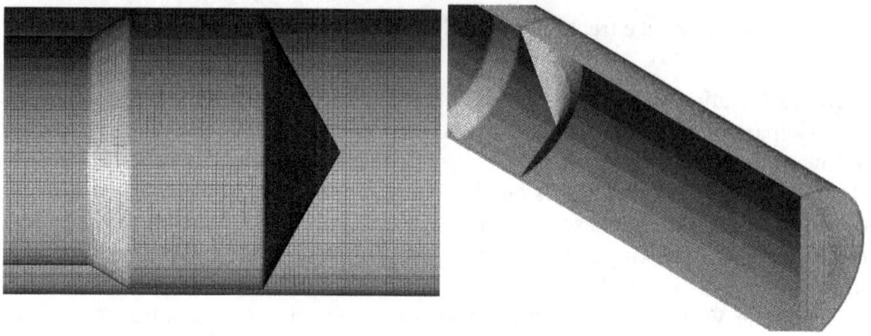

Figure 4: 3D axisymmetric contactor view of (top) mixing zone and gap region near the end of the contactor (bottom) outlet and divergence region.

4 COMPUTATIONAL DETAILS

4.1 Mesh

The simulations to measure primary breakup are set up using a CLSVOF model in ANSYS Fluent for a shorter 254 mm section of the full 3D contactor. The stable droplet diameter for this problem is calculated to be 143 μm, which requires that the maximum grid size to be equal to that value. As a result, a hexahedral-dominant mesh is employed here, which consists of approximately 3×10^6 grid points. On the other hand, for the secondary phase change, the entire contactor length is used to evaluate and track particle locations. Similar to the mesh for primary breakup, a hexahedral dominant mesh is used to improve accuracy and convergence. For this research, a mesh size of approximately 725,000 elements and 795,000 nodes is used and is shown in Fig. 4. Earlier research shows, through mesh dependence tests, that this mesh size is sufficient to accurately capture the parameters of interest [3], [6], [7].

4.2 Numerical methods

The CLSVOF model is a transient simulation that simultaneously solves for both the continuous and discrete phases (steam and cement, respectively). Listed below are the schemes used to solve the model based on research and ANSYS guidance [13], [18], [19].

- Pressure-Velocity Coupling: SIMPLE
- Pressure: PRESTO! (PREssure STaggering Option)
- Momentum, Density, Energy, Turbulence: First Order Upwind
- Gradient: Least Squares Cell Based
- Level-set Function: First Order Upwind
- Time: First Order Upwind Implicit
- Volume Fraction: Geo-Reconstruct

The Geo-Reconstruct scheme is particularly useful for time-dependent solutions with sharp interface between phases like those in this research.

The DPM simulation benefits from the accuracy of a second-order upwind discretization. A maximum number of trajectory time steps is set to 42,000 with a default step length factor of 5. The time step necessary to progress the DPM injections and solve for particle trajectories is chosen to be 0.0001 s. The choice of the current parameters is based on several numerical tests carried out previously [3]. All boundaries in the model are set as reflective, except for

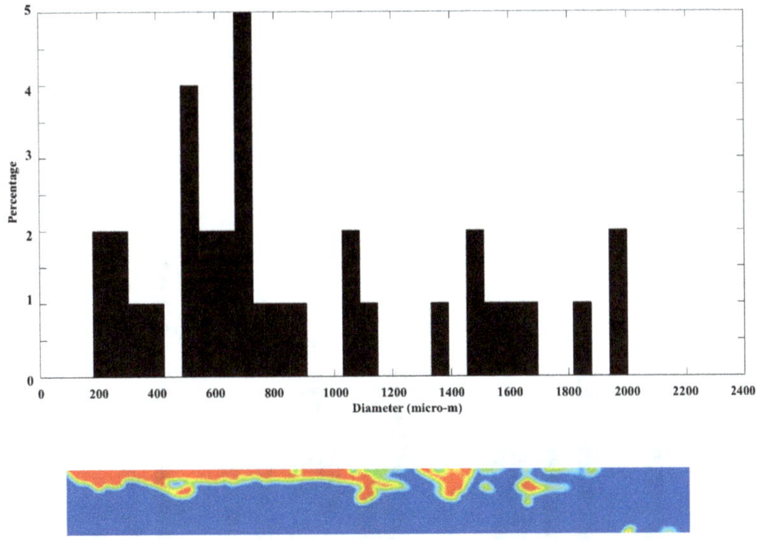

Figure 5: Upper plane view of original geometry CLSVOF results and associated particle diameter distribution from image analysis.

the outlet. For the purposes of this simulation, the particle interaction with the walls of the contactor is assumed to be completely elastic, meaning the particle retains all of its normal and tangential momentum after the collision [13]. The outlet is set to an escape boundary condition so that it is the only exit for the particles from the contactor, except in cases of reverse flow at the inlets.

5 RESULTS AND DISCUSSIONS

5.1 Primary breakup

The simulations to determine primary breakup are set up using a CLSVOF model in ANSYS Fluent for a shorter 254 mm section of the full 3D contactor. This length has been determined to be sufficient for viewing the initial liquid atomization from previous studies, which included validation with experiments as well [3]. The intent is to capture just the initial droplet formation without going through the computational expense of seeing further phase change, which will be modeled in a separate 3D axisymmetric DPM model.

So once the inlet and outlet flow conditions reach a relative equilibrium and uniformity, images of the breakup were analysed in an image analysis formulation in MATLAB. The diameter distributions are plotted in the form of a histogram form, anything above 3000 μm is neglected as being either a part of the initial liquid stream or captured against the wall without separation, leaving the total histogram percentage less than 100%. Because of this limitation, little consideration is afforded towards the amount of cement sticking to the wall and its impacts on the final polymer crumb quality. A comparison of the minimum, maximum, and average diameters resulting from the CLSVOF results for each geometry is provided in Table 2. The original geometry's distribution, visually depicted in Fig. 5, appears "favorable" (towards a better quality polymer crumb) with a majority of particles being less than 1000

Table 2: Minimum, average, and maximum droplet diameter in μm for each geometry.

Geometry	Min	Avg	Max
Original	181	944	1998
Modified	127	1662	1941

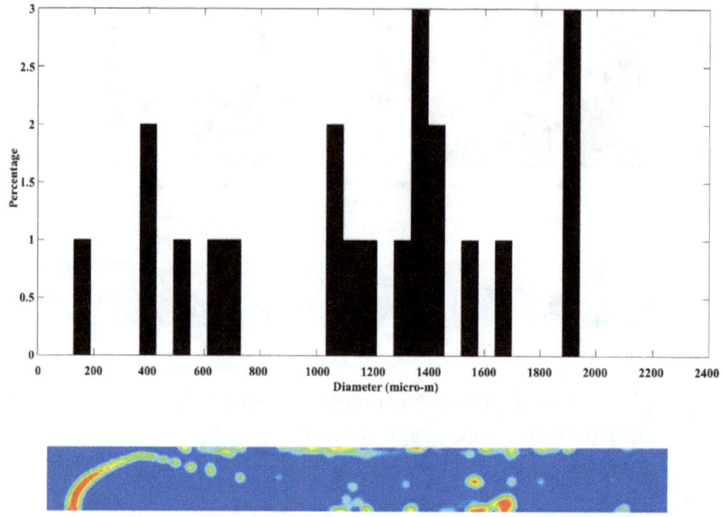

Figure 6: Lower plane view of Dual Steam – 90° CLSVOF results and associated particle diameter distribution from image analysis.

μm in diameter. It also has some of the smallest droplets created, which are only at 181 μm in diameter. This, in large part, is due to separation that occurs as the cement droplets "shed" from the column of fluid moving along the wall. This primary breakup is different than in any of the other geometries considered and reflects more of a bag-type breakup described earlier.

In the modified geometry, the downstream steam inlet injects superheated steam normal to the wall of the contactor. This downstream steam inlet forces the column of cement liquid to become more vertical, which leads to a reduced flowrate of steam coming from the upstream inlet as it comes into contact with the dense cement mixture. This interaction between the two steam flows leads to higher aerodynamic forces, shearing the cement unevenly and nearer to the opposite wall. This can be observed in the image and histogram shown in Fig. 6.

5.2 Secondary phase change

The second stage of the simulation effort involves the DPM calculations in ANSYS Fluent. In these calculations the entire contactor length is considered in order to evaluate relevant parameters of the outlet particles prior to entering the coagulation process. The transition from the CLSVOF model to the DPM model is accomplished by determining the particle diameters from the CLSVOF calculations, fitting them to a Rosin-Rammler distribution, and then employing them as inlet conditions for the particle injections in the DPM calculations. When the number of particles exiting equal the number being injected and when the flow is

Table 3: Comparison of outlet particle parameters.

	Diam (μm)	Res Time (s)	Cyclohexane MF	Temp (K)
Original (Min)	84.7	0.021	0.000	283.7
Dual steam (Min)	59.9	0.019	0.000	271.9
Original (Avg)	727.8	0.046	0.537	322.2
Dual steam (Avg)	699.8	0.031	0.484	321.5
Original (Max)	1687.4	0.069	0.815	474.7
Dual steam (Max)	1589.3	0.038	0.768	493.7

considered to be reasonably uniform, and at this stage, a sample of the particles at the outlet is picked in order to determine diameter, cyclohexane mass fraction, residence time, and temperature. The comparison of those results is shown in Table 3. The results of the original geometry have been validated with experimental data measured in the manufacturing facility of the industrial partner. From this table and from the corresponding histogram plots of the same parameters in Fig. 7, a couple of observations can be made. Firstly, while the effects on the minimum and maximum diameters are relatively equal between the models, the average particle diameter in the dual steam geometry is reduced by 30% more than in the original geometry. Secondly, an interesting feature of the results that seems counter-intuitive to the first observation is that the maximum particle residence time is 45% lower than in the original geometry. It is possible that one could erroneously assume that the average and maximum particle diameters would be larger due to the lesser residence time in the contactor and therefore lesser solvent evaporation. As mentioned previously, the results clearly show lower minimum, maximum, and average particle diameters than those from the original geometry calculation. Also, as it relates to the cyclohexane mass fraction, Fig. 8 show that the content is noticeably lower before the converging-diverging region of the contactor due to elevated particle temperatures.

The reason for this behaviour lies in the particle velocity predictions shown in Fig. 9. It can be seen here that at the outer edges of the original contactor, there are exceedingly slow moving particles (indicated by the dark blue-colored particles) depicting very high levels of cyclohexane mass fraction (as seen in Fig. 8) and also exceedingly low values of temperature (not shown here). This is increasingly true at the bottom of the contactor where it is likely that large stagnate particles have settled to the bottom of the contactor due to gravity. This is further supported by the residence time predictions (not shown here), where the particles with the longest residence time are coming from the bottom of the contactor. These particles would represent larger, stagnate cement adhering to the wall.

In the dual steam design, this occurs to a much lesser extent, because the figures depict a more uniform set of properties. This is a result of mixing with the superheated steam. Where there is insufficient contact with the steam, as in the case of geometries like the original contactor where much of the cement is adhering to the wall, then there is poor solvent removal regardless of residence time. When the amount of cement sticking to the wall is reduced, cyclohexane removal is enhanced, and the particles flow through the contactor as intended.

6 CONCLUSIONS

The intent of this research is to accurately model the polymer devolatilization process and the effects of contactor design changes on the final polymer product quality. Because of the complexity of the devolatilization process, measurements in the production facility are cumbersome and expensive. Therefore, the use of CFD modeling is particularly effective at

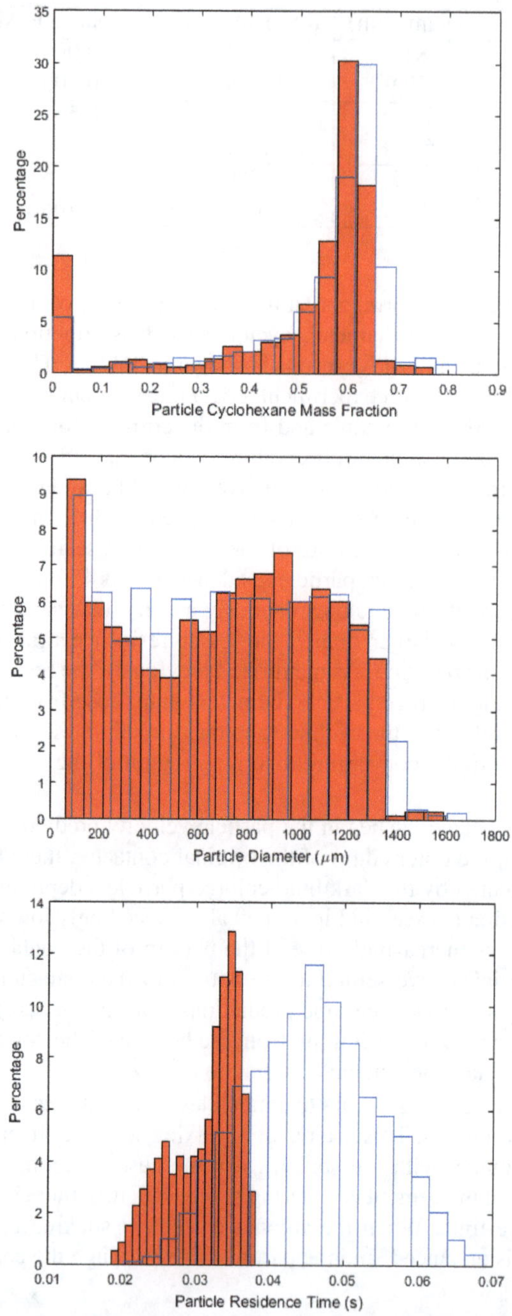

Figure 7: Histograms of cyclohexane mass fraction (top), diameter (middle) and residence time (bottom) of the particles at the outlet; open bars – original geometry, closed bars – modified geometry.

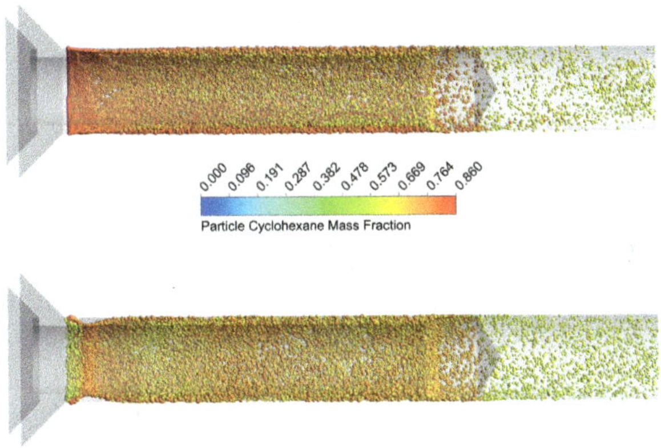

Figure 8: Particle cyclohexane mass fraction tracks for (top) original geometry and (bottom) dual steam inlets – 90°.

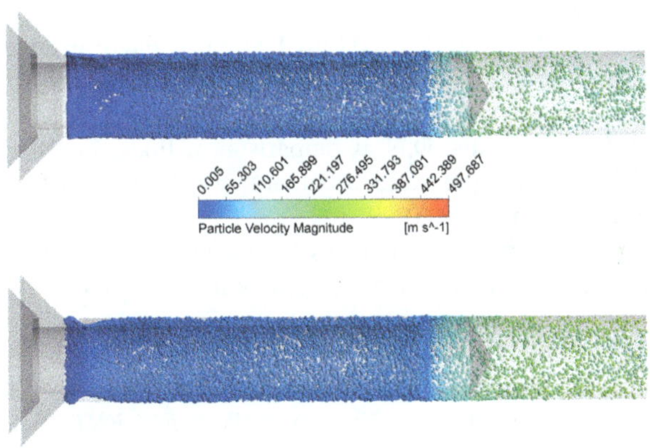

Figure 9: Particle velocity magnitude tracks for (top) original Geometry and (bottom) dual steam inlets – 90°.

providing a high confidence, lower cost solution. Towards this effort, 3D CFD calculations of polymer devolatilization in a steam contactor are conducted in two stages. The first stage develops a multiphase model for the initial or primary breakup of the liquid polymer mixture by steam. This stage employs the CLSVOF-based multiphase model to analyse the droplet diameter distribution in two different designs, where a polymer mixture sheet is stripped into droplets by superheated steam. The second stage then uses the resulting diameter distribution to model the secondary phase change of the polymer mixture. In this stage, the Eulerian–Lagrangian DPM approach is employed to model the phase change involving evaporation of polymer mixture droplets into polymer crumb. Results showed that slight geometric modifications resulted in ideal droplet diameter distributions and furthermore in increasing heat transfer and cyclohexane evaporation in the body the contactor at the stage of phase change.

REFERENCES

[1] Albalak, R., *Polymer Devolatilization*, Vol. 33, CRC Press, 1996.

[2] Tadmor, Z. & Gogos, C.G., *Principles of Polymer Processing*, John Wiley & Sons, 2013.

[3] Gabor, K.M., *Computational investigations of polymer devolatilization processes in steam contactors*. PhD thesis, The University of Akron, 2016.

[4] Astarita, G. & Maffettone, P.L., Devolatilization of polymers. *Adv. Transport Proc.*, **9**, 1993.

[5] Shindle, B., Gabor, K.M. & Chandy, A.J., Modeling polymer sheet breakup for devolatilization processes in steam contactors. *2nd Thermal and Fluids Engineering Conference (TFEC)*, ASTFE: Las Vegas, NV, USA, Paper # 18457, 2017.

[6] Gabor, K.M., Shindle, B. & Chandy, A.J., CFD investigation of control parameters of a steam contactor used in polymer devolatilization. *International Journal of Thermal Sciences*, **120**, pp. 19–30, 2017.

[7] Gabor, K.M., Shindle, B. & Chandy, A.J., CFD simulations of polymer devolatilization in steam contactors. *Engineering Applications of Computational Fluid Mechanics*, **11**(1), pp. 273–292, 2017.

[8] Walters, D.K. & Cokljat, D., A three-equation eddy-viscosity model for Reynolds-averaged Navier-Stokes simulations of transitional flow. *Journal of Fluids Engineering*, **130**, pp. 1–14, 2008.

[9] Burns, A.D., Frank, T., Hamill, I. & Shi, J.M., The Favre averaged drag model for turbulent dispersion in Eulerian multi-phase flows. *International Conference on Liquid Atomization and Spray Systems*, 2006.

[10] Papageorgakis, G. & Assanis, D.N., Comparison of linear and nonlinear RNG-based k-epsilon models for incompressible turbulent flows. *Numerical Heat Transfer: Part B: Fundamentals*, **35**(1), pp. 1–22, 1999.

[11] Hou, Q. & Zou, Z., Comparison between standard and renormalization group k-epsilon. models in numerical simulation of swirling flow tundish. *ISIJ International*, **45**(3), pp. 325–330, 2005.

[12] Jawarneh, A.M., Tlilan, H., Al-Shyyab, A. & Ababneh, A., Strongly swirling flows in a cylindrical separator. *Minerals Engineering*, **21**(5), pp. 366–372, 2008.

[13] ANSYS Academic Research, Help System, *ANSYS FLUENT Theory Guide*. Release 14.5 edition, 2012.

[14] Sussman, M. & Puckett, E.G., A coupled level set and volume-of-fluid method for computing 3d and axisymmetric incompressible two-phase flows. *Journal of Computational Physics*, **162**(2), pp. 301–337, 2000.

[15] Li, X., Massarotti, N., Nithiarasu, P., Ray, B., Biswas, G., Sharma, A. & Welch, S.W., CLSVOF method to study consecutive drop impact on liquid pool. *International Journal of Numerical Methods for Heat & Fluid Flow*, **23**(1), pp. 143–158, 2013.

[16] Shindle, B.W., *Computational investigations of polymer sheet breakup For optimization of devolatilization processes in steam Contactors*. PhD thesis, The University of Akron, 2017.

[17] Ajilkumar, A., Sundararajan, T. & Shet, U., Numerical modeling of a steam-assisted turbular coal gasifier. *International Journal of Thermal Sciences*, **48**, pp. 308–321, 2008.

[18] Andreassi, L., Baudille, R. & Biancolini, M.E., Spew formation in a single lap joint. *International Journal of Adhesion and Adhesives*, **27**(6), pp. 458–468, 2007.

[19] Cloete, S.W., Eksteen, J.J. & Bradshaw, S.M., A mathematical modelling study of fluid flow and mixing in full-scale gas-stirred ladles. *Progress in Computational Fluid Dynamics, An International Journal*, **9**(6-7), pp. 345–356, 2009.

SECTION 3
MULTIPHASE FLOWS

NUMERICAL INVESTIGATION OF LIQUID FILM INSTABILITIES AND EVAPORATION IN CONFINED OSCILLATING SLUG-PLUG FLOWS

MANOLIA ANDREDAKI, ANASTASIOS GEORGOULAS, NICOLAS MICHE & MARCO MARENGO
Advanced Engineering Centre, School of Computing Engineering and Mathematics, University of Brighton, UK

ABSTRACT

An enhanced volume of fluid (VOF)-based numerical simulation framework that accounts for conjugate heat transfer between solid and two-phase flow regions and phase-change due to boiling/condensation, is utilised in order to investigate the effect of flow oscillation amplitude and frequency on the liquid film evaporation and instability formation in slug-plug flows within heated channels, in saturated flow boiling conditions. Various series of parametric numerical simulations are performed, for different values of flow oscillation amplitude and frequency for a variety of working fluids. For one of the working fluids two different channel diameters are also tested. The oscillations in each case are induced by applying an oscillating pressure boundary condition at the inlet of the channel, keeping the pressure constant at the outlet, after an initial period of constant pressure drop between the inlet and the outlet. Capillary ridges that are initiated at the liquid film, in the vicinity of the leading edge of the considered vapour slugs, are identified as a result of the imposed oscillations, which are translated in the form of capillary waves towards the rear end of the bubbles. It is shown that the formation frequency as well as the geometric characteristics of the generated ridges, are directly related to the corresponding frequency and amplitude of the induced flow oscillations. Furthermore, it is shown that in the initial stages of the bubble fate after the application of the oscillations liquid film evaporation is enhanced with the increase of the oscillation amplitude while it degrades as the frequency of the oscillation becomes higher. However, for large oscillation amplitudes and channel diameters, liquid jets penetrate into the elongated bubbles leading in a lot of cases to bubble break-up.
Keywords: volume of fluid, conjugate heat transfer, oscillating vapour slug, micro-channel.

1 INTRODUCTION

The demand for increasingly higher performances of electronic equipment, has pushed in the development of a new generation of heat transfer systems based on the local phase-change of a working fluid. Efficient thermal control, of densely packed electronic components, has become of crucial importance. Flow boiling within micro-channels, constitute one of the most promising cooling technologies to dissipate high heat fluxes from micro-scale electronic components. Two-phase cooling devices utilising evaporators consisting of micro-channels that are in direct contact with a micro-chip, can remove heat fluxes up to 300 W/cm^2 [1]. However, the flow complexity in cases of multi-microchannel two-phase flow evaporators, can lead to high-amplitude and high-frequency temperature and pressure oscillations [2]. Such oscillations, in the slug flow regime, may generate significant instabilities in the liquid film that surrounds the elongated vapour slugs. Such flow oscillations and instabilities are also present in the case of oscillating, closed loop, two-phase cooling devices, such as Pulsating Heat Pipes that might consist of micro- (e.g. [3]) or mini-channels (e.g. [4]).

In the present paper, an enhanced volume of fluid (VOF) based numerical simulation framework that accounts for conjugate heat transfer between a solid region and a two-phase flow region with phase-change due to evaporation and/or condensation, is applied in order to investigate the effect of global flow oscillation frequency and amplitude on the liquid film evaporation and in the liquid film instability generation, in the case of confined isolated, elongated vapour slugs, in saturated flow boiling conditions.

WIT Transactions on Engineering Sciences, Vol 123, © 2019 WIT Press
www.witpress.com, ISSN 1743-3533 (on-line)
doi:10.2495/MPF190121

The utilised, enhanced, VOF-based numerical model, has been extensively validated in the past (by the same research group as the authors of the present work) against existing analytical solutions and experimental data on adiabatic and diabatic bubble and droplet dynamics [8], [10]–[12]. Some, of these validation results are reproduced in Fig. 1. As it can be observed, the proposed VOF-based numerical modelling framework, accurately predicts adiabatic and diabatic bubble and droplet dynamics as well as conjugate heat transfer between solid and fluid domains. Therefore, since all underpinned mechanisms have been validated in the past (interfacial dynamics, phase-change, conjugate heat transfer) it can safely be used for the purposes of the present investigation.

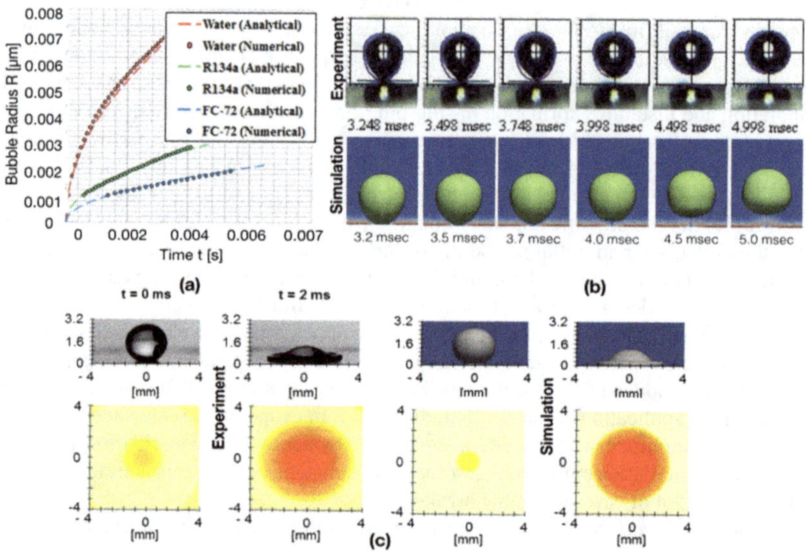

Figure 1: (a) Comparison between numerical simulation and analytical solution results for a bubble evaporating in a superheated domain [10]; (b) Numerical simulation results and experimental results in pool boiling [10]; and (c) Conjugate heat transfer numerical simulation results and experimental measurements on droplet impact cooling [11].

2 INVESTIGATION METHODOLOGY

2.1 Numerical method

2.1.1 Governing equations

In this section, the governing equations for mass, momentum, energy, and volume fraction are presented. It should be mentioned that liquid and vapour phases are both treated as incompressible, Newtonian fluids. The mass conservation equation is given as:

$$\nabla \cdot \left(\rho \vec{U}\right) = \dot{\rho}, \tag{1}$$

where U is the fluid velocity and ρ is the bulk density. The source term on the right-hand side accounts for the phase change. The conservation of momentum is given by the following equation:

$$\frac{\partial}{\partial t}(\rho\vec{U}) + \nabla\cdot(\rho\vec{U}\vec{U}) - \nabla\cdot\left\{\mu\left[\nabla\vec{U} + (\nabla\vec{U})^{T}\right]\right\} = -\nabla p + \vec{f}_{ST} + \vec{f}_{g}, \tag{2}$$

where p is the pressure and μ is the bulk dynamic viscosity. The momentum source terms on the right-hand side of the equation account for the effects of surface tension and gravity, respectively. The surface tension term is modelled according to the classical approach of Brackbill et al. [5]. The conservation of energy balance is given by the following equation:

$$\frac{\partial}{\partial t}(\rho c_p T) + \nabla\cdot(\vec{U}\rho c_p T) - \nabla\cdot(\lambda\nabla T) = \dot{h}, \tag{3}$$

where c_p is the bulk heat capacity, T the temperature field, and λ is the bulk thermal conductivity. The source term on the right-hand side of the equation represents the contribution of the enthalpy of evaporation/condensation or else the cooling/heating associated with the latent heat of the phase-change. The volume fraction α is advected by the flow field by the following equation:

$$\frac{\partial \alpha}{\partial t} + \nabla\cdot(\alpha\vec{U}) - \nabla\cdot(\alpha(1-\alpha)U_r) = \frac{\dot{\rho}}{\rho}\alpha. \tag{4}$$

Interface sharpening is very important in simulating two-phase flows of two immiscible fluids. In OpenFOAM the sharpening of the interface is achieved artificially by introducing the extra compression term $\nabla\cdot(\alpha(1-\alpha)U_r)$ in eqn (4). U_r is an artificial compression velocity. The divergence of the compression velocity U_r, ensures the conservation of the volume fraction α, while the term α(1-α) limits this artificial compression approach only in the vicinity of the interface, where $0 < \alpha < 1$ [6]. The level of compression depends on the value of C_γ [6], [7]. For the simulations of the present investigation, initial, trial simulations indicated that a value of $C_\gamma = 1$ should be used, in order to maintain a quite sharp interface without at the same time having unphysical results. The source term on the right-hand side of the eqn (4) is needed because, due to the local mass source terms, the velocity field is not free of divergence. Finally, the bulk fluid properties γ are computed as the averages over the liquid (γ_l) and vapour (γ_v) phases, weighted with the volume fraction α. The VOF-based solver that is used in the present investigation has been modified accordingly in order to account for an adequate level of spurious currents suppression. More details on the proposed development and validation as well as on the proposed improved VOF method can be found in the paper by Georgoulas et al. [8].

The conservation of energy equation in the solid domains is defined as:

$$\frac{\partial}{\partial t}(\rho_s c_{ps} T) = \nabla\cdot(\lambda_s \nabla T). \tag{5}$$

The coupling at the interface between the solid region and fluid region is achieved iteratively through the following conditions:

$$T_f = T_s, \tag{6}$$

$$\lambda_f \frac{\partial T_f}{\partial n} = \lambda_s \frac{\partial T_s}{\partial n}. \tag{7}$$

2.1.2 Phase change model

The utilized phase change model that was implemented in the improved OpenFOAM VOF solver that is used in the present investigation, will be described briefly in this section. In the case of evaporation, the evaporating mass flux at the liquid–vapour interface j_{evap} is calculated from the following equation:

$$j_{evap} = \frac{T_{int} - T_{sat}}{R_{int} h_{lv}}, \tag{8}$$

where T_{int} is the temperature of the interface, T_{sat} is the saturation temperature, R_{int} is the interfacial heat resistance and h_{lv} is the latent heat of evaporation at the saturation temperature. The amount of liquid that evaporates is calculated locally and the resulting source term field is smeared over a few cells in order to avoid numerical instabilities. The evaporating mass is taken away on the liquid side of the interface and reappears on the vapour side, following the process originally suggested in the work of Hardt and Wondra [9]. Further details on the proposed development can be found in the work of Georgoulas et al. [10].

3 APPLICATION OF NUMERICAL METHOD

3.1 Set-up of numerical simulations

In the current section of the present work the previously described, VOF-based, numerical simulation framework is further applied for the conduction of parametric numerical simulations, aiming to identify and quantify the effect of global flow oscillation frequency and amplitude on the liquid film evaporation and instability formation, around isolated, elongated vapour slugs, within a heated micro-channel, in saturated flow boiling conditions. In more detail, various series of parametric numerical experiments are conducted, applying a constant heat flux value of $\dot{q}$ = 5,000W/m^2 in the heated section of the two in total considered circular micro-channels with external and internal diameter values of D_o = 0.69 and 1.19 mm and D_i = 0.5 and 1.00 mm, respectively. In each series of parametric numerical simulations, the imposed global flow oscillation amplitude and frequency are varied. The flow oscillations are generated inducing an oscillating pressure boundary condition at the inlet of the channel while keeping the pressure constant at the outlet. The same values of oscillation amplitude and frequency are tested for four different working fluids: R245fa, FC72, ethanol and water for the D_i = 0.5mm channel (Series A, C, D and E in Table 1) while R245fa only is tested for the D_i = 1.0mm channel (Series B in Table 1). Each simulation is conducted in three stages. At the first stage a single-phase simulation is run with only liquid flowing into the channel, applying a constant pressure drop of ΔP = 500 Pa. In the second stage, after the single-phase flow has reached a quasi-steady state and the hydrodynamic and thermal boundary layers have been developed (t = 0.2 s), an elongated vapour slug is patched upstream of the heated section of the channel at a certain distance from the channel inlet, having an initial length of 3xD_i and an initial liquid film thickness of $h_{film(init)}$ = 20 μm. This initial vapor slug is then left to be carried away downstream by the previously developed liquid flow, towards the heated section of the circular micro-channel. At the third and final stage, after approximately 0.02 s (depending on the working fluid and diameter) that the vapour slug front has reached approximately the middle of the channel in each case (ensuring that it is almost completely contained within the heated region of the channel), an oscillating relative pressure boundary condition is imposed at the inlet of the channel, keeping the relative pressure at the outlet constant at 0 Pa. The resulting pressure value for the inlet with respect to time, can be described by the following equation

$$P(t) = (1 + \alpha \sin(\pi f t))P_{ref} + P_0, \tag{9}$$

where α and f are the amplitude and frequency of the oscillation, while P_0 and P_{ref} are the initial pressure and a reference pressure, respectively. It should be mentioned that for all the simulations presented here, the reference pressure is taken as unity and the initial pressure as zero. For each channel, an axisymmetric computational domain is constructed with the axis of symmetry coinciding with the central/longitudinal axis of the considered circular micro-channels. The computational domain, mesh and the applied boundary conditions are illustrated in Fig. 2, for each of the considered channel diameters.

Table 1: Inlet pressure oscillation characteristics, working fluids and internal channel diameters for the considered cases.

Case	a (Pa)	f (Hz)	Fluid	D_i (mm)	Case	a (Pa)	f (Hz)	Fluid	D_i (mm)
A1	100,000	500	R245fa	0.5	**D1**	100,000	500	Ethanol	0.5
A2	100,000	750	R245fa	0.5	**D2**	100,000	750	Ethanol	0.5
A3	100,000	1000	R245fa	0.5	**D3**	100,000	1000	Ethanol	0.5
A4	50,000	500	R245fa	0.5	**D4**	50,000	500	Ethanol	0.5
A5	75,000	500	R245fa	0.5	**D5**	75,000	500	Ethanol	0.5
A6	200,000	500	R245fa	0.5	**D6**	200,000	500	Ethanol	0.5
A7	300,000	500	R245fa	0.5	**D7**	300,000	500	Ethanol	0.5
B1	100,000	500	R245fa	1.0	**E1**	100,000	500	Water	0.5
B2	100,000	750	R245fa	1.0	**E2**	100,000	750	Water	0.5
B3	100,000	1000	R245fa	1.0	**E3**	100,000	1000	Water	0.5
B4	50,000	500	R245fa	1.0	**E4**	50,000	500	Water	0.5
B5	75,000	500	R245fa	1.0	**E5**	75,000	500	Water	0.5
B6	200,000	500	R245fa	1.0	**E6**	200,000	500	Water	0.5
B7	300,000	500	R245fa	1.0	**E7**	300,000	500	Water	0.5
C1	100,000	500	FC72	0.5	–	–	–	–	–
C2	100,000	750	FC72	0.5	–	–	–	–	–
C3	100,000	1000	FC72	0.5	–	–	–	–	–
C4	50,000	500	FC72	0.5	–	–	–	–	–
C5	75,000	500	FC72	0.5	–	–	–	–	–
C6	200,000	500	FC72	0.5	–	–	–	–	–
C7	300,000	500	FC72	0.5	–	–	–	–	–

The computational domain consists of a solid and a fluid region that are coupled at their interface. The fluid and solid domains consist of a total number of 143,750 and 115,000 computational cells for the 0.5 mm internal diameter channel and 287,500 and 115,00 computational cells for the 1.0 mm internal diameter channel, respectively. The mesh density in each case was selected after an appropriate mesh independency study. A structured mesh, consisting of hexahedral and prismatic elements is used, with a grid clustering towards the coupling interface between the solid and fluid regions. The overall computation domain constitutes a 5° wedge of the circular micro-channel under consideration in each of the two cases, since this is the way that 2D axisymmetric simulations are conducted in OpenFOAM, the open-source CFD platform that is utilized for the overall computations. Table 1 summarizes the imposed oscillation characteristics, the working fluid and the channel diameter for each of the examined cases, in the present numerical investigation. The initial state of the bubble just before the application of the oscillation for each series of parametric numerical simulations is depicted in Fig. 3.

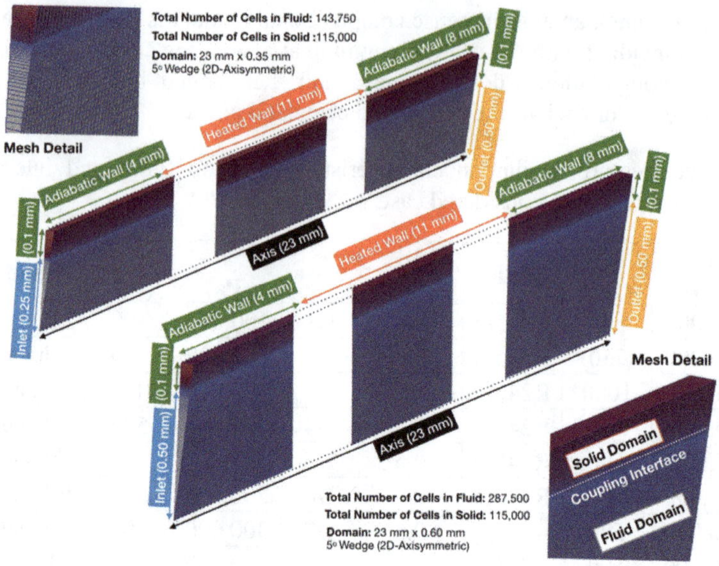

Figure 2: Geometry, Mesh and Boundary conditions, for each of the considered microchannels (D_i = 0.5 in the top-left part and D_i = 1.0 mm in the bottom-right part of the figure).

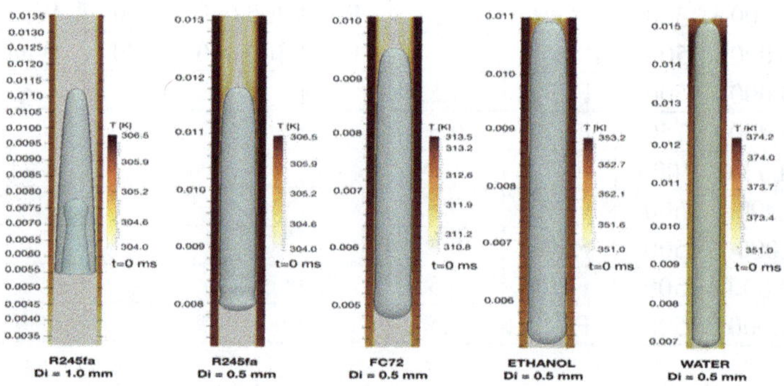

Figure 3: Initial state for all series of parametric numerical simulation (Series A, B. C, D and E), before the application of the fluctuating pressure at the inlet boundary.

3.2 Effect of flow oscillation frequency

The spatial and temporal evolution of the considered vapour slugs in the case of R245fa, under different pressure oscillation frequencies at the inlet of the considered micro-channel, are depicted in Figs 4 and 5 for the D_i = 0.5 mm and D_i = 1.0 mm channels respectively. Fig. 4 depicts the numerical simulation results for cases A1–A3, while Fig. 5 illustrates the corresponding results for cases B1–B3. In more detail, in both cases the amplitude of the pressure oscillation is kept constant to 100 kPa and the frequency is increased progressively taking the values of 500, 750 and 1000 Hz. As it can be observed, the imposed flow

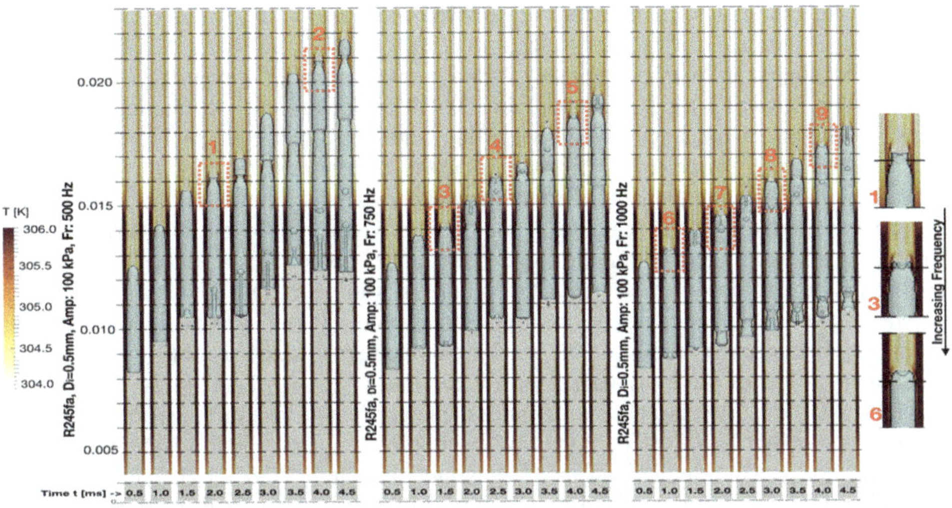

Figure 4: Spatial and temporal evolution of the considered vapour slugs in the case of R245fa, under different pressure oscillation frequencies at the inlet of the considered micro-channel (D_i = 0.5 mm). Cases A1 (left), A2 (middle) and A3 (right), from Table 1. The red numbered squares mark the initiation of capillary ridges. A zoomed version of the first instance in each case is also provided.

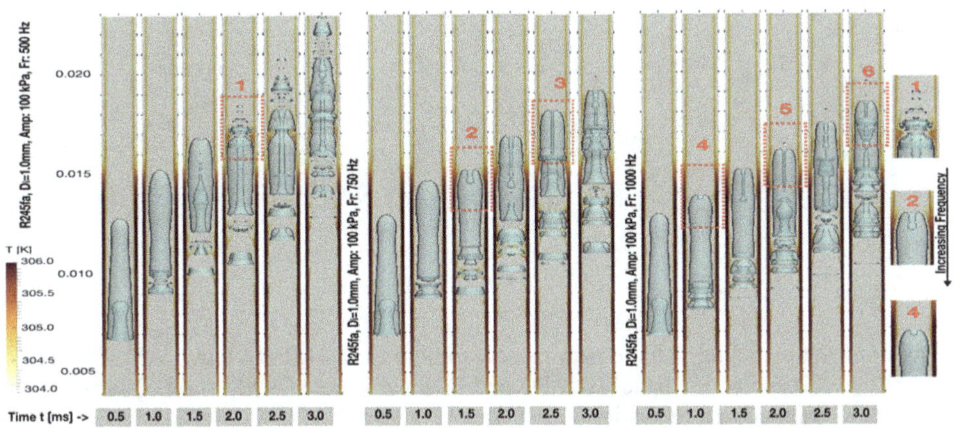

Figure 5: Spatial and temporal evolution of the considered vapour slugs in the case of R245fa, under different pressure oscillation frequencies at the inlet of the considered micro-channel (D_i = 1.0 mm). Cases B1 (left), B2 (middle) and A3 (right), from Table 1. The red numbered squares mark the initiation of capillary ridges. A zoomed version of the first instance in each case is also provided.

oscillations are responsible for the periodic generation of a capillary ridges in the vicinity of the leading edge of the evaporating vapour slug. These capillary ridges, after their formation (see for example the numbered red dotted squares in Figs 4 and 5) are then traveling downstream towards the trailing edge of the bubble, gradually attenuating (reducing their maximum height and increasing their overall length).It is characteristic that for both channel diameters, the period of the capillary ridge formation, coincides with the period of a complete pressure oscillation at the inlet (i.e. every 2 ms for cases A1 and B1, every 1.33 ms for cases A2 and B2 and every 1 ms for cases A3 and B3). It is also evident that, the increase in the pressure oscillation frequency at the inlet of the considered micro-channel in each case, is accompanied by a corresponding reduction in the maximum height of the periodically generated capillary ridge. Some interesting inertial phenomena are also evident, that result in liquid jets penetrating at the rear end of the bubbles leading to the encapsulation of small liquid droplets and fingers within the elongated vapour bubbles and/or in the detachment of small satellite bubbles from its leading and/or its trailing edges. The proposed inertia generated interfacial phenomena are more enhanced in the 1.0 mm internal diameter channel, that in most cases lead to the complete break-up of elongated vapour bubbles into 2 or more parts. Similar, trends were also observed for the rest of the working fluids, however the visualisation results from these working fluids are not presented here due to space limitations.

In order to quantify the effect of the flow oscillation frequency, in the evaporation of the liquid film between the elongating vapour slugs and the heated wall of the channel, the vapour slug volume increase with time is plotted for each of the considered working fluids in Fig. 6.

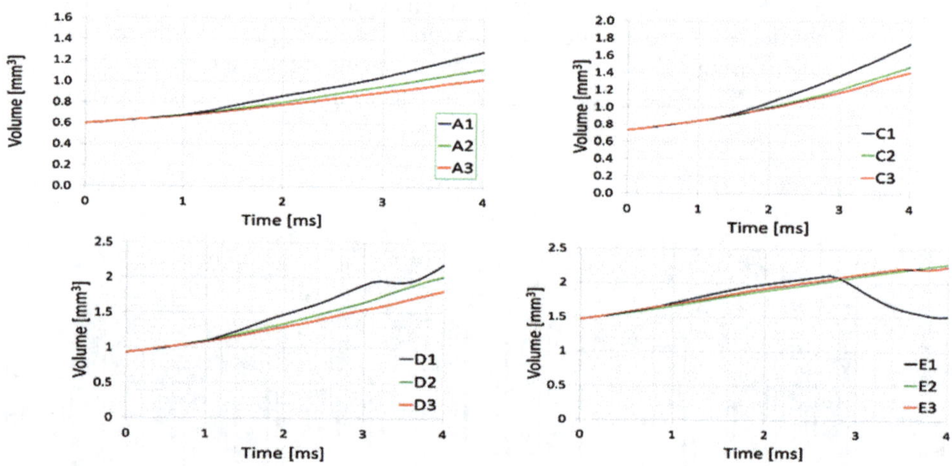

Figure 6: Vapour slug volume with respect to time for R245fa (A1, A2, A3), FC72 (C1, C2 and C3), ethanol (D1, D2 and D3) and water (E1, E2 and E3) for different oscillation frequencies.

As it can be observed, for all four simulated working fluids, the increase in flow oscillation frequency causes a significant decrease in the liquid film evaporation. It seems however that as the frequency increases further that the proposed effect becomes progressively less intense (e.g. C2 to C3 in comparison to C1 to C2 transition). In the case of ethanol and water and for flow oscillation frequency of 500 Hz (D1 and E1), just after around t = 3 ms in both cases, a decrease in the vapour volume is observed. This can be attributed to the significant amount of liquid that has been encapsulated in the form of liquid droplets, jets and fingers, within the elongating vapour slugs in these cases.

3.3 Effect of flow oscillation amplitude

The spatial and temporal evolution of the considered vapour slugs in the case of R245fa, under different pressure oscillation amplitudes at the inlet of the considered micro-channel, are depicted in Figs 7 and 8 for the D_i = 0.5 mm and D_i = 1.0 mm channels respectively. Fig. 7 depicts the numerical simulation results for cases A4–A7, while Fig. 8 illustrates the corresponding results for cases B4–B7. In more detail, in both cases the frequency of the pressure oscillation is kept constant to 500 Hz and the amplitude is increased progressively taking the values of 50, 75, 100, 200 and 300 kPa. In this case it is evident that, the increase in the pressure oscillation amplitude at the inlet of the considered micro-channel in each case, is accompanied by a corresponding increase in the maximum height of the periodically generated capillary ridge. In some cases, the height of the formed capillary ridge riches the channel radius causing the bubble to break up into 2 parts at this point (see for example highlighted red square number 5 in Fig. 7 and squares number 3, 4 and 5 in Fig. 8). Also, here inertial phenomena are also evident, that result in liquid jets penetrating at the rear end of the bubbles leading to the encapsulation of small liquid droplets jets and fingers within the elongated vapour bubbles and/or in the detachment of small satellite bubbles from its leading and/or its trailing edges.

The proposed inertia generated interfacial phenomena are more enhanced in the 1.0 mm internal diameter channel, that in most cases lead to the complete break-up of elongated vapour bubbles into two or more parts. Moreover, these become more intense as the amplitude of the flow oscillation increases. Similar, trends were also observed for the rest of the working fluids, however the visualisation results from these working fluids are not

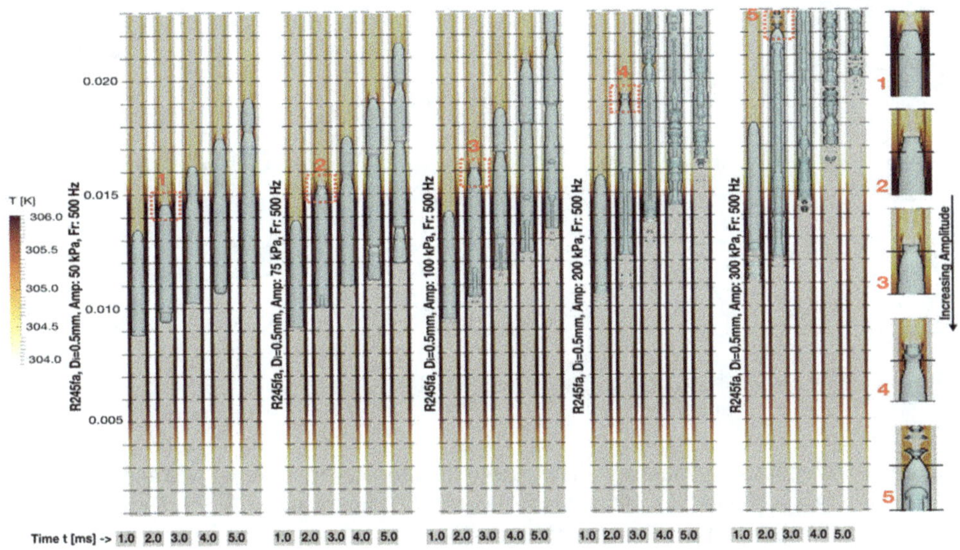

Figure 7: Spatial and temporal evolution of the considered vapour slugs in the case of R245fa, under different pressure oscillation amplitudes at the inlet of the considered micro-channel (D_i = 0.5 mm). Cases A4, A5, A1, A6 and A7 (left to right), from Table 1. The red numbered squares mark the initiation of the first capillary ridge in each case. A zoomed version of these instances for each case is also provided.

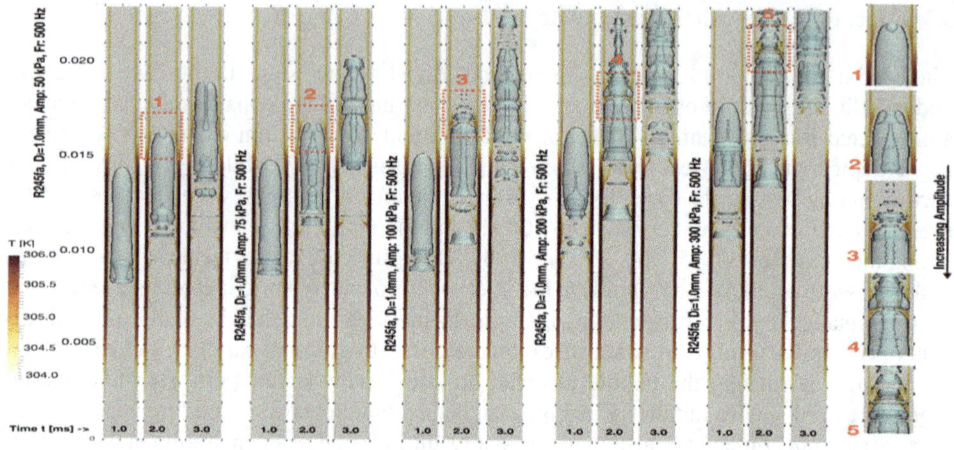

Figure 8: Spatial and temporal evolution of the considered vapour slugs in the case of R245fa, under different pressure oscillation amplitudes at the inlet of the considered micro-channel (D_i = 1.0 mm). Cases B4, B5, B1, B6 and B7 (left to right), from Table 1. The red numbered squares mark the initiation of the first capillary ridge in each case. A zoomed version of these instances for each case is also provided.

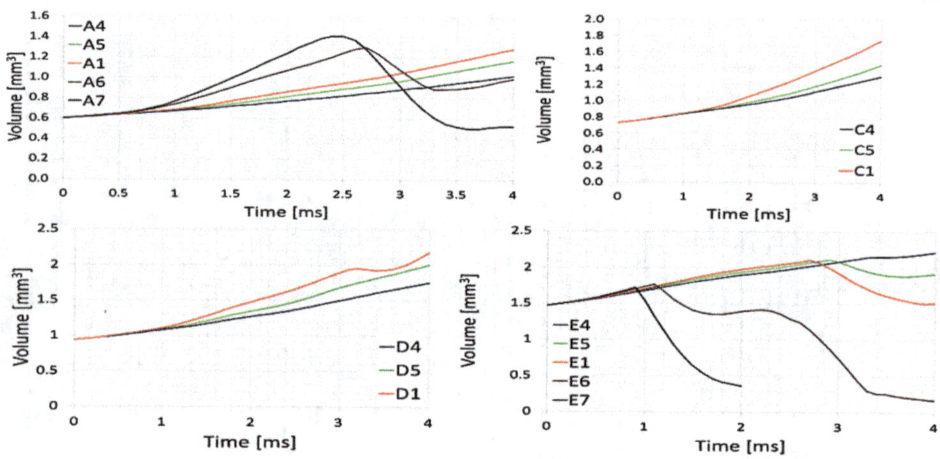

Figure 9: Vapour slug volume with respect to time for R245fa (A4, A5, A6 and A7), FC72 (C4, C5 and C1), ethanol (D4, D5 and D1) and water (E4, E5, E1, E6 and E7) for different oscillation amplitudes.

presented here due to space limitations. In order to quantify the effect of the flow oscillation amplitude, in the evaporation of the liquid film between the elongating vapour slugs and the heated wall of the channel, the vapour slug volume increase with time is plotted for each of the considered working fluids in Fig. 9.

As it can be observed, for all four simulated working fluids, the increase in flow oscillation amplitude, for the initial stages of the bubble growth, causes a corresponding increase in the liquid film evaporation. It seems in most cases that as the amplitude of the oscillation increases further that the proposed effect becomes progressively more intense (e.g. C4 to C5 in comparison to C5 to C1 transition). In all working fluid cases apart from FC72 and for specific oscillation amplitudes, at a certain in each case time instance, which varies in relation to both the oscillation amplitude and the working fluid, a reduction in the vapour volume with respect to time is observed. This can be attributed to the significant amount of liquid that has been encapsulated in the form of liquid droplets, jets and fingers, within the elongating vapour slugs in all these cases, and also to the fact that in some cases the bubble has already exited the channel outlet.

4 CONCLUSIONS

In the present paper, a previously validated and optimized, enhanced, VOF-based numerical modelling framework is applied in order to identify and quantify the effects of global flow oscillation frequency and amplitude on the liquid film evaporation and instabilities formation, around isolated, elongated vapour slugs, within heated micro-channels, in saturated flow boiling conditions. Four different working fluids are considered. From the post-processing and analysis of the results it is obvious that after the application of the induced flow oscillation, capillary ridges are periodically formed in the liquid film between the vapour slug and the heated wall of the channel, in the vicinity of the trailing edge of the elongating bubbles, as they preferentially move towards the outlet. It is shown that the period of formation coincides with the period of the induced flow oscillations for all working fluids tested and that the increase in the flow oscillation frequency, is accompanied by a reduction in the maximum height of the periodically generated capillary ridges, while the increase in the flow oscillation amplitude, causes a corresponding increase in their maximum height that for high oscillation amplitudes can eventually lead to the bubble break-up into two or more parts. Finally, it is found that the liquid film evaporation rate is significantly enhanced by the corresponding increase of the flow oscillation amplitude, while it degrades by the increase of the flow oscillation frequency. For future investigations the content of the present paper can be further extended by performing similar parametric numerical investigations for longer channels and for different applied heat fluxes in order to examine the effect of flow oscillation characteristics in cases of different initial bubble lengths.

ACKNOWLEDGEMENTS

The authors would like to acknowledge the UK's Engineering and Physical Science Research Council support, through the grant EP/P013112/1 as well as the financial support from ESA MAP Projects INWIP and ENCOM 3.

REFERENCES

[1] Magnini, M., Pulvirenti, B. & Thome, J.R., Numerical investigation of hydrodynamics and heat transfer of elongated bubbles during flow boiling in a microchannel. *Int. J. Heat Mass Transf.*, **59**, pp. 451–471, 2013.

[2] Szczukiewicz, S., Magnini, M. & Thome, J.R., Proposed models, ongoing experiments, and latest numerical simulations of microchannel two-phase flow boiling. *Int. J. Multiph. Flow*, **59**, pp. 84–101, 2014.

[3] Lee, J. & Kim, S.J., Effect of channel geometry on the operating limit of micro pulsating heat pipes. *Int. J. Heat Mass Transf.*, **107**, pp. 204–212, 2017.

[4] Mangini, D., Mameli, M., Georgoulas, A., Araneo, L., Filippeschi, S. & Marengo, M., A pulsating heat pipe for space applications: Ground and microgravity experiments. *Int. J. Therm. Sci.*, **95**, 2015.

[5] Brackbill, J., Kothe, D. & Zemach, C., A continuum method for modeling surface tension. *J. Comput. Phys.*, **100**(2), pp. 335–354, 1992.

[6] Hoang, D.A., van Steijn, V., Portela, L.M., Kreutzer, M.T. & Kleijn, C.R., Benchmark numerical simulations of segmented two-phase flows in microchannels using the Volume of Fluid method. *Comput. Fluids*, **86**, pp. 28–36, 2013.

[7] Deshpande, S.S., Anumolu, L. & Trujillo, M.F., Evaluating the performance of the two-phase flow solver interFoam. *Comput. Sci. Discov.*, **5**(1), p. 14016, 2012.

[8] Georgoulas, A., Koukouvinis, P., Gavaises, M. & Marengo, M., Numerical investigation of quasi-static bubble growth and detachment from submerged orifices in isothermal liquid pools: The effect of varying fluid properties and gravity levels. *Int. J. Multiph. Flow*, **74**, pp. 59–78, 2015.

[9] Hardt, S. & Wondra, F., Evaporation model for interfacial flows based on a continuum-field representation of the source terms. *J. Comput. Phys.*, **227**(11), pp. 5871–5895, 2008.

[10] Georgoulas, A., Andredaki, M. & Marengo, M., An enhanced VOF method coupled with heat transfer and phase change to characterise bubble detachment in saturated pool boiling. *Energies*, **10**(3), 2017.

[11] Teodori, E., Pontes, P., Moita, A., Georgoulas, A., Marengo, M. & Moreira, A., Sensible heat transfer during droplet cooling: Experimental and numerical analysis. *Energies*, **10**(6), 2017.

[12] Andredaki, M., Georgoulas, A., Miché, N. & Marengo, M., Break-up mechanisms and conditions for vapour slugs within mini-channels. *15th UK Heat Transfer Conference, UKHTC2017,* Brunel University, London, 4–5 September, 2017.

INTERACTION OF A CAVITATION BUBBLE WITH AN ANTIBUBBLE

BAI LIXIN[1], XU WEILIN[2], YAN JIUCHUN[3] & ZENG ZHIJIE[1]
[1]State Key Laboratory of Acoustics, Institute of Acoustics, Chinese Academy of Sciences, China
[2]State Key Laboratory of Hydraulics and Mountain River Engineering, Sichuan University, China
[3]State Key Laboratory of Advanced Welding Production Technology, Harbin Institute of Technology, China

ABSTRACT

An antibubble is the opposite of a bubble: a thin spherical gas shell containing liquid inside and surrounded by liquid outside. Although antibubbles are rarely used in industry, they do have some potential applications, such as cleaning, lubrication, filtration and material transport. The dynamic behaviour of antibubbles and spark-induced cavitation bubbles is experimentally studied by using high-speed photography. It is found that when spark discharges occur between the electrodes and the cavitation bubble begin to expand, the antibubble splits into upper and lower parts. The upper parts with more gas produces annular liquid jet under the action of shock wave, and split into two bubbles in the next few cycles under the action of inertial force. The lower parts are composed of semi-spherical shell-like gas film. When the compression wave arrives, the gas film becomes thinner and contacts between the two sides of the gas film appear at some points. When the expansion wave arrives, the gas film becomes thicker and the contact point becomes larger to form holes. As the holes become larger and the gas film shrinks, many scattered microbubbles are formed. The secondary shock wave caused by the collapse of the cavitation bubble which occur a few milliseconds after the first shock wave (spark discharge) was observed by schlieren photography. The physical mechanism of interaction between cavitation bubble and antibubble is analysed.
Keywords: cavitation bubble, antibubble, film, shock wave, high-speed photography.

1 INTRODUCTION

Cavitation refers to the formation and subsequent dynamic life of bubbles in liquids subjected to a sufficiently low pressure [1]. Extreme high temperature and pressure could be generated inside the cavitation bubble during the collapse, and shock waves and microjets could be generated in the liquid outside the cavitation bubbles. Cavitation can concentrate the diffuse energy to produce many effects with many applications. Applications of cavitation span many industrial sectors, from ultrasonic brazing, through ultrasonic lithotripsy, sonochemistry, ultrasonic cleaning, and wastewater treatment, to jet cutting. These applications of cavitation occur through the interaction between cavitation bubbles and other objects. Many investigation have been carried out. Lindau and Lauterborn [2], Shaw et al. [3], Brujan et al. [4] and Tomita et al. [5] investigated the interaction between a cavitation bubble and a rigid boundary. Robinson et al. [6] analysed the motion of two cavitation bubbles in a line perpendicular to the free surface. Bai et al. [7] investigated the interaction of an air bubble (isolated in water or attached to a boundary) with shock waves induced by electric sparks. Kodama et al. [8] investigated the interaction of a cavitation bubble with two air bubbles attached to a boundary. Poulain et al. [9] investigated the interaction of a cavitation bubble with a distant and small tethered spherical particle. Until recently, the interaction of a cavitation bubble with a gas film (antibubble) has not been studied.

An antibubble is the opposite of a soap bubble: a thin spherical gas shell containing liquid inside and surrounded by liquid outside. Although antibubbles are rarely used in industry, they do have some potential applications, such as cleaning, lubrication, filtration and material transport. Many investigation have been carried out on formation [10]–[14], aging [15]–[18],

WIT Transactions on Engineering Sciences, Vol 123, © 2019 WIT Press
www.witpress.com, ISSN 1743-3533 (on-line)
doi:10.2495/MPF190131

collapse [19], [20] and stabilization [21]–[23] of antibubbles. However, no investigation on the interaction between antibubbles and other fluid objects has been carried out.

In this paper, the interaction of a cavitation bubble with an antibubble is experimentally studied by using high-speed photography. The dynamic behaviour of antibubbles and spark-induced cavitation bubbles is analysed. The present work is, from the authors' knowledge, the first analysis on the interaction of a cavitation bubble with an antibubble.

2 EXPERIMENT

The experimental setup consists of a cavitation bubble generation system, an antibubble generation system, and imaging and illumination devices (Fig. 1). Cavitation bubbles are generated by using high-voltage spark discharge. The discharge time is of special importance, and is accurately controlled. Fine copper wires are used as electrodes in the experiment. The dripping device can produce jets of different velocities and droplets of different sizes. A rectangular plexiglass container (220 mm × 150 mm × 170 mm) is used to hold a mixture of tap water and linear alkylbenzenesulfonate (LAS) (about 10 times the critical micellar concentration (2.2 mM). The behaviour of cavitation bubbles and antibubbles is recorded with a high-speed camera (Photron Fastcam SA-1, Photron Ltd., Japan) equipped with two long distance microscopes (Zoom 6000, Navitar, USA; LM50JCM, Kowa, Japan) respectively. The frames are illuminated with PI-LUMINOR high-light LED lamp (150W), Cree XHP70 white high power LED, HALOGEN lamp (2600W). In the experiment, the liquid in the dripping device is the same as that in the rectangular glass tank. The gas in the antibubble film is the air in the environment (temperature: 20°C, pressure: 1 atm).

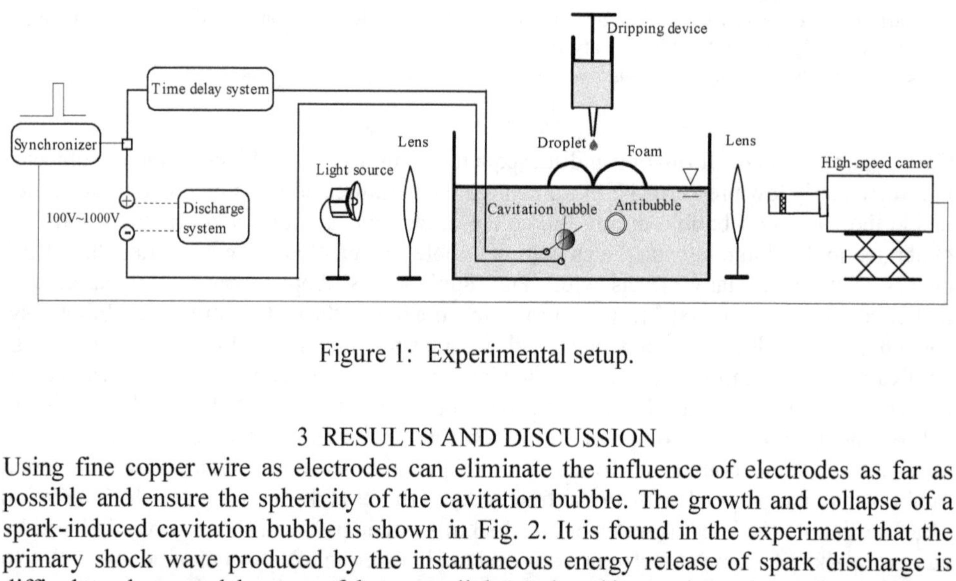

Figure 1: Experimental setup.

3 RESULTS AND DISCUSSION

Using fine copper wire as electrodes can eliminate the influence of electrodes as far as possible and ensure the sphericity of the cavitation bubble. The growth and collapse of a spark-induced cavitation bubble is shown in Fig. 2. It is found in the experiment that the primary shock wave produced by the instantaneous energy release of spark discharge is difficult to photograph because of the strong light produced by spark (as shown the schematic diagram in Fig. 2). The secondary shock wave produced by the collapse of cavitation bubble is visualized clearly by schlieren photography (as shown the subfigure on the right in Fig. 2). The primary shock wave and the secondary shock wave will affect the deformation of antibubbles.

It is found in the experiment that antibubbles will split horizontally into two parts (upper part and lower part) during the growth of the cavitation bubble (Fig. 3). The splitting line is always near the top of the antibubble and always horizontal. Through many experiments, it is found that there is a stable phase relationship between the cavitation bubble and the antibubble (Fig. 4). A large number of experiments show that the antibubble splitting characteristics are independent of azimuth (Fig. 5). Experiments show that if the primary shock wave is too strong, the antibubble may completely break into microbubbles without splitting (Fig. 6). In this paper, only moderate- and low-intensity shock waves are discussed.

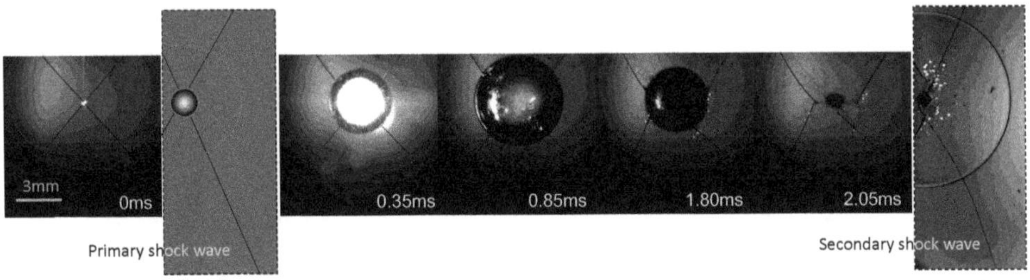

Figure 2: The growth and collapse of a spark-induced cavitation bubble.

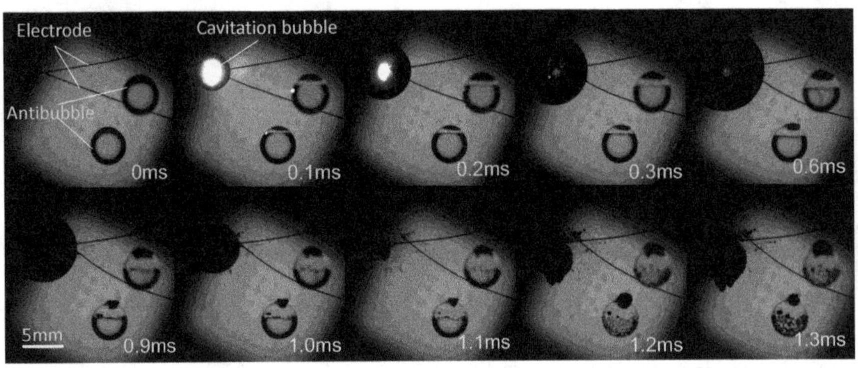

Figure 3: Interaction of a cavitation bubble with two antibubbles (medium-intensity shock wave).

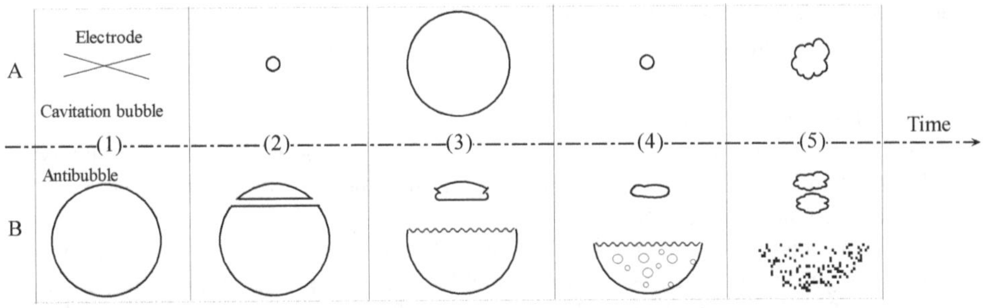

Figure 4: Schematic diagram of phase relation between cavitation bubble deformation and antibubble deformation.

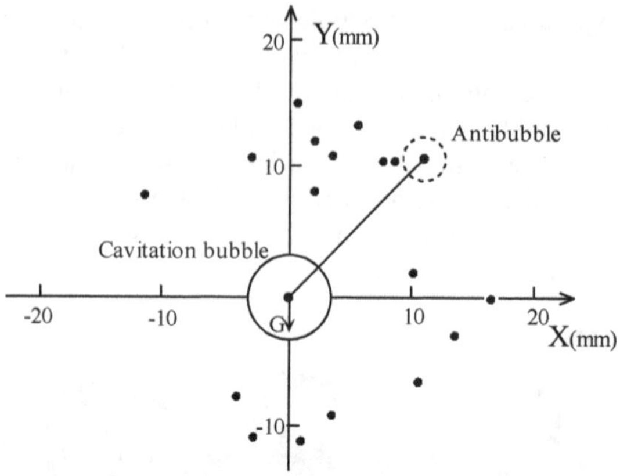

Figure 5: Relative positions of cavitation bubble and antibubble in the experiments.

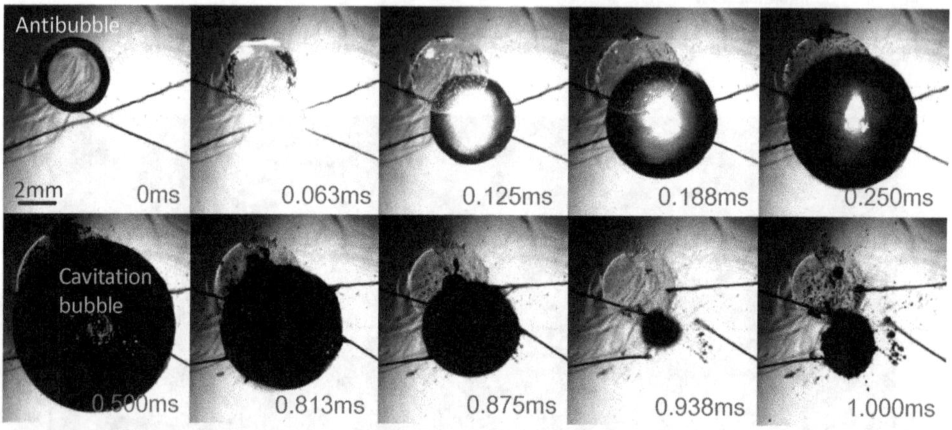

Figure 6: Interaction of a cavitation bubble with a antibubble (high-intensity shock wave).

The detailed process of antibubble deformation can be seen from Fig. 7. The lower edge of the upper part of the antibubble shrinks inward and produces a liquid jet (Fig. 7(a) (13)). When the secondary shock wave arrives, the upper part of the antibubble appears as a flat disc (Fig. 7(a) (25) and Fig. 7(b) (τ = 0.8033 ms)). Then, the upper part of the antibubble splits vertically into two air bubbles (Fig. 7(a) (36) and Fig. 7(b) (τ = 1.2 ms)). The upper edge of the lower part of the antibubble shrinks downward until the secondary shock wave arrives (Fig. 7(a) (1)–(24) and Fig. 7(b) (τ = 0–0.8 ms)). There are some holes in the gas film (Fig. 7(a) (25) and Fig. 7(b) (τ = 0.8033 ms)). Then the holes became larger and the gas film fragmented into many microbubbles.

Because the gas density is smaller than that of water, under the action of gravity and buoyancy, the gas in the lower part of the antibubble film will gather up to the upper part, so the upper gas film is thicker than the lower one (Fig. 8 (1)). When the primary shock wave arrives, the gas film in the antibubble is compressed. Because the upper part of antibubble contains more gas, so the upper part shrinks more obviously. The suction effect of the upper

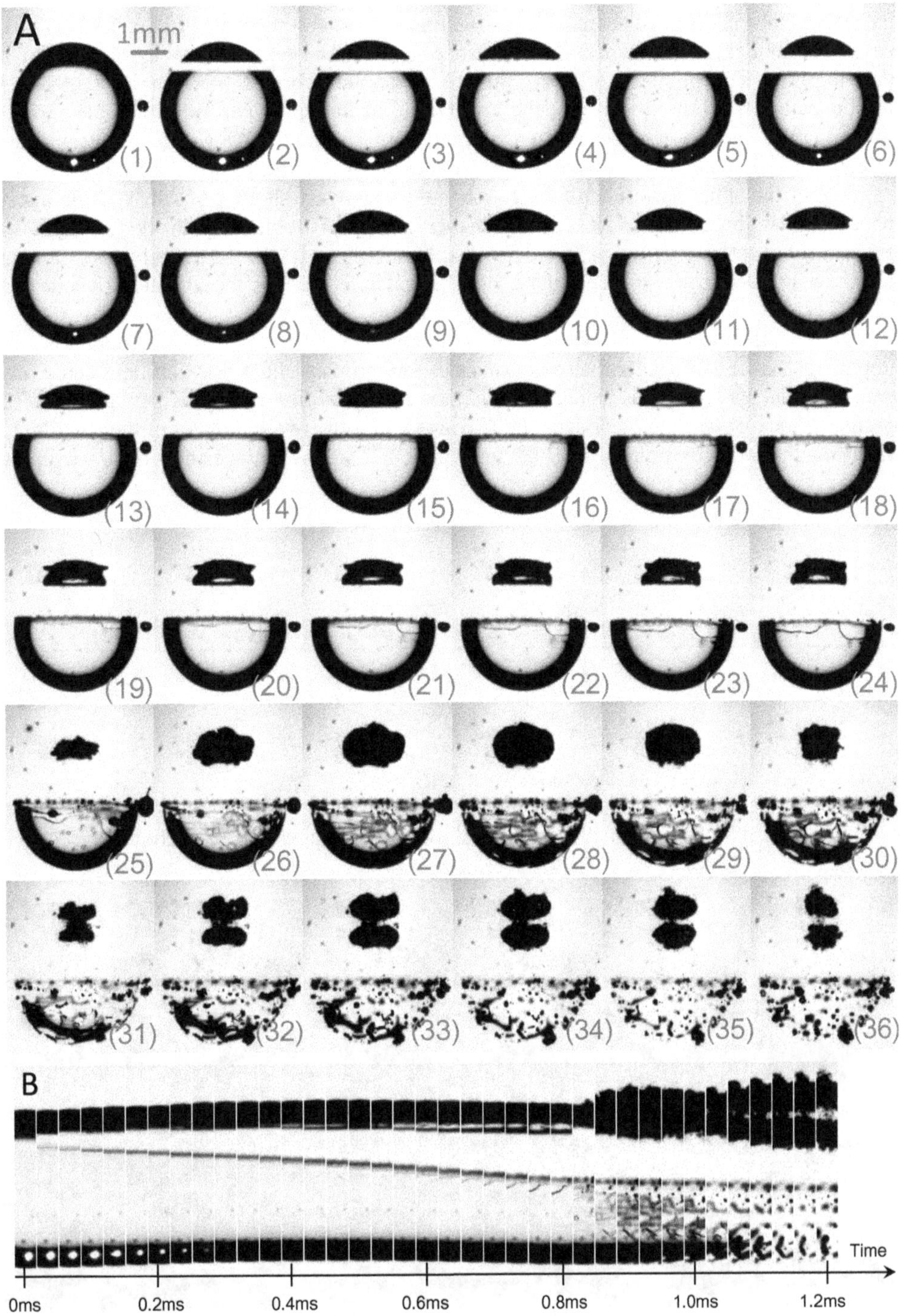

Figure 7: Deformation of antibubble. (a) High-speed photos (Frame rate 30 000 fps, exposure time 1/184000 s); and (b) Streak plot of antibubble deformation.

gas causes the film to break somewhere. The fracture location is related to the thickness (d) of the gas film at that location and the distance (L) from the center of gravity of gas film to that location. The points that meets this condition will form a horizontal circle. After disconnecting from the lower part, the lower edge of the upper part of antibubble shrinks inward under the action of surface tension, and form a liquid jet (Fig. 8 (3)). Because the direction of the jet is closer to the tangent direction of the arc, the jet does not cause the splitting of the upper part of antibubble. When the secondary shock wave arrives, the upper part of the antibubble appears as a flat disc (Fig. 8 (5, 6)). After the shock wave and its subsequent negative pressure pass, the flat disc air bubble shrinks inward and forms a circular jet under the action of surface tension. The circular jet divides the air bubbles into two (Fig. 8 (7, 8)).

Although the gas film of lower part of antibubble is very thin (submicron to micron), its impact resistance is stronger than we expected. The primary shock wave did not cause any damage to the gas film. It is well known that the secondary shock wave is stronger than the primary shock wave. The secondary shock waves cause the two interfaces of the gas film to stick together at some point (Fig. 9 (2) and Fig. 10 (2)). The stronger the shock wave, the

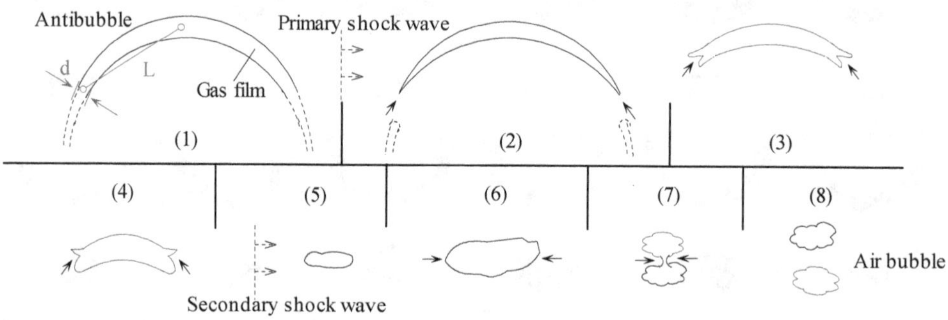

Figure 8: Schematic diagram of deformation of upper part of antibubble.

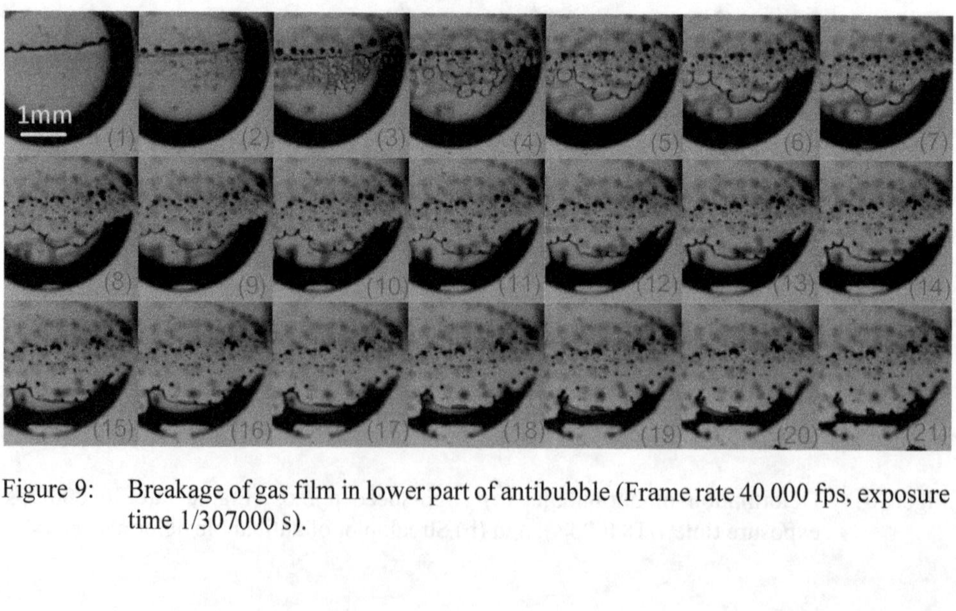

Figure 9: Breakage of gas film in lower part of antibubble (Frame rate 40 000 fps, exposure time 1/307000 s).

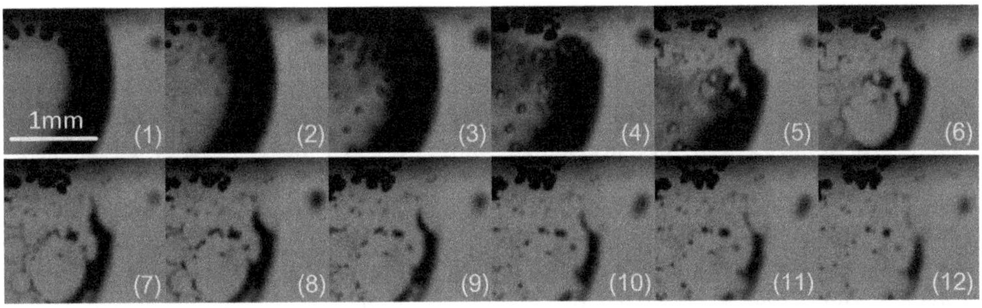

Figure 10: Details of gas film rupture (Frame rate 100 000 fps. Exposure time 1/297000 s).

more points of contact. Under the action of surface tension, the contact points become holes. The gas gathers around the hole (thicker than the gas film), forming a ring-shaped gas bubble connected with the gas film (Fig. 9 (3) and Fig. 10 (3–5)). As the holes become larger, they connect with each other and break to form many microbubbles.

The total reflection caused by the gas–liquid interface on both sides of the gas film makes the antibubble show a thick black edge on the outer edge in the transmission light. Under the impact of secondary shock wave, the inner edge of the black edge blurs into gray (Fig. 11(a) (2) and Fig. 11(b) (2)). This indicates that a small amount of light passes through the original total reflection region. The contact point of the film is probably the channel through which the small amount of light may pass (Fig. 11(b) (3)).

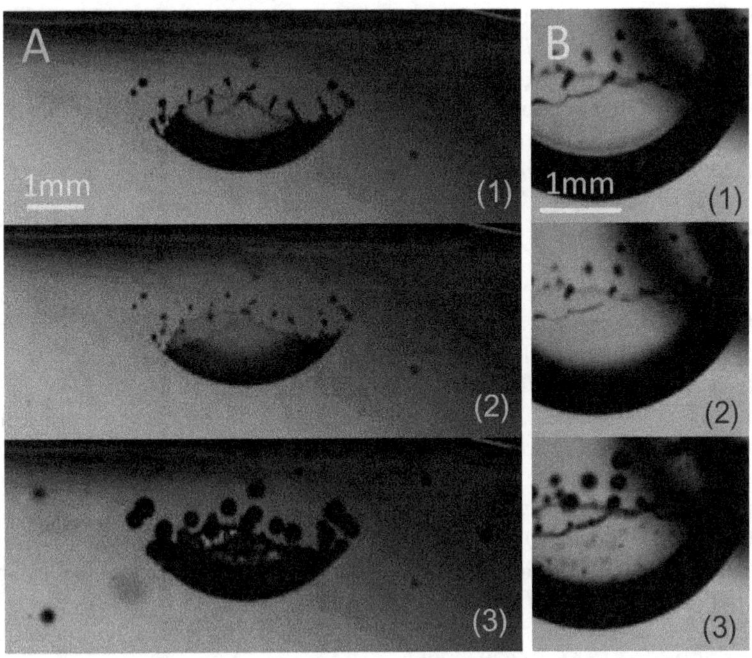

Figure 11: Black edge caused by total reflection (Frame rate 100 000 fps, exposure time 1/297000 s).

4 SUMMARY AND CONCLUSIONS

The interaction of a spark-induced cavitation bubble with an antibubble is experimentally investigated by using high-speed photography. It is found that there is a stable phase relationship between the cavitation bubble deformation and the antibubble deformation. Antibubbles are always divided horizontally into two parts, independent of azimuth. The suction effect of the upper gas causes the gas film to break. The fracture location is related to the thickness of the gas film at that location and the distance from the center of gravity of gas film to that location. The formation of the liquid jet in the upper part of the antibubble and the splitting of the disc-shaped air bubble are all related to the surface tension. The secondary shock waves cause the two interfaces of the gas film in the lower part of antibubbles to stick together at some point. These defects eventually lead to partial loss of total reflection and the fragmentation of the gas film.

ACKNOWLEDGEMENT

This work was supported by the National Natural Science Foundation of China (No. 11874062, No. 11674350).

REFERENCES

[1] Bai, L., Xu, W., Tian, Z. & Li, N., A high-speed photographic study of ultrasonic cavitation near rigid boundary. *Journal of Hydrodynamics*, **20**, pp. 637–644, 2008.

[2] Lindau, O. & Lauterborn, W., Cinematographic observation of the collapse and rebound of a laser-produced cavitation bubble near a wall. *Journal of Fluid Mechanics,* **479**, pp. 327–348, 2003.

[3] Shaw, S.J., Schiffers, W.P. & Emmony, D.C., Experimental observations of the stress experienced by a solid surface when a laser-created bubble oscillates in its vicinity. *Journal of the Acoustical Society of America*, **110**, pp. 1822–1827, 2001.

[4] Brujan, E.A., Keen, G.S., Vogel, A. & Blake, J.R., The final stage of the collapse of a cavitation bubble close to a rigid boundary. *Physics of Fluids*, **14**, pp. 85–92, 2002.

[5] Tomita, Y. et al., Growth and collapse of cavitation bubbles near a curved rigid boundary. *Journal of Fluid Mechanics,* **466**, pp. 259–283, 2002.

[6] Robinson, P.B., Blake, J.R., Kodama, T., Shima, A. & Tomita, Y., Interaction of cavitation bubbles with a free surface. *Journal of Applied Physics*, **89**, pp. 8225–8237, 2001.

[7] Bai, L., Xu, W., Li, C. & Gao, Y., The counter jet formation in an air bubble induced by the impact of shock waves. *Journal of Hydrodynamics*, **23**(5), pp. 562–569, 2011.

[8] Kodama, T., Takayama, K. & Nagayasu, N., The dynamics of two air bubbles loaded by an underwater shock wave. *Journal of Applied Physics*, **80**, pp. 5587–5592, 1996.

[9] Poulain, S., Guenoun, G., Gart, S., Crowe, W. & Jung, S., Particle motion induced by bubble cavitation. *Physical Review Letters*, **114**(21), 214501, 2015.

[10] Bai, L., Xu, W., Wu, P., Lin, W. & Li, C., Formation of antibubbles and multilayer antibubbles. *Colloids and Surfaces A: Physicochemical and Engineering Aspects*, **509**(20), pp. 334–340, 2016.

[11] Ganan-Calvo, M. & Gordillo, J.M., Perfectly monodisperse microbubbling by capillary flow focusing. *Physical Review Letters*, **87**, 274501, 2001.

[12] Tufaile, A. & Sartorelli, J.C., Bubble and spherical air shell formation dynamics. *Physical Review E*, **66**, 056204, 2002.

[13] Kim, P.G. & Stone, A.H., Dynamics of the formation of antibubbles. *Europhysics Letters*, **83**, 54001, 2008.

[14] Postema, M., Ten Cate, F.J., Schmitz, G., De Jong, N. & Van Wamel, A., Generation of a droplet inside a microbubble with the aid of an ultrasound contrast agent: First result. *Letters in Drug Design and Discovery*, **4**, pp. 74–77, 2007.
[15] Dorbolo, S., Reyssat, E., Vandewalle, N. & Quéré, D., Aging of an antibubble. *Europhysics Letters*, **69**, pp. 966–970, 2005.
[16] Dorbolo, S. et al., Antibubble lifetime: Influence of the bulk viscosity and of the surface modulus of the mixture. *Colloids and Surfaces A*, **365**, pp. 43–45, 2010.
[17] Scheid, B., Dorbolo, S., Arriaga, L.R. & Rio, E., The drainage of an air film with viscous interfaces. *Physical Review Letters*, **109**, 264502, 2012.
[18] Scheid, B., Zawala, J. & Dorbolo, S., Gas dissolution in antibubble dynamics. *Soft Matter*, **10**, pp. 7096–7102, 2014.
[19] Sob'yanin, D.N., Theory of the antibubble collapse. *Physical Review Letters*, **114**, 104501, 2015.
[20] Zou, J., Ji, C., Yuan, B.G., Ruan, X.D. & Fu, X., Collapse of an antibubble. *Physical Review E*, **87**, 061002(R), 2013.
[21] Dorbolo, S., Caps, H. & Vandewalle, N., Fluid instabilities in the birth and death of antibubbles. *New Journal of Physics*, **5**, pp. 161.1–161.9, 2003.
[22] Kim, P.G. & Vogel, J., Antibubbles: Factors that affect their stability. *Colloids and Surfaces A*, **289**, pp. 237–244, 2006.
[23] Poortinga, A.T., Long-lived antibubbles: Stable antibubbles through Pickering stabilization. *Langmuir*, **27**, pp. 2138–2141, 2011.

FULLY DEVELOPED BUBBLY TWO-PHASE FLOW THROUGH A PIPE: AN ANALYTICAL SOLUTION

WAQAS ALI[1], MUBBASHAR NAZEER[2] & AHMED ZEESHAN[3]
[1]Faculty of Engineering Technology, University of Twente, The Netherlands
[2]Department of Mathematics, Riphah International University, Pakistan
[3]Department of Mathematics and Statistics, International Islamic University Islamabad, Pakistan

ABSTRACT

In this work, a novel homotopy analysis method for the solution of two-fluid models is presented. A fully developed bubbly through a pipe is considered. Different physical and mathematical properties of the two-fluid model were identified. The problem was solved in the central region of the pipe where the wall forces were neglected, which implies that void fractions and velocity profiles are not affected by the wall. The system of equations was reduced to a single equation without parameters with an intrinsic length scale L. Away from the wall of the pipe, the flat void fraction depends on the applied pressure gradient, the density of different phases and gravity. It was also found that the effective specific weight of the fluid column was cancelled by the pressure gradient.

Keywords: two-phase flow, laminar flow, bubbly flow, two-fluid model, fully developed flow.

1 INTRODUCTION

For the simulation of two-phase flows, the two-fluid model proposed by Ishii [1] is widely used. This model is based on the time averaging of two-phase field equations. The averaging process used in Ishii model needs to be accompanied with several closure approximations, which lack practical understanding. For example, lift term that models transversal force influencing particles or bubbles in a velocity field is not completely understood. In the current work, we analyse the two-fluid model by considering a fully developed bubbly flow in a circular pipe. Our main objective is to look into the generic properties of the model.

We first recall the two-fluid model in brief, with algebraic manipulations of the equations. The solution of the system is first provided away from the wall which follows the solution of the complete system.

2 MATHEMATICAL FORMULATION

We consider the mass and momentum conservation equations with no temperature effects similar to Drew and Passman [2], assumption.

$$\frac{\partial(\epsilon_k \rho_k)}{\partial t} + \nabla.(\epsilon_k \rho_k \overrightarrow{v_k}) = 0, \tag{1}$$

$$\frac{\partial(\epsilon_k \rho_k \overrightarrow{v_k})}{\partial t} + \nabla.(\epsilon_k \rho_k \overrightarrow{v_k}\,\overrightarrow{v_k}) = \nabla.[\epsilon_k(T_k + \tau_k^{Re})] + \epsilon_k \rho_k \vec{g} + M_k, \tag{2}$$

in which k is a phase (L for liquid, G for gas), ϵ_k is the volumetric fraction of phase k, $\overrightarrow{v_k}$ is the corresponding velocity, ρ_k is the density, T_k is the stress tensor, defined as

$$T_k = -p_k I + \mu_k[\nabla\overrightarrow{v_k} + (\nabla\overrightarrow{v_k})^T], \tag{3}$$

where p_k is the pressure, and μ_k is the viscosity τ_k^{Re} is the Reynolds like stress due to statistical fluctuations, modelled by Nigmatulin [3]

WIT Transactions on Engineering Sciences, Vol 123, © 2019 WIT Press
www.witpress.com, ISSN 1743-3533 (on-line)
doi:10.2495/MPF190141

$$\tau_k^{Re} = -\epsilon_G \rho_L [A|\overrightarrow{v_G} - \overrightarrow{v_L}|^2 I + B(\overrightarrow{v_G} - \overrightarrow{v_L})(\overrightarrow{v_G} - \overrightarrow{v_L})], \tag{4}$$

where values for A and B are $\frac{3}{20}$ and $\frac{1}{20}$, respectively. Assume $\vec{v}_G - \vec{v}_L = \vec{v}_r$ and define $v_r = |\vec{v}_r|$. In eqn (2), M_k represents the interfacial momentum exchange, which relates to the actual force between phases (M_k') according to

$$M_k = p_{ki} \nabla \epsilon_k - \tau_{ki} . \nabla \epsilon_k + M_k', \tag{5}$$

where values at the interphase are represented by subscript *i*. Different models exist for M_k, in particular for the liquid phase ($k = L$) Antal et al. [4] used

$$M_L = p_{Li} \nabla \epsilon_L - \mu_L (\nabla v_L + \nabla v_L^T) + \rho_L (A|\vec{v}_r|^2 I + B(\vec{v}_r \otimes \vec{v}_r)) + M_L', \tag{6}$$

whereas on the other hand Drew and Passman [2], used

$$M_L = p_{Li} \nabla \epsilon_L - [\rho_L (ab|\vec{v}_r|^2 I + a^2 (\vec{v}_r . \vec{v}_r))] \nabla \epsilon_L - \rho_L \epsilon_L \nabla . (\vec{v}_r \otimes \vec{v}_r) + M_L', \tag{7}$$

where $a = -\frac{9}{20}$ and $b = \frac{3}{20}$. The pressure at the interphase (p_{Li}) can be derived from the expression by Stuhmiller [5],

$$p_L - p_{Li} = C_{\rho_L} \epsilon_L |\vec{v}_r|^2, C = \frac{1}{4}. \tag{8}$$

We will consider drag, lift and wall forces,

$$M_L' = M_K'^D + M_K'^L + M_K'^W, \tag{9}$$

with

$$M_G'^D = -M_L'^D = -\frac{3}{8} \frac{\epsilon_G}{R_b} C_D \rho_L \vec{v}_r |\vec{v}_r|, \tag{10}$$

$$C_D = \frac{24}{Re} (1 + 0.1 Re^{0.75}), \tag{11}$$

$$Re = \frac{2 R_b \rho_L |\vec{v}_r|}{\mu_m}, \mu_m = \frac{\mu_L}{(1 - \epsilon_G)}, \tag{12}$$

$$M_L'^L = -M_L'^L = -C_L \epsilon_G \rho_L \vec{v}_r \wedge (\nabla \wedge \vec{v}_L), \tag{13}$$

$$M_G'^W = -M_L'^W = \begin{cases} \epsilon_L \rho_L |u_\parallel|^2 \left[C_{w1} + C_{w2} \left(\frac{R_b}{y_0}\right)\right] \widetilde{n_b} & if \ \frac{\epsilon_G \rho_L |u_\parallel|^2}{R_b} \left[C_{w1} + C_{w2} \left(\frac{R_b}{y_0}\right)\right] > 0, \\ 0 & if \ \frac{\epsilon_G \rho_L |u_\parallel|^2}{R_b} \left[C_{w1} + C_{w2} \left(\frac{R_b}{y_0}\right)\right] \leq 0 \end{cases}, \tag{14}$$

$$u_\parallel = \vec{v}_r - [\widetilde{n_w} . \vec{v}_r], \tag{15}$$

$$C_{w1} = -0.1, C_{w2} = 0.147, \tag{16}$$

where, y_0 is the distance to the wall, $\breve{n}_w$ the exterior unit normal and R_b the bubble radius.

Considering incompressible unidirectional laminar fully developed flow of Newtonian fluid with bubble floating through a vertical cylindrical duct. The set of five differential equations corresponding to five unknown functions, i.e., $v_L(r), v_G(r), p_L(r,z), p_G(r,z)$ and $\epsilon_G(r,z)$, as follows.

Momentum conservation equation of the gas phase in the radial direction $\breve{r}$

$$\epsilon_G\left[\frac{\partial p_G}{\partial r}+C_L\rho_L v_r\frac{\partial v_L}{\partial r}+\frac{\rho_L v_r^2}{R_b}\left(C_{w1}+C_{w2}\frac{R_b}{(R-r)}\right)\right]=0, \tag{17}$$

where R is the internal radius of the duct.

Momentum conservation equation of the gas phase in the axial direction $\breve{z}$

$$\frac{\partial p_G}{\partial r}=-\rho_G g-\frac{3}{8}\frac{C_D}{R_b}\rho_L v_r|v_r|. \tag{18}$$

Relative velocity v_r is calculated by using the above equation, which is considered uniform. Momentum conservation equation of the liquid phase in the radial direction $\breve{r}$

$$\epsilon_L\frac{\partial p_L}{\partial r}=-\frac{\partial}{\partial r}[A\rho_L\epsilon_L(1-\epsilon_L)v_r^2]+C_L\epsilon_G\rho_L v_r\frac{\partial v_L}{\partial r}+C_{\rho_L}(1-\epsilon_G)v_r^2\frac{\partial \epsilon_G}{\partial r}$$
$$+\frac{\epsilon_G\rho_L v_r^2}{R_b}\left(C_{w1}+C_{w2}\frac{R_b}{(R-r)}\right)-C_1\rho_L v_r^2\frac{\partial\epsilon_G}{\partial r}. \tag{19}$$

Momentum conservation equation of the liquid phase in the axial direction $\breve{z}$

$$\epsilon_L\frac{\partial p_L}{\partial r}=\frac{1}{r}\frac{\partial}{\partial r}\left[r\epsilon_L\mu_L\frac{\partial v_L}{\partial r}\right]+\frac{3}{8}\frac{\epsilon_G}{R_b}C_D\rho_L v_r|v_r|-\epsilon_L\rho_L g+C_2\mu_L\frac{\partial v_L}{\partial r}\frac{\partial \epsilon_G}{\partial r}. \tag{20}$$

Jump condition at the interphase is given by

$$p_G-p_L=\frac{2\sigma}{R_b}-C_{\rho_L}(1-\epsilon_G)v_r^2. \tag{21}$$

3 PROBLEM SOLUTION

3.1 Region away from the wall

We define

$$\rho\equiv\rho_{eff}=\epsilon_L\rho_L+\epsilon_G\rho_G, \bar{\rho}\equiv\frac{1}{S}\int\rho.ds, \qquad \frac{\partial P}{\partial z}\equiv\frac{\partial p}{\partial z}+\bar{\rho}g, \tag{22}$$

where S is the cross-sectional area of the pipe.

Eliminating the drag force from eqns (18) and (20), and using $\epsilon_L=\frac{(\rho-\rho_G)}{(\rho_L-\rho_G)}$ we get

$$\frac{\partial p}{\partial z}+(\rho-\bar{\rho})g=\frac{(\rho-\rho_G)}{(\rho_L-\rho_G)}\frac{\mu_L}{r}\frac{\partial}{\partial r}\left(r\frac{\partial v_L}{\partial r}\right)+(C_2-1)\mu_L\frac{\partial v_L}{\partial r}\frac{\partial\epsilon_G}{\partial r}. \tag{23}$$

On the other hand, replacing eqns (19) and (21) into eqn (17) we get, after multiplication by $\frac{\epsilon_L}{(\rho_L v_r)}$,

$$\frac{(\rho-\rho_G)}{(\rho_L-\rho_G)}\frac{1}{(\rho_L-\rho_G)}\frac{\partial p}{\partial z}=\frac{C_L}{(2C-A)v_r}\frac{\partial v_L}{\partial r}+\frac{1}{(2C-A)R_b}\left(C_{w1}+C_{w2}\frac{R_b}{(R-r)}\right)$$
$$-\frac{(C_1+A\epsilon_G)}{(2C-A)}\frac{1}{(\rho_L-\rho_G)}\frac{\partial p}{\partial r}. \tag{24}$$

As a result, the system is reduced to only two equations. Introducing non-dimensional variables:

$$\frac{\partial p^*}{\partial z}=\frac{1}{(\rho_L-\rho_G)}\frac{\partial p}{\partial z}, L^2=\frac{2\mu_L v_L(C-A)}{g(\rho_L-\rho_G)C_L}, v^*=\frac{v_L}{v_r}\frac{C_L}{2(C-A)}, r^*=\frac{r}{L}, R^*=\frac{R}{L}, R_L^*=\frac{R_b}{L}, E=\frac{C_{w_1}}{2(C-A)R_b^*}, D=\frac{C_{w_2}}{2(C-A)}. \tag{25}$$

From eqns (23) and (24), using the scaled variables in eqn (25) and assuming the wall-force terms are negligible, the system is reduced to following ordinary differential equation

$$(\epsilon_L)^2\frac{\partial^2\epsilon_L}{\partial r^{*2}}+(2-C_2)\epsilon_L\left(\frac{\partial\epsilon_L}{\partial r^*}\right)^2+\frac{(\epsilon_L)^2}{r^*}\frac{\partial\epsilon_L}{\partial r^*}-\epsilon_L+\frac{(C_1+A)}{2(C-A)}\left[\epsilon_L\left(\frac{1}{r^*}\frac{\partial\epsilon_L}{\partial r^*}+\frac{\partial^2\epsilon_L}{\partial r^{*2}}\right)-(C_2-1)\left(\frac{\partial\epsilon_L}{\partial r^*}\right)^2\right]=\frac{\partial p^*}{\partial z}-\bar{\epsilon_L}, \tag{26}$$

where $\bar{\epsilon_L}=\frac{(\bar{\rho}-\rho_G)}{(\rho_L-\rho_G)}$. This equation is valid where the wall force is not active $\left(R-r>\frac{DL}{E}\right)$, It can be seen that the right-hand side is a constant, independent of r^*. The values of the constants A and B are fixed ($A=\frac{3}{20}, B=\frac{1}{20}$), and considered following two set of values for C_1 and C_2, with $C_1=-\frac{3}{20}$ and $C_2=1$, taken for Antal et al. [4] (Model I), and $C_1=\frac{37}{100}$ and $C_2=0$, for Drew and Passman [2] (Model II).

In the case of Model I, eqn (26) is reduced to

$$(\epsilon_L)^2\frac{\partial^2\epsilon_L}{\partial r^{*2}}+\epsilon_L\left(\frac{\partial\epsilon_L}{\partial r^*}\right)^2+\frac{(\epsilon_L)^2}{r^*}\frac{\partial\epsilon_L}{\partial r^*}-\epsilon_L=\frac{\partial p^*}{\partial z}-\bar{\epsilon_L}, \tag{27}$$

which is the equation of laminar fully developed bubbly flow far from the wall and contains no parameter except constant on the right-hand side. Eqn (27) can be solved to get a different solution for pressure gradient and the amount of gas in the flow. By using $r^*=0$, $\varepsilon_L(0)$ eqn (27) can be rewritten as

$$\frac{\partial P^*}{\partial z}-\bar{\epsilon_L}=-\epsilon_L(0)(1+\lambda), \tag{28}$$

finally resulting in

$$(\epsilon_L)^2\frac{\partial^2\epsilon_L}{\partial r^{*2}}+\epsilon_L\left(\frac{\partial\epsilon_L}{\partial r^*}\right)^2+\frac{(\epsilon_L)^2}{r^*}\frac{\partial\epsilon_L}{\partial r^*}-\epsilon_L=-\epsilon_L(0)(1+\lambda). \tag{29}$$

If the model of Drew and Passman is used we following equation is obtained

$$(\epsilon_L)^2\frac{\partial^2\epsilon_L}{\partial r^{*2}}+2\epsilon_L\left(\frac{\partial\epsilon_L}{\partial r^*}\right)^2+\frac{(\epsilon_L)^2}{r^*}\frac{\partial\epsilon_L}{\partial r^*}-\epsilon_L+\frac{33}{80}\left[\epsilon_L\left(\frac{1}{r^*}\frac{\partial\epsilon_L}{\partial r^*}+\frac{\partial^2\epsilon_L}{\partial r^{*2}}\right)+\left(\frac{\partial\epsilon_L}{\partial r^*}\right)^2\right]=-\epsilon_L(0)(1+\lambda), \tag{30}$$

where λ is a measure of the difference between the applied pressure gradient and the effective weight of the mixture at the centre of the pipe. With the given initial conditions $\epsilon_L(0)$ and $\epsilon_L'(0)$ we then solve eqns (29) and (30).

4 SOLUTION OF THE PROBLEM

In the previous section problem in the region where the wall forces are not active. In addition, all applied pressure gradients were not considered as when $-\partial p_L / \partial z$ is greater than $\rho_L \bar{g}$, at the centre by adjusting the void fraction, the pressure gradient cannot be balanced. This is valid even if the void fraction is assumed zero.

Eqns (17)–(21) can be solved for $v_L(r)$ and $\varepsilon_G(r)$, which after mathematical manipulation and elimination of drag coefficient results in

$$\frac{\partial p_L}{\partial z} + (1-\epsilon_G)\rho_L g + \epsilon_G \rho_G g = \frac{1}{r}\frac{\partial}{\partial r}\left[r(1-\epsilon_G)\mu_L \frac{\partial v_L}{\partial r}\right] + C_2 \mu_L \frac{\partial v_L}{\partial r}\frac{\partial \epsilon_G}{\partial r}, \tag{31}$$

$$\epsilon_G \frac{\partial \epsilon_G}{\partial r} v_r^2 \left[\left(\frac{1}{2} - A + C_1\right) - \frac{1}{5}\epsilon_G\right] = -\epsilon_G C_L v_r \frac{\partial v_r}{\partial r} - \epsilon_G \left(C_{w1} + C_{w2}\frac{R_b}{Rr}\right)\frac{v_r^2}{R_b}, \tag{32}$$

with following boundary conditions

$$\frac{\partial v_L}{\partial r}(0) = 0, v_L(R) = 0, \frac{\partial \epsilon_G}{\partial r}(0) = 0. \tag{33}$$

Using the non-dimensional parameter defined in eqn (25) we get the following form

$$\frac{\partial P}{\partial z} + \frac{\rho_L}{\rho_L - \rho_G} - \epsilon_G = C_2 \frac{\partial v}{\partial r}\frac{\partial \epsilon_G}{\partial r} + \frac{1}{r}(1-\epsilon_G)\frac{\partial v}{\partial r} + \frac{\partial v}{\partial r}\frac{\partial \epsilon_G}{\partial r} + (1-\epsilon_G)\frac{\partial^2 v}{\partial r^2}, \tag{34}$$

$$\epsilon_G \frac{\partial \epsilon_G}{\partial r}\frac{\left(\frac{1}{2} - A + C_1\right)}{2(C-A)} - \epsilon_G^2 \frac{\partial \epsilon_G}{\partial r}\left(\frac{1}{10(C-A)}\right) = -\epsilon_G \frac{\partial v}{\partial r} - E\epsilon_G - \frac{D\epsilon_G}{R-r}, \tag{35}$$

subject to boundary conditions

$$\frac{\partial v_L}{\partial r}(0) = 0, \qquad v_L(R) = 0, \qquad \frac{\partial \epsilon_G}{\partial r}(0) = 0,$$
$$\text{where } P = \frac{\partial P}{\partial z} + \frac{\rho_L}{\rho_L - \rho_G}. \tag{36}$$

Eqns (34)–(36) are solved in *Mathematica* [6], for a wide set of values by using the homotopy analysis method (HAM). Zero-order deformation of the system of equations, the linear operator L and the non-linear operator N are defined as

$$L(f) = \frac{d^2 f}{dr^2}, \tag{37}$$

$$L(g) = \frac{dg}{dr}, \tag{38}$$

$$N(f) = \frac{\partial p}{\partial z} + \frac{\rho_L}{\rho_L - \rho_G} - \epsilon_G - C_2 \frac{\partial v}{\partial r}\frac{\partial \epsilon_G}{\partial r} - \frac{1}{r}(1-\epsilon_G)\frac{\partial v}{\partial r} - \frac{\partial v}{\partial r}\frac{\partial \epsilon_G}{\partial r}$$
$$-(1-\epsilon_G)\frac{\partial^2 v}{\partial r^2}, \tag{39}$$

$$N(g) = \epsilon_G \frac{\partial \epsilon_G}{\partial r}\frac{\left(\frac{1}{2} - A + C_1\right)}{2(C-A)} - \epsilon_G^2 \frac{\partial \epsilon_G}{\partial r}\left(\frac{1}{10(C-A)}\right) + \epsilon_G \frac{\partial v}{\partial r} + E_{\epsilon_G} + \frac{D\epsilon_G}{R-r}. \tag{40}$$

In the view of boundary condition, the initial guess and boundary condition are

$$f(0) = \tfrac{1}{4}(Pr^2 - PR^2), \tag{41}$$

$$g(0) = \tfrac{(r^2-R^2)}{2}, \tag{42}$$

$$f(R,h) = 0; f(0,h) = 0; g(0,h) = 0, \tag{43}$$

$$\begin{aligned} L\{f_m(x)\} - (1 - x_m)L\{f_0\} &= X_m R_{1m} \\ L\{g_m(x)\} - (1 - x_m)L\{g_0\} &= X_m R_{2m}, \end{aligned} \tag{44}$$

and the mth order deformation is

$$\begin{aligned} R_{1m}(V_{m-1}, \epsilon_{m-1}) = \epsilon_{Gm-1}(r) + C_2 \sum_{j=0}^{m-1} V_j'(r)\epsilon_{Gm-1}'(r) + \frac{1}{r}V_{m-1}'(r) \\ - \sum_{j=0}^{m-1} \epsilon_{Gj}(r)V_{m-1-j}'(r) + \sum_{j=0}^{m-1} V_j'(r)\epsilon_{Gm-1-j}'(r) \\ + V_{m-1}'' + \sum_{j=0}^{m-1} \epsilon_{Gj}(r)V_{m-1-j}''(r) - X_m\lambda', \end{aligned} \tag{45}$$

$$\begin{aligned} R_{2m}(V_{m-1}, \epsilon_{m-1}) = \frac{(\frac{1}{2}A + C_1)}{2(C-)} \sum_{j=0}^{m-1} \epsilon_{Gj}(r)\epsilon_{Gm-1-j}'(r) \\ - \frac{1}{10(C-A)} \sum_{k=0}^{j}\sum_{j=0}^{m-1} \epsilon_{Gk}(r)\,\epsilon_{Gj-k}(r)\epsilon_{Gm-1-j}'(r) \\ + \sum_{j=0}^{m-1} \epsilon_{Gj}(r)V_{m-1-j}'(r) + E_{\epsilon Gm-1}(r) + \frac{D}{R-r}\epsilon_{Gm-1}(r), \end{aligned} \tag{46}$$

where

$$X_m = \begin{cases} 0 & when\, m \le 1. \\ 0 & otherwise \end{cases}$$

Zero-order and first-order solution by using HAM are

$$f_0 = \frac{R^4}{12} - \frac{PR^4}{30} - \frac{R^{2x^2}}{10} + \frac{x^4}{60} + \frac{Px^4}{30} + \frac{1}{4}(-PR^2 + Px^2) \tag{47}$$

$$\begin{aligned} g_0 = \frac{147x^2}{2000} - \frac{R^{2x^2}}{10} - \frac{PR^{2x^2}}{20} - \frac{R^{4x^2}}{20} + \frac{49Lx^3}{500R_b} + \frac{3x^4}{20} + \frac{3Px^4}{40} - \frac{3R^{2x^4}}{20} - \frac{x^6}{12} \\ + \frac{1}{2}(-R^2 + x^2) \end{aligned} \tag{48}$$

$$f_1 = \frac{43241R^4}{180000} - \frac{7159PR^4}{90000} + \frac{16R^6}{1125} - \frac{7PR^6}{4500} - \frac{P^2R^6}{1500} + \frac{11R^8}{42000} - \frac{23PR^8}{21000} - \frac{49LR^5}{25000R_b} - \frac{147LPR^5}{25000R_b} - \frac{7R^4x^2}{25} + \frac{7159x^4}{180000} + \frac{7159Px^4}{90000} - \frac{17R^2x^4}{900} - \frac{23PR^2x^4}{1800} - \frac{P^2R^4x^4}{300} - \frac{R^4x^4}{600} + \frac{49Lx^5}{25000R_b} + \frac{147LPx^5}{25000R_b} + \frac{7x^6}{1500} + \frac{43Px^6}{3000} + \frac{P^2x^6}{250} + \frac{R^2x^6}{600} + \frac{PR^2x^6}{300} - \frac{x^8}{1680} - \frac{Px^8}{280} + \frac{(-PR^2 + Px^2)}{4}, \tag{49}$$

$$g_1 = -\frac{272391x^2}{2000000} - \frac{3853R^2x^2}{20000} - \frac{4147PR^2x^2}{40000} - \frac{551R^4x^2}{10000} + \frac{PR^4x^2}{1000} + \frac{R^6x^2}{50} + \frac{PR^6x^2}{200} + \frac{R^8x^2}{200} - \frac{7203LRx^2}{500000R_b} - \frac{147Rx^3}{5000} - \frac{147PRx^3}{10000} - \frac{49R^3x^3}{3750} + \frac{81193Lx^3}{500000R_b} - \frac{147LR^2x^3}{2500R_b} - \frac{49LPR^2x^3}{5000R_b} - \frac{147LPR^4x^3}{5000R_b} + \frac{8913x^4}{40000} + \frac{1959Px^4}{16000} + \frac{32R^2x^4}{625} - \frac{11PR^2x^4}{100} - \frac{3P^2R^2x^4}{400} - \frac{9R^4x^4}{40} - \frac{3PR^4x^4}{50} - \frac{3R^6x^4}{40} + \frac{21609L^2x^4}{1000000R_b} + \frac{49Rx^5}{2500} + \frac{1421Lx^5}{12500R_b} + \frac{833LPx^5}{25000R_b} + \frac{1421LR^2x^5}{12500R_b} + \frac{929x^6}{18000} + \frac{11Px^6}{90} + \frac{P^2x^6}{80} + \frac{13R^2x^6}{30} + \frac{PR^2x^6}{8} + \frac{13R^4x^6}{60} - \frac{399Lx^7}{5000R_b} - \frac{133x^8}{600} - \frac{161Px^8}{2400} - \frac{161Px^8}{2400} - \frac{133R^2x^8}{600} + \frac{3x^{10}}{40} + \frac{21609R^2x}{1000000(R+x)} - \frac{147R^4x}{10000(R+x)} - \frac{147PR^4x}{20000(R+x)} - \frac{7203LR^3x}{250000R_b(R+x)} + \frac{(-R^2+x^2)}{2} - \frac{21609R^2Log[-R+x]}{1000000} + \frac{147R^4Log[-R+x]}{10000} + \frac{147PR^4Log[-R+x]}{20000} + \frac{49R^6Log[-R+x]}{10000} + \frac{7203LR^3Log[-R+x]}{250000R_b}, \tag{50}$$

and so on.

5 RESULTS AND DISCUSSION

Analysis of the two-fluid model in case of upward bubbly flows shows that there exists a so-called peak in the void fraction in the close vicinity of the wall. It is evident that, for the gas–liquid mixture to flow upward, the pressure gradient must overcome the horizontally averaged weight $\bar{\rho}\,\bar{g}$. It is clear that the bubble should vanish on the wall.

Figs 1 and 2 show upward flow which is away from the wall for the different values of λ. The greater value of λ indicates, greater pressure gradient as compared to effective weight. Graphs for model I and model II are shown and upward motion is rapid for large values of λ. Fig. 3 shows a comparison of the model I and model II. It can be seen that both models converge approximately to the same value.

Figs 4–6, show the downward flow of void space for model I and model II for different values of $\epsilon_G(0)$. The gravity effect dominates pressure gradient and bubble moves towards the lower wall. Figs 7–9 shows the variation of initial coordinate $\epsilon_G(0)$, on both model I and model II. For upward flow a greater value of $\epsilon_G(0)$, show a quick rise of the bubble. Close to the wall, forces are effective, eqns (33)–(35) in non-dimensional form are simulated by using homotopy analysis method. The results shown in Figs 10–12 are for different values of bubble radius R_b, intrinsic length L and λ'. It is seen that void fraction profile decreases as the absolute value of applied pressure gradient increases, peak enlarges and radical position of maximum moves towards the wall. This is due to the result of the rise in lift force that comply to increasing liquid velocity gradient at the region close to the wall.

In fully developed laminar flow the solutions of the two-fluid model are such that the effective weight at the centre of the pipe (i.e., $\rho_{eff}\vec{g}$) practically balances the applied pressure gradient. Also at some particular limiting value of $r = r^*$called critical radius at which solution is unreal, i.e., the value of liquid fraction either become zero or greater than one.

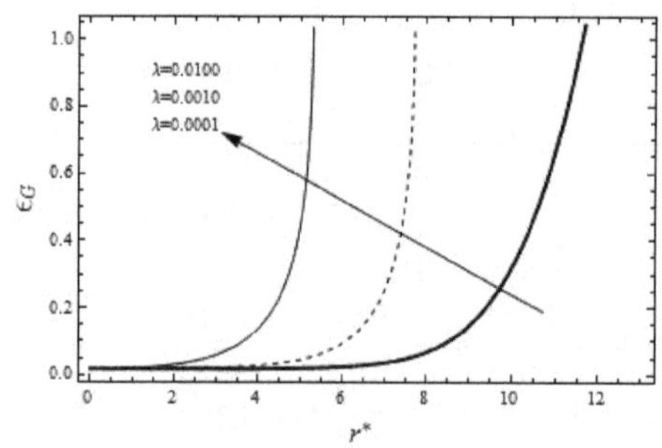

Figure 1: Upward flow for model I.

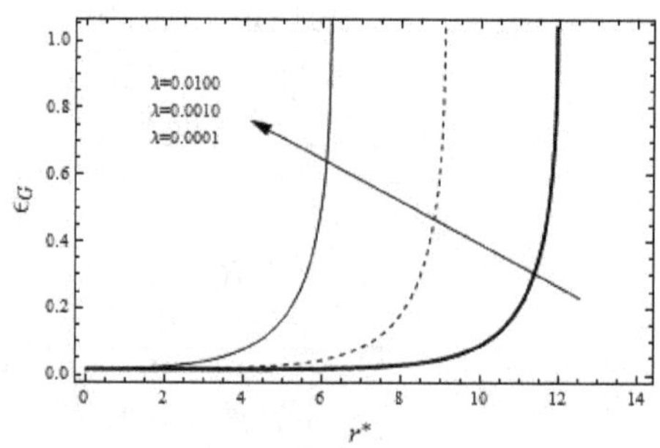

Figure 2: Upward flow for model II.

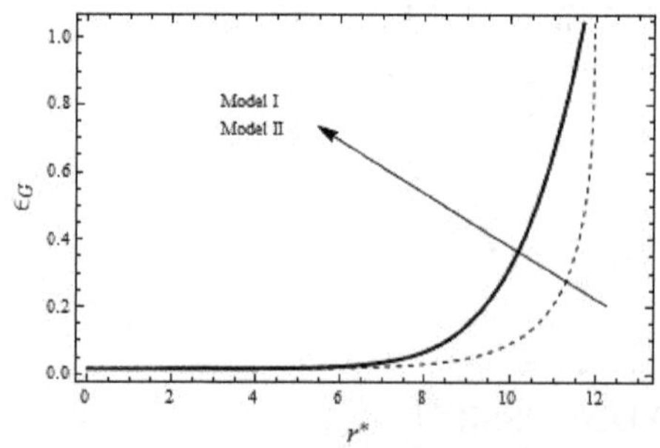

Figure 3: Upward flow for λ= 0:0001.

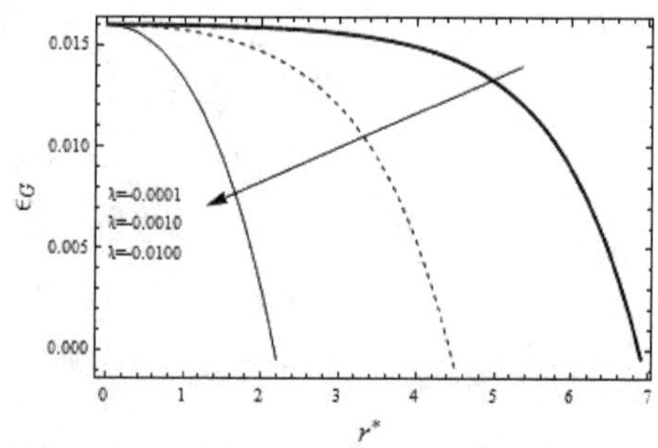

Figure 4: Downward flow for model I.

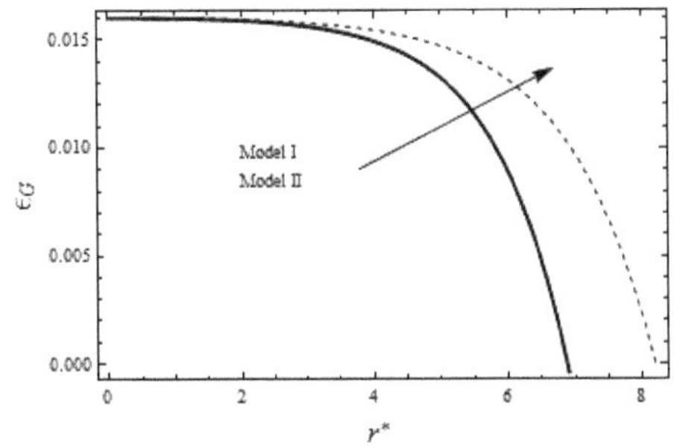

Figure 5: Downward flow for model II.

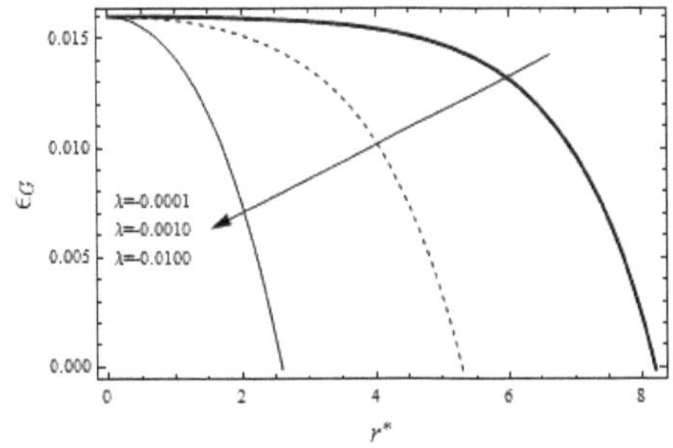

Figure 6: Comparison of downward flow for model I and model II for $\lambda = 0.0001$.

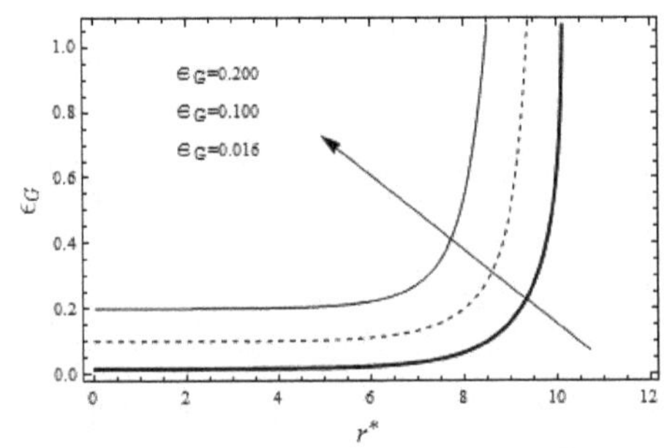

Figure 7: $\varepsilon_G\left(r^*\right)$ for different $\varepsilon_G\left(0\right)$ model I.

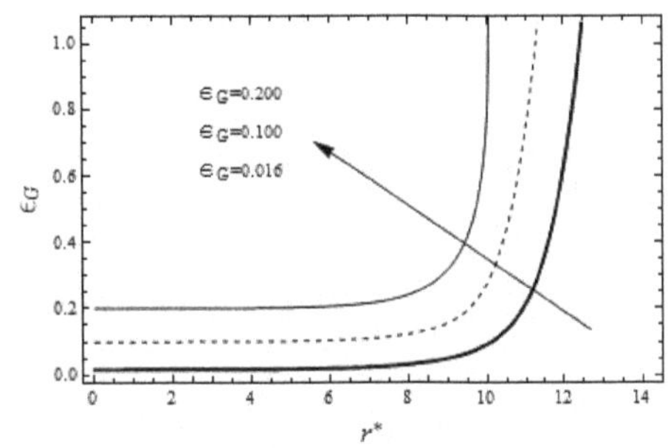

Figure 8: $\varepsilon_G\left(r^*\right)$ for different $\varepsilon_G\left(0\right)$ model II.

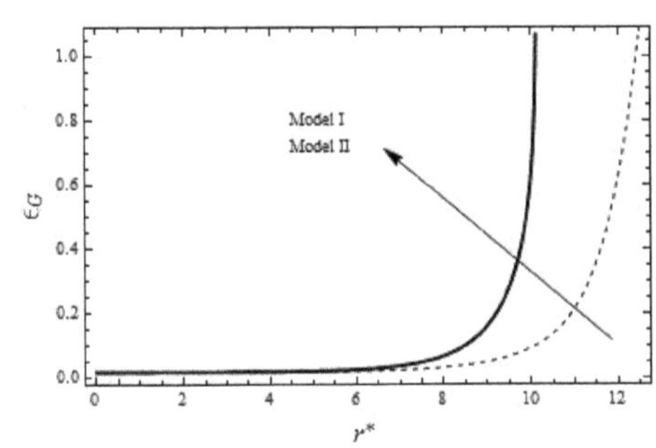

Figure 9: Comparison of $\varepsilon_G\left(r*\right)$ for model I and model II for $\varepsilon_G\left(0\right) = 0.016$.

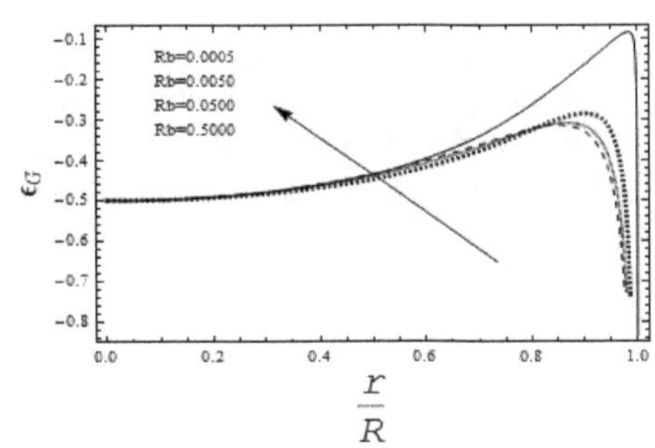

Figure 10: Void fraction profile variation in R_b.

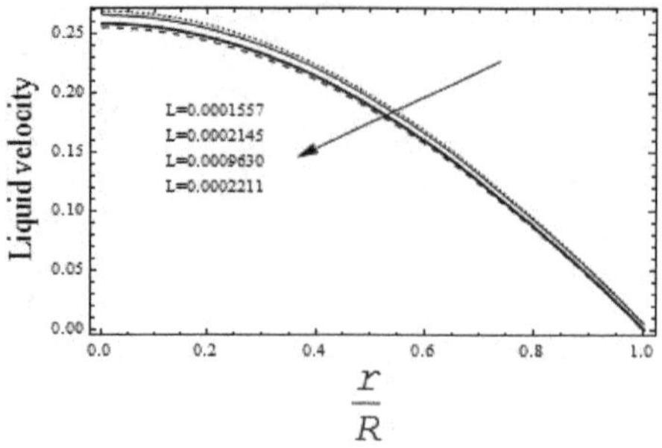

Figure 11: Velocity profile for variation of L.

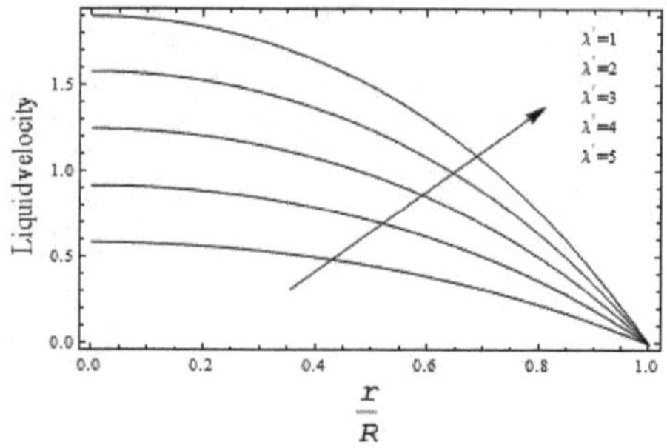

Figure 12: Liquid velocity profile for variation of upward flow.

In current work several physical and mathematical properties of the two-fluid model in the simple case of a laminar fully developed bubbly flow in case of a circular pipe have been identified. In vertical two-phase flow, it is found that the pressure gradient is cancelled with the effective specific weight of the fluid column. It is also found that the lift force balances the applied pressure gradient. For reduced gravity values, a prominent effect of lift force was observed. In future work, the effect of different boundary conditions on the model will be studied.

REFERENCES

[1] Ishii, M., *Thermo-Fluid Dynamic Theory of Two-Phase Flow*, Direction des Etudes et Recherches d'ElectricitXe de France: Eyrolles, 1975.
[2] Drew, D. & Passman, S., *Theory of Multicomponent Fluids*, Springer: Berlin, 1998.
[3] Nigmatulin, R., Spatial averaging in the mechanics of heterogeneous and dispersed systems. *International Journal of Multiphase Flow*, **4**, pp. 353–385, 1979.
[4] Antal, S., Lahey, R. Jr. & Flaherty, J., Analysis of phase distribution in fully developed laminar bubbly two-phase flow. *International Journal of Multiphase Flow*, **17**(5), pp. 635–652, 1991.
[5] Stuhmiller, J., The influence of interfacial pressure on the character of two-phase flow model equations. *ASME Symposium on Computational Techniques for Non-Equilibrium Two-Phase Phenomena*, p. 118, 1977.
[6] Wolfram, S., *Mathematica*, Addison-Wesley: Reading, MA, 1991.

EXPERIMENTAL INVESTIGATION OF PRESSURE AND VOID FRACTION CHANGES FOR TWO-PHASE FLOW THROUGH FLOW-RESTRICTING ORIFICES

NAIEF ALMALKI & WAEL H. AHMED
School of Engineering, University of Guelph, Canada

ABSTRACT

In this work, the change of pressure and void fraction of adiabatic air–water two-phase flow through orifices are experimentally investigated. Horizontal pipes with an internal diameter of 25.4 mm with multiple orifices with area ratios of 0.062, 0.14, 0.25 and 0.54 are considered. Both pressure and void fraction distributions upstream and downstream of the orifice are obtained for intermittent flow patterns and are compared with a straight pipe without the restriction for gas superficial velocity of 0.657 m/s and liquid superficial velocity of 0.523 m/s. The flow redistribution across the orifices is also recorded using a high-speed imaging camera at a frame rate of up to 3 kHz. The effect of the area ratio on the local pressure and void fraction upstream and downstream of the restriction is investigated. The results show that the fully developed void fraction upstream of the orifice increased with the increase in the pressure-drop across the orifice. Far from the orifice, the values of the average pressure gradient and the time average void fraction of the piping with orifices approached the fully developed values similar to the case of the straight pipe without restriction. The flow pattern changes across the orifice are found to significantly depend on the area ratio.

Keywords: two-phase flow, orifices, pressure drop, void fraction.

1 INTRODUCTION

Two-phase flow through flow restricting orifices is present in many applications including power generation, oil and gas industries, and food and chemical industries. These type of piping components are used to control the pressure of the system or the direction of the flow or both. Also, they are used for metering purposes to measure flow rate of the two-phase flow or mass quality [1], [2]. They also can be used to relive excessive pressure as it is the case in rupture disc in nuclear and chemical reactors [3], [4]. Moreover, multiple orifice valves (MOV) are used in the oil and gas wellhead applications to control the pressure and flow of the multiphase flow from the well [5]. Differences in the fluid properties of the two phases when passing through orifices makes the dynamic characteristics of the two-phase flow very complex. Further, piping degradation and failures downstream of orifices could lead to catastrophic failures and series loss [6]. Also, the hydrodynamic behaviour and phases redistribution downstream of the orifice is significant to safely operate and reasonably design such systems.

Research on two-phase flow through restricting orifices has been reported by many researchers. Many of these studies were concerned about predicting the pressure drop across the orifices of different sizes and thickness [9]–[14]. Also, extensive work to determine the effect of the orifice geometry as well as its shape on the two-phase flow was performed [4], [5], [15]–[17]. However, limited research has been carried out to evaluate the effect of orifices on void fraction and flow pattern development. For example, Fossa and Guglielmini [18] focused on the effect of the thin and thick orifices of area ratio of 0.54 and 0.73 on the local pressure drop and void fraction for intermittent flows. They reported that the void fraction increases and reaches a maximum value just downstream of the orifice regardless of its thickness. The slip ratio was also calculated using the void fraction measurements and

WIT Transactions on Engineering Sciences, Vol 123, © 2019 WIT Press
www.witpress.com, ISSN 1743-3533 (on-line)
doi:10.2495/MPF190151

compared to the available slip ratio correlations. They concluded that the measured slip ratio just downstream of the orifice is less than the homogenous model while far from the orifice, the slip ratio is predicted by Chisholm [19] and the Armand and Treschev [20] correlations.

A study by Fossa et al. [21] focused on the effect of orifices of area ratio 0.54 and 0.73 on intermittent two-phase flow. Using the impedance technique, the local time averaged void fraction in 2D upstream and 4D downstream were reported. The results showed that the void fraction reaches maximum value at 1D downstream the orifice. This void fraction value could be double of the upstream one due to change in slip between phases. The void fraction in the fully developed region downstream of the restriction in general increases compare to the fully developed upstream.

Zeghloul et al. [3] performed experiments on two-phase flow through orifice in a vertical pipe with area ratio of 0.54 and 0.73. In this study, void fraction measurements were conducted in nine locations using conductance method for bubbly, slug and churn flow patterns observed upstream. The axial distribution of the void fraction was investigated and indicated recovery length of 20, 10 and 7 pipe diameters downstream of the orifice for bubbly, slug and churn flow patterns respectively. They also recorded that the frequencies of the bubbles both upstream and downstream of the orifice and found to be similar for bubbly and slug flow.

From the above literatures, it can be concluded that not much work has been performed on the effect of orifice area ratio on the two-phase flow structure for area ratio less than 0.5. Also, not much work has been reported to show the relation between the pressure and the void fraction distribution in horizontal piping structures with an orifice. Therefore, this paper aims to investigate the effect of area ratio less than 0.5 on adiabatic air-water two-phase flow for horizontal pipe. Measurements of pressure and instantaneous void fraction change upstream and downstream of the orifice were carried out for gas superficial velocity of 0.657 m/s and liquid superficial velocity of 0.523 m/s. Pressure and void fraction were recorded upstream and downstream of orifice with area ratio of 0.62, 0.14, 0.25 and 0.56. For the same liquid and gas superficial velocities, the experiments also were carried out for horizontal straight pipe. High speed imaging camera was also used to show the effect of the orifices on the flow pattern development across the orifice.

2 EXPERIMENTAL FACILITY

The experimental facility was constructed to simulate air-water mixture flows in a closed loop at the ambient temperature as shown in Fig. 1. Deionized water was pumped from a 310-L tank by a centrifugal pump that was controlled by a variable frequency drive. The liquid flow rate was measured using an Omega turbine flow meter with reading accuracy of ±1.5% full scale. The air was supplied from main laboratory supply line and controlled by four rotameters of total range of to 100 L/min with accuracy of ±2% full scale. The air and water were mixed in an annular two-phase mixer at a distance of 3 m upstream of the orifice. The mixer is made of a concentric aluminium tubing with total length of 80 mm and 25.4 mm inner diameter. The internal piece of the mixer is made of perforated tube with a total number of 64 holes distributed equally at eight sets spaced 10mm. Air is injected radially through these holes while water is introduced axially. The two-phase flow mixture was then transported through a horizontal straight pipe with inner diameter of 25.4 mm and a total length of 2.4 m both upstream and downstream of the orifice. The pipe is made of clear polycarbonate in order to allow for flow visualization.

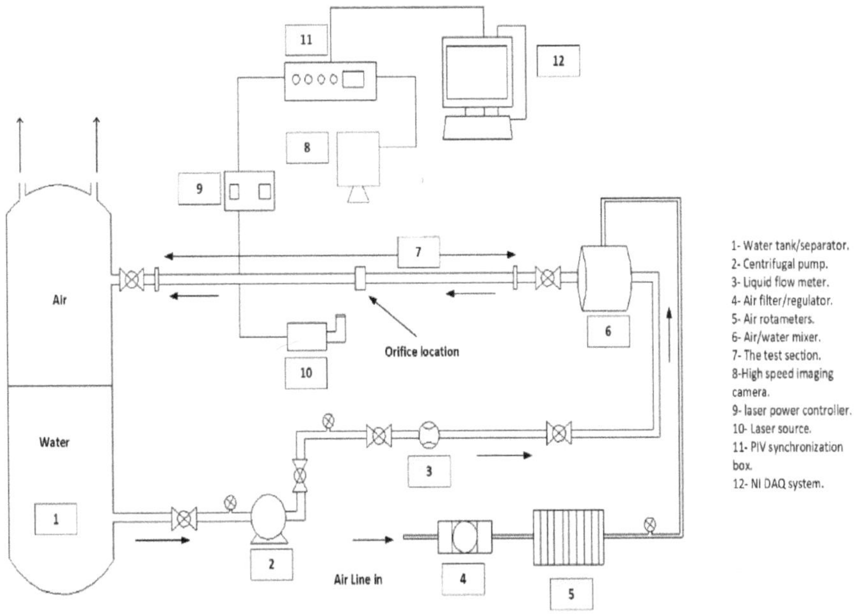

Figure 1: Experimental set-up.

The pressure was measured at nine and thirteen locations upstream and downstream of the orifice using pressure taps made of clear tubing to insure no air was trapped inside which may affect the measurements. These pressure taps were connected to a mechanical pressure scanner from Scannivalve. The output of the pressure signal from the pressure scanner was then divided to two lines and connected to negative ports of two DP15 differential pressure sensors from Validyne with accuracy of ±0.25% full scale. The two differential pressure sensors have two pressure ranges namely; high pressure range of 350 kPa and low pressure range of 35 kPa. The positive ports of the two differential pressure sensors then combined in one line and connected to the first pressure tap after the mixer to represent the reference pressure in the test section. A pressure transducer from Omega (model: PX481A) with accuracy of ±0.3% full scale and range of 414 kPa was also installed to measure the reference pressure.

The void fraction was measured at total of twelve locations distributed equally upstream and downstream of the orifice. The void fraction was measured using a concave capacitance sensor consists of four electrodes made of copper of 0.12 mm thickness with 50 mm length and 24 mm width. The four electrodes are connected through coaxial cables to a Boonton 7200 DAQ system with sampling frequency up to 3 kHz. Detail of the capacitance sensor design adopted in this work was presented by Ahmed [22] and Elsaftawy et al. [23]. Repeatability test for the void fraction measurements indicates that the measured data is within ±5% full scale. Details on the pressure taps and void fraction sensor measurement locations is shown in Fig. 2.

The flow pattern was visualized using high speed imaging camera from Speedsense with maximum resolution of 1920×1200 pixels and trigger rate up to 3000 frame per second.

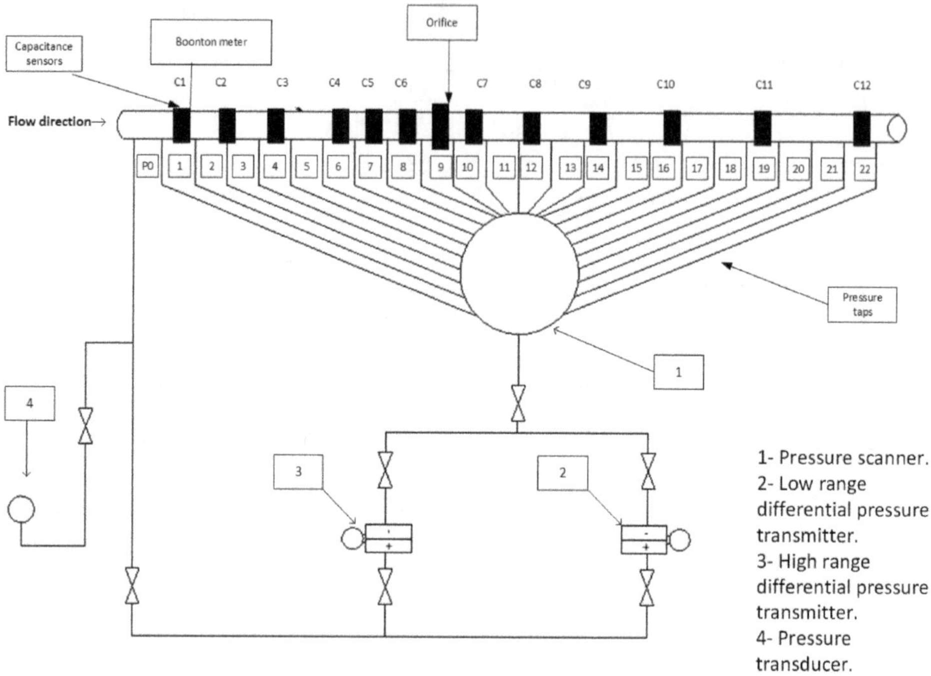

Figure 2: The test section instrumentations.

3 RESULT AND DISCUSSION

3.1 Single-phase pressure drop in orifice

Single-phase pressure drop across orifices experiments were performed to validate the experimental facility, instrumentation and the experimental procedure. Three thin orifices were used to evaluate the single-phase pressure drop. The total pressure drop caused by the orifice is found by calculating the difference between the upstream best fit equation and the downstream one.

Therefore, the single phase pressure drop caused by an orifice can be calculated as [18]:

$$\Delta P_{sp} = \frac{2\rho V^2}{2}\left[\left(\frac{1}{\sigma\sigma_c}\right)^2 - 1\right], \tag{1}$$

where σ is the area ratio and σ_c is contraction coefficient. The relationship between the coefficient of discharge and the contraction coefficient is:

$$\sigma_c = \frac{1}{\sigma + \frac{\sqrt{1-\sigma^2}}{C_d}}, \tag{2}$$

where C_d is the coefficient of discharge. It should be noted that the above relations are valid only for thin orifice in which the ratio of the orifice thickness to its diameter (s/d) is less than 0.5 according to Chisholm [19]. On the other hand, the contraction coefficient can be calculated using Chisholm correlation [19] in term of area ratio only as:

$$\sigma_c = \frac{1}{[0.639(1-\sigma)^{0.5}+1]} \tag{3}$$

Summary of the RMSD of the experimental contraction coefficient as well as the ones calculated using Chisholm correlation [19] are presented in Table 1.

Table 1: Summary of the single-phase orifice contraction coefficients.

Area ratio	Average contraction coefficient	RMSD (Contraction coefficient vs Chisholm eqn)	RMSD (Ideal vs experimental pressure drop)
0.062	± 0.58	± 4.61	± 10.38
0.25	± 0.62	± 3.33	± 8.33
0.54	± 0.69	± 1.64	± 5.64

3.2 The effect of the orifice on the flow structure

The flow structure upstream and downstream of the orifice was investigated using high speed imaging camera as shown in Fig. 3. It should be mention that the discussion is limited to the presented cases at liquid superficial velocity of 0.523 m/s with gas superficial velocity of 0.657 m/s. In these cases, mainly slug flow upstream of the orifice was observed. The images represent the flow at a distance of 4D upstream and 10D downstream from the orifice. It should be mention here that the flow pattern just downstream of the orifice encounters a considerable fluctuation in the pressure and consequently the density of the gas phase. Therefore, a sequence of flow patterns downstream in the developing region may occur for the same liquid and gas superficial velocities and area ratio used. Three observations can be seen in these images: first; the shape of the gas pocket in the stratified region in the slug flow just upstream of the orifice is affected by the area ratio. It was noticed that the bubbles coalescence and the gas pocket length increase as the area ratio increases. This is mainly due to the decrease in the system pressure. For instance, the pressure upstream in the case of area ratio of 0.62 is about 92 kPa and therefore smaller gas pocket is expected. On the other hand, the upstream pressure in case of area ratio of 0.56 is about 6 kPa which is close to the unrestricted pipe case and consequently a longer gas pocket was observed. Second: once the gas pocket in the slug unit starts to pass the orifice, a separation between the phases occurs. This separation results in a liquid jet surrounded by gas voids as well as thin liquid layers on the pipe wall due to the considerable change in the gas density. When the gas pocket passes completely, the liquid jet with less liquid film on the perimeter of the pipe and gas voids are formed. The length of the liquid jet in the axial direction increases as the area ratio decreases due to momentum increase. Third: the mixture downstream flows in reverse direction of the liquid jet due the lower pressure in the area just downstream of the orifice. Dispersed bubbly flow with high vortices was observed at the location where the liquid jet merges with the back flow. The cycle is repeated as another slug hits the orifice upstream. Also, the fluctuating in the flow pattern in the developing region downstream decreases as the area ratio increases.

3.3 The effect of the area ratio

The pressure and averaged void fraction distributions along the test section are shown in Figs 3 and 4 for liquid superficial velocity of 0.523 with gas superficial velocity of 0.657 m/s. It can be seen that the pressure upstream of the orifice increases as the area ratio decreases due to the restriction. Subsequently, the fully developed averaged void fraction upstream ($Z/D = -17.5$) decreases as the pressure of the system increases due to the compressibility of the gas phase. Far downstream of the orifice, the pressure and the averaged void fraction in the fully developed region ($Z/D = 90.5$) reach values close to the one in the case of the pipe without restriction. This trend was also observed for other cases not reported in this paper. A closer look at the pressure distribution shows that the pressure decreases as the two-phase flow approaches the orifice due to the frictional losses. Once the two-phase flow passes through the orifice, the pressure drops dramatically due to the increase in the flow turbulence downstream of the orifice. The magnitude of this pressure drop increases as the area ratio decreases for the same liquid and gas superficial velocities. The pressure then recovers at some distance downstream of the orifice due to the deceleration of the phases.

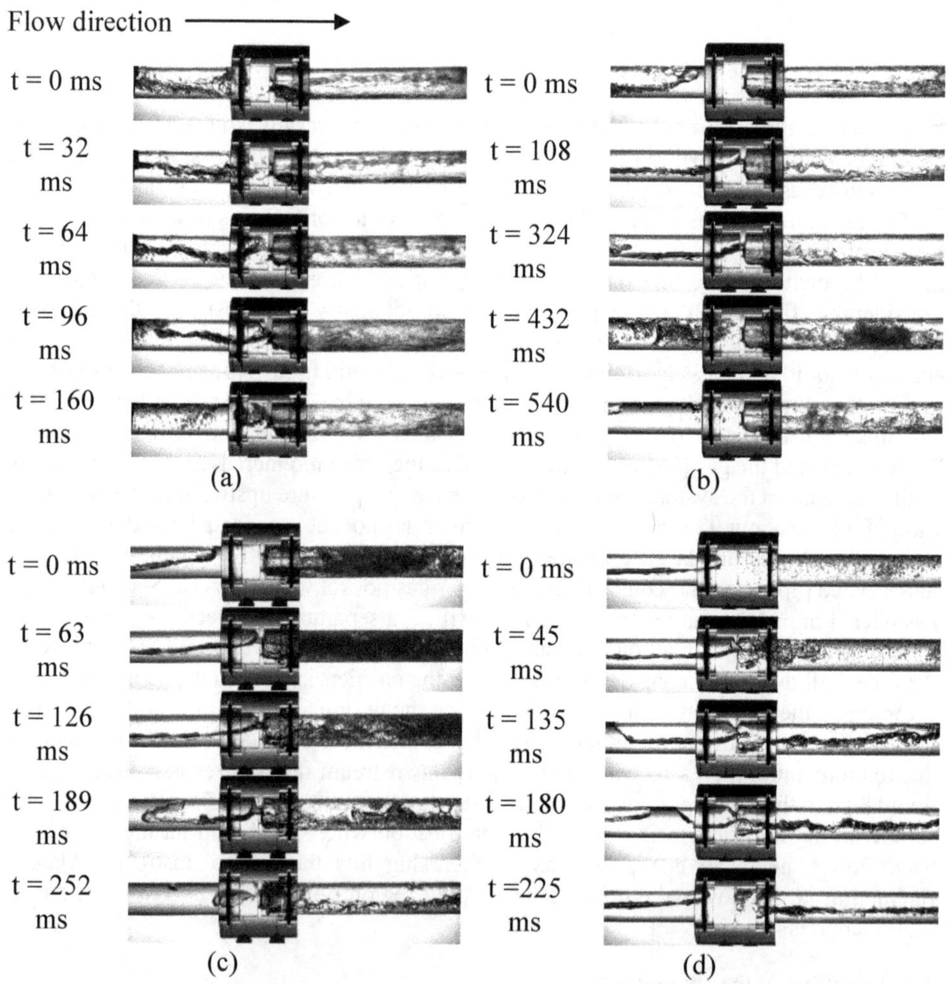

Figure 3: Time sequence of images showing the phase redistribution downstream of the orifice: (a) $\sigma = 0.62$; (b) $\sigma = 0.14$; (c) $\sigma = 0.25$; (d) $\sigma = 0.56$.

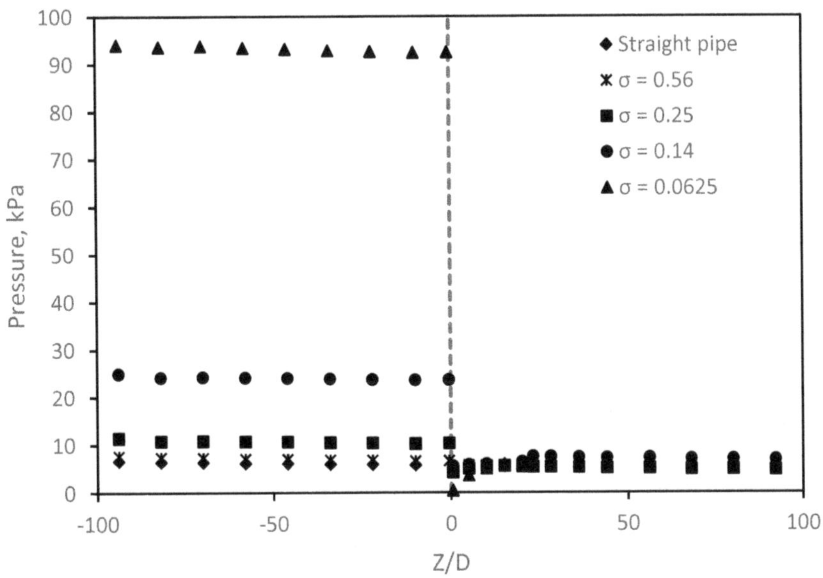

Figure 4: Pressure distribution along the test section.

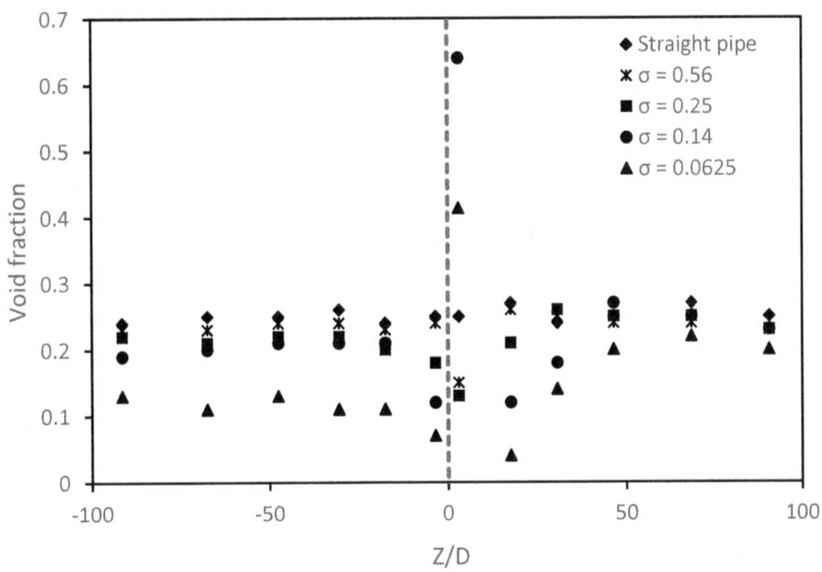

Figure 5: Averaged void fraction along the test section.

This pressure recovery length decreases as the area ratio increases for the same liquid and gas superficial velocities. On the other hand, the averaged void fraction increases as the two-phase flow approaches the orifice due the frictional pressure drop. Before the two-phase flow passes through the orifice, a slight increase in the pressure and a subsequent decrease in the void fraction just before the orifice ($Z/D = -3.5$) was noticed due to the presence of the restriction in the flow path. The ratio of the void fraction downstream of the orifice ($Z/D = 3$) to the upstream one ($Z/D = -3.5$) can be calculated from the presented data as 0.63, 0.72, 5.33 and 5.91 for area ratios 0.56, 0.25, 0.14 and 0.62 respectively. This clearly indicates that the void fraction just downstream of the orifice increases as the area ratio decreases compare to its initial values upstream. Similar to the pressure downstream, the void fraction downstream starts to recover as the velocities of the phases restore their initial ones upstream. However, the void fraction continues to develop with lesser rate due to bubbles coalescence as well as the frictional pressure drop. Therefore, it is expected that the fully developed void fraction downstream of the orifice to increase compared to the upstream value. This is clear from the void fraction distribution for the case of area ratio of 0.62 in which it increases from 0.11 at $Z/D = -17.5$ upstream to 0.2 at $Z/D = 90.5$ downstream.

3.4 Spatial development of the flow pattern across the orifice

The development of the flow pattern across the orifice is discussed using the time series of the void fraction for the studied area ratios. An example of the time series of the void fraction for area ratio of 0.62 is shown in Fig. 7. This graph represents the time series of the void fraction at one location upstream ($Z/D = -3.5$) and six locations downstream ($Z/D = 3, 17.5, 30.5, 46.5, 68.5$ and 90.5). As expected, the gas pocket of the slug unit upstream increases as the area ratio increases ($Z/D = -3.5$) as discussed in the previous sections. At the location just downstream of the orifice ($Z/D = 3$), the instantaneous void fraction changes as the two-phase flow passes the orifice for all ratios. The magnitude of this change is more obvious as the area ratio decreases. Also, despite the area ratio used, the flow starts to reform downstream. The distance at which flow pattern changes is however dependant on the area ratio as it decreases as the area ratio increases. Therefore, a fully developed void fraction was observed at $Z/D =$ 17.5, 30.5, 46.5 and 68.5 for area ratio of 0.56, 0.25, 0.14 and 0.62 respectively. It was also observed that the fully developed void fraction downstream is around 0.4 for all cases. This is because the fact that the pressure downstream of the orifice in the fully developed reign is almost the same for all cases as discussed in the pressure distribution results.

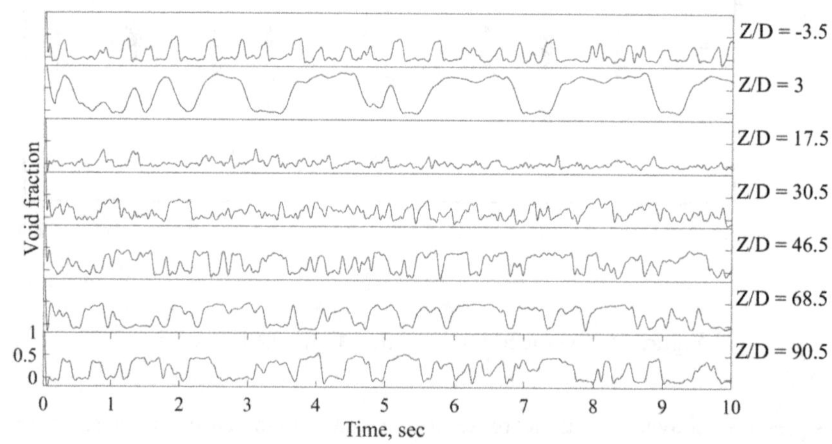

Figure 6: The effect of area ratio on flow pattern development for $\sigma = 0.62$.

4 CONCLUSION

A new set of experiments were carried out to study the effect of the area ratio of the orifices on adiabatic air-water two phase flow in horizontal pipe. Four area ratios of 0.62, 0.14, 0.25, and 0.56 as well the straight pipe without restriction were used. The two-phase flow was characterized throughout the investigation by pressure and void fraction measurements and high-speed camera. Cases for gas superficial velocity of 0.657 m/s and liquid superficial velocity of 0.523 m/s were presented. The investigation shows that the fully developed void fraction upstream of the orifice decreases as the area ratio decreases due to the increase in the local pressure upstream. The void fraction just downstream of the orifice increases as the area ratio decreases. For the same liquid and gas superficial velocities, similar pressure and void fraction and subsequently flow pattern were observed in the fully developed region downstream of the orifice regardless of the area ratio used.

ACKNOWLEDGEMENT

The authors would like to acknowledge the support of Natural Sciences and Engineering Research Council of Canada (NSERC) under the discovery grant program.

REFERENCES

[1] Lin, Z.H., Two-phase flow measurements with sharp-edged orifices. *International Journal of Multiphase Flow*, **8**(6), pp. 683–693, 1982.
[2] Oliveira, J.L.G., Passos, J.C., Verschaeren, R. & Geld, C., Mass flow rate measurements in gas–liquid flows by means of a venturi orifice plate coupled to a void fraction sensor. *Experimental Thermal and Fluid Science*, **33**(2), pp. 253–260, 2009.
[3] Zeghloul, A., Azzi, A., Saidj, F., Azzopardi, B.J. & Hewakandamby, B., Interrogating the effect of an orifice on the upward two-phase gas–liquid flow behaviour. *International Journal of Multiphase Flow*, **74**, pp. 96–105, 2015.
[4] Shannak, B., Friedel, L., Alhusein, M. & Azzi, A., Experimental investigation of contraction in single- and two-phase flow through sharp-edged short orifice. *Forschung im Ingenieurwesen*, **11**(64), pp. 291–295, 1999.
[5] Alimonti, C., Falcone, G. & Bello, O., Two-phase flow characteristics in multiple orifice valves. *Experimental Thermal and Fluid Science*, **34**(8), pp. 1324–1333, 2010.
[6] Ahmed, W.H., Bello, M.M., El Nakla, M. & Al Sarkhi, A., Flow and mass transfer downstream of an orifice under flow accelerated corrosion conditions. *Nuclear Engineering and Design*, **252**, pp. 52–67, 2012.
[7] Ahmed, W.H., Bello, M.M., El Nakla, M., Al Sarkhi, A. & Badr, H., Experimental investigation of flow accelerated corrosion under two-phase flow conditions. *Nuclear Engineering and Design*, **267**, pp. 34–43, 2014.
[8] Bamidele, O.E., Ahmed, W.H. & Hassan, M., Two-phase flow induced vibration of piping structure with flow restricting orifices. *International Journal of Multiphase Flow*, **113**, pp. 59–70, 2019.
[9] Sharp, R.M., Two-phase pressure losses in valves and fittings. Master's thesis, Georgia Institute of Technology, 1953.
[10] Chisholm, D., Pressure losses in bends and tees during steam-water flow. NEL Report No. 318, 1967.
[11] Chisholm, D. & Sutherland, L.A., Prediction of pressure gradients in pipeline systems during two-phase flow. *SAGE Journals*, **184**(3), pp. 24–32, 1969.
[12] Sookprasong, P., Brill, J.P. & Schmidt, Z., Two-phase flow in piping components. *Journal of Energy Resources Technology*, **108**(3), pp. 197–201, 1986.

[13] Saadawi, A.A., Grattan, E. & Dempster, W.M., Two phase pressure loss in orifice plates and gate valves in large diameter pipes. *Proceedings of the 2nd Symposium on Two-Phase Flow Modelling and Experimentation*, Pisa, Italy, 1999.

[14] Morris, S.D., Two phase pressure drop across valves and orifice plates. Technical Report, European Two Phase Flow Group Meeting, 1985.

[15] Salcudean, M., Groeneveld, D.C. & Leung, L., Effect of flow-obstruction geometry on pressure drops in horizontal air-water flow. *International Journal of Multiphase Flow*, **9**(1), pp. 73–85, 1983.

[16] Salcudean, M.E. & Leung, L.K.H., Two-phase pressure drop through obstructions. *Nuclear Engineering and Design*, **105**(3), pp. 349–361, 1988.

[17] Klausner, J. & Warren, B.A., Developing lengths in horizontal two-phase flow. *Journal of Fluids Engineering*, **117**(3), pp. 512–518, 1995.

[18] Fossa, M. & Guglielmini, G., Pressure drop and void fraction profiles during horizontal flow through thin and thick orifices. *Experimental Thermal and Fluid Science*, **26**(5), pp. 513–523, 2002.

[19] Chisholm, D., *Two-Phase Flow in Pipelines and Heat Exchangers*, Longman Group Ed: London, 1983.

[20] Armand, A.A. & Treschev, G., Investigation of the resistance during movement of the steam–water mixtures in heated boiler pipe at high pressure. *Izvestia User. Teplo. Inst. AERE Lib/Trans 816*, **4**, pp. 1–5, 1947.

[21] Fossa, M., Guglielmini, G. & Marchitto, A., Two-phase flow structure close to orifice contractions during horizontal intermittent flows. *International Communications in Heat and Mass Transfer*, **33**(6), pp. 698–708, 2006.

[22] Ahmed, W., Capacitance sensors for void-fraction measurements and flow-pattern identification in air–oil two-phase flow. *IEEE Sensors Journal*, **6**(5), pp. 1153–1163, 2006.

[23] Elsaftawy, A., Potts, J. & Ahmed, W., Oscillation frequency LC-based sensor for characterizing two-phase flows in energy systems. *IEEE Sensors*, **19**(1), 2019.

[24] Roul, M.K. & Dash, S.K., Single-phase and two-phase flow through thin and thick orifices in horizontal pipes. *Journal of Fluids Engineering*, **134**(9), p. 14, 2012.

STUDY OF POLYMER ELASTIC BEHAVIOR IN THE DISPLACEMENT OF OIL DROPS AT PORE SCALE

LUIS PRADA, ARLEX CHAVES, JULIO PEDRAZA, JOSE GÓMEZ & SAMUEL MUÑOZ
Universidad Industrial de Santander, Colombia

ABSTRACT

Polymeric liquids have been used in the oil industry, especially at enhanced oil recovery (EOR). From the rheological point of view, polymers have the particularity of being viscoelastic liquids. One of the most common and useful models to describe that behavior is the upper convected Maxwell model (UCM). The main characteristic of the polymer used in the EOR process is the increase in viscosity which pushes the oil outside of the reservoir. The elasticity could contribute to the drag of the oil that stays in the reservoir. Studying the elastic effect on the oil drop at the pore scale, brings an explanation if the addition of elastic force could mobilize the oil. This research explores if the contraction and expansion of the polymer in the pore-scale may increase the elastic behavior of this kind of fluid. For that reason, this work simplified the pore geometry and built two simple geometries with micrometer lengths. Using source terms with the user defined function this work introduces the UCM model in the ANSYS Fluent simulator with the purpose of evaluating the elastic effect of the polymer in a contraction and expansion geometry. Also, using the Eulerian multiphase model this research considers the possibility that extra elastic force will show a deformation effect on the oil, for that reason, this work considers an oil drop on the upper wall of the geometry. Finally, all the simulations exhibit that at pore scale conditions extra vortices exist in the UCM model but it is not possible to deform the oil completely and push it outside of the restrictions.

Keywords: ANSYS Fluent, interfacial fluids mechanics, polymers, pore scale, viscoelasticity.

1 INTRODUCTION

Polymerflooding is an enhanced oil recovery (EOR) mechanism that improves the sweep efficiency of the water injected, to increase oil production. However, for decades have been considered that even with the viscosity of the polymer is not possible to decrease the oil trapped at the pore. Some research as is mention in this paper have been probing that the elastic behavior of the polymer is the main factor to consider in these phenomena.

Wreath [1] was the first scientist to consider the polymer effect at the reduction of the residual oil because of the viscoelastic potential. He compares the polymerflooding between two types of rocks, where at the more heterogeneity rock (Antolini) get a reduction of the residual oil. Wang et al. [2] studied many configurations of the residual oil trapped at the pore. He found that polymer could deform the oil volume that is trapped on the concavity of the rock or thinning the oil film that sticks on the wall of the pore. He explains that polymers are large and flexible entangled chains, therefore, when they move, they interact between each other creating elongation and movement of the other molecules, connecting with the oil that stay on the concavity. Researchers such as Urbissinova et al. [3], Seright [4] and Qi et al. [5] found in their experiments a reduction in at least 5% of the oil saturation with a polymerflooding process, increasing that value when they used more viscoelastic polymer.

For the simulation aspect, Yin et al. [6] use the upper convective Maxwell model to describe the viscoelastic behavior of the polymer solution. They show the contour of the stream function and the velocities for different Weissenberg and Reynold numbers in simplified pore geometries. With the increase of the elasticity of the polymer Yin et al. [6], found that the velocity of the polymer increases at the corner where the oil could be trapped, for that reason there is a chance for the polymer to decrease the oil saturation in that places.

WIT Transactions on Engineering Sciences, Vol 123, © 2019 WIT Press
www.witpress.com, ISSN 1743-3533 (on-line)
doi:10.2495/MPF190161

Afsharpoor and Balhoff [7] studied four pore geometries and evaluated the effects of the pressure, velocity and normal stress around the oil drop during the polymerflooding. This research indicated that the increase in elasticity could change the oil drop shape. Finally, Fan et al. [8] use the upper convective Maxwell (UCM) to evaluate the effect of high Weissenberg number in the polymerflooding at pore scale. Fan conclude that at higher Weissenberg number the normal stress increase, creating favorable condition to the deformation of the oil that stay on the reservoir. In this paper, the pore-scale flooding processes were investigated with the volume of fluid method for enhanced oil recovery. A single pore body/throat was established to investigate the wettability and interfacial tension, and viscoelastic effects on the polymer/oil flow interaction in pore scale. The results can provide some insights into the basic understanding of the polymer flooding process for enhanced oil recovery.

2 MATHEMATICAL MODEL

In this study, the computational fluid dynamic (CFD) technique was used, specifically the ANSYS Fluent software to evaluate the effect of the viscoelastic component of the polymer on the phenomena of the polymerflooding at the pore scale. The CFD modeling is applied here for viscoelastic flow around stationary oil droplets. First, as a problem in 2d geometry governing equations: momentum and continuity can be written as follows:

$$\frac{\partial u}{\partial x} + \frac{\partial v}{\partial y} = 0, \tag{1}$$

$$\rho\left(\frac{\partial u}{\partial t} + u\frac{\partial u}{\partial x} + v\frac{\partial u}{\partial y}\right) = -\frac{\partial p}{\partial x} + \frac{\partial \tau_{xx}}{\partial x} + \frac{\partial \tau_{xy}}{\partial y}, \rho\left(\frac{\partial v}{\partial t} + u\frac{\partial v}{\partial x} + v\frac{\partial v}{\partial y}\right) = -\frac{\partial p}{\partial y} + \frac{\partial \tau_{xy}}{\partial x} + \frac{\partial \tau_{yy}}{\partial y}, \tag{2}$$

where u, v are the x and y component of the velocity, ρ is density, τ_{xx}, τ_{xy} and τ_{yy} are the stresses.

2.1 Upper convected Maxwell model

In this study the UCM model (eqn (3)) is used as the constitutive equation for the non-Newtonian fluid. The motivation of using this model is because it was developed to generalize the linear viscoelastic model to be valid under varying conditions of stress and motion, and suggested that describes the rheology of HPAM, one of the polymers most used in enhanced oil recovery. The UCM is given by:

$$\boldsymbol{\tau} + \lambda \overset{\nabla}{\boldsymbol{\tau}} = 2\mu \boldsymbol{D}, \tag{3}$$

where $\boldsymbol{\tau}$ is the stress tensor, λ is the relaxation time, $\boldsymbol{D}$ is the rate of deformation tensor, μ is the viscosity of the polymeric solution and $\overset{\nabla}{\boldsymbol{\tau}}$ is the upper convected derivative of the stress tensor defined as:

$$\overset{\nabla}{\boldsymbol{\tau}} = \frac{D\tau}{Dt} - (\nabla v)^T \cdot \tau - \tau \cdot \nabla v. \tag{4}$$

Eqn (3) could be written in Einstein notation as:

$$\frac{\partial \tau_{ij}}{\partial t} + V_k \frac{\partial \tau_{ij}}{x_k} = \boldsymbol{\tau}_{kj}\frac{\partial v_i}{\partial x_k} + \boldsymbol{\tau}_{ik}\frac{\partial v_j}{\partial x_k} + \frac{\mu_p}{\lambda}\boldsymbol{\gamma}_{ij} - \frac{1}{\lambda}\boldsymbol{\tau}_{ij}, \tag{5}$$

where γ_{ij} is the rate of deformation tensor.

2.1.1 Implementation in ANSYS Fluent software

Eqn (5) represents the viscoelastic constitutive equation and is embedded into the Fluent code through UDF functionality. In UDF, the general user-defined scalar transport as follows:

$$\frac{\partial(\rho\phi)}{\partial t} + \frac{\partial}{\partial x_i}\left(F_i\phi - \Gamma\frac{\partial\phi}{\partial x_i}\right) = S_\phi. \tag{6}$$

In eqn (6), there are four terms to be customized: transient term, convection term, diffusion term and source term, where ϕ is the scalar, corresponding to components of the stress tensor τ_{ij} in eqn (5). Γ is the diffusion coefficient, and is set equal to zero in the present calculations. The transient term corresponds to the first term on the left side in eqn (5), the convection term is the second term on the left side; the remaining term in eqn (5) were grouped in the source term $S_\phi$as shown in the following expression for each one the four polymer stress components

$$S_{\tau_{ij}} = \tau_{kj}\frac{\partial v_i}{\partial x_k} + \tau_{ik}\frac{\partial v_j}{\partial x_k} + \frac{\mu_p}{\lambda}\gamma_{ij} - \frac{1}{\lambda}\tau_{ij}. \tag{7}$$

2.1.2 Two phase model

Finally, to model the two-phase fluid (water–oil and oil–polymer), the volume of fluid (VOF) model was used to model the two immiscible liquids by solving a single set of momentum equation and tracking the volume fraction of each of the fluids throughout the domain. The tracking of the interface(s) between the phases is accomplished by the solution of a continuity equation for the volume fraction of one (or more) of the phases. For the qth phase, this equation has the following form:

$$\frac{1}{\rho_q}\left[\frac{\partial}{\partial t}(\alpha_q\rho_q) + \vec{\nabla}\cdot(\alpha_q\rho_q\vec{v_q}) = S_{\alpha_q} + \sum_{p=1}^{n}(\dot{m}_{pq} - \dot{m}_{qp})\right], \tag{8}$$

where $\dot{m}_{pq}$ is the mass transfer from phase q to phase p and $\dot{m}_{qp}$ is the mass transfer from phase p to phase q. In the problem that we solve S_{α_q} is the source term we added before. The volume fraction equation will not be solved for the primary phase, the primary-phase will be computed based on the following constraint:

$$\sum_{q=1}^{n}\alpha_q = 1. \tag{9}$$

The fraction equation for this simulation was solver through implicit time discretization [9].

3 VALIDATION

Validation of the model is an important step in a simulation. In this paper, we present two kinds of validation: quantitative and qualitative validation.

3.1 Quantitative validation

The quantitative validation was performed comparing Yin et al. [6] velocities profiles to ANSYS Fluent simulation results, for the particular case of expansion geometry. Yin et al. used a relation of 4:1 height of the geometry and differences of Newtonian and non-Newtonian flows as Fig. 1 shows. For the case of Fig. 1(b), the Weissenberg number represent the ratio of a fluid time constant to a characteristic time of flow and is defined as a fluid time constant λ multiplied with a deformation rate describing the flow.

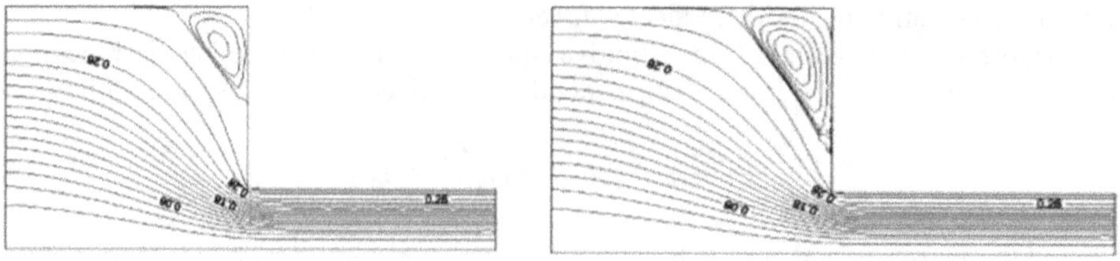

(a) (b)

Figure 1: Streamlines in [6]. (a) We = 0; and (b) We = 0.6.

To validate Yin's results, a geometry of micrometer size was built because representing the pore scale, with the same 4:1 geometry relation. Properties of the mesh and other conditions are presented in Table 1. Using this condition for the geometry and fluid in ANSYS simulation, it is easy to observe that an increase in the viscoelasticity of the fluid creates more vorticity on the right corner of the geometry as Fig. 2 illustrated, that is in agreement with the results obtained by Yin.

Table 1: Properties of the fluid simulated.

Properties	Value
Flow velocity (m/s)	$1 * 10^{-5}$
Viscosity (kg/m-s)	0.0001
Density (kg/m^3)	1000
We	0 and 6
Reynold	$1 * 10^{-5}$

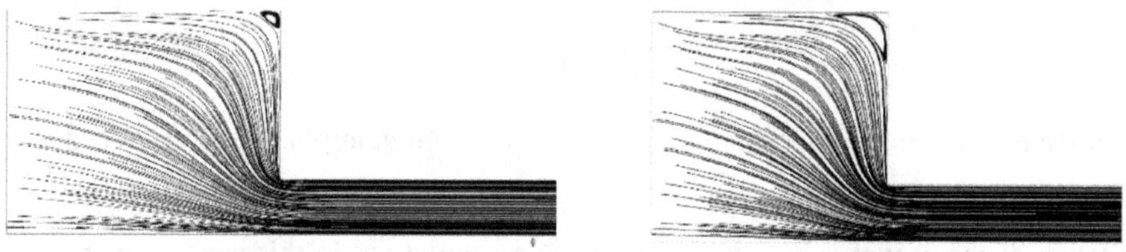

(a) (b)

Figure 2: Streamlines in ANSYS Fluent. (a) We = 0; and (b) We = 0.6.

3.2 Qualitative validation

The qualitative validation was performed comparing results of velocities for the Rayleigh problem of a viscoelastic fluid. The Rayleigh problem consider a fluid initially at rest ($t = 0$) on an infinitely long flat plate. For $t > 0$, the plate moves in the right direction, and the liquid is putting in motion as shown in Fig. 3.

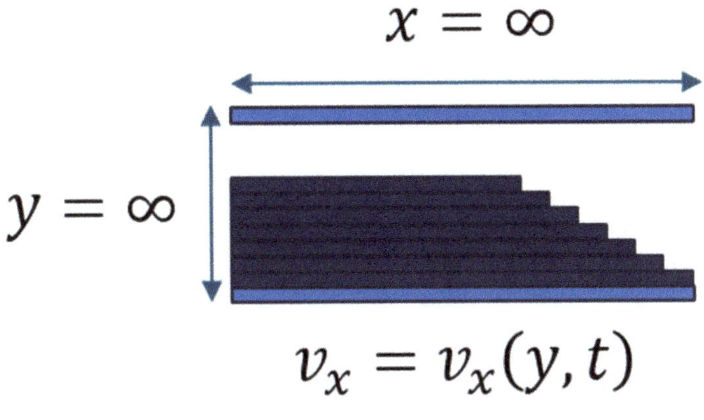

Figure 3: Rayleigh-type problem.

The analytical solution for the Rayleigh-type problem for a non-Newtonian fluid was developed by Fetecau and Fetecau [10] using the Upper Convective Maxwell model and is given by the following expression:

$$v(y,t) = V - \frac{2V}{\pi}\int_0^{1/(2\sqrt{v\lambda})} \frac{r_3 e^{(r_4 t)} - r_4 e^{(r_3 t)}}{r_3 - r_4} \times \frac{\sin(y\xi)}{\xi}\, d\xi - \frac{2V}{\pi} e^{\left(-\frac{t}{2\lambda}\right)} \int_{1/(2\sqrt{v\lambda})}^{\infty} \left[\cos\left(\frac{\gamma t}{2\lambda}\right) + \frac{1}{\gamma}\sin\left(\frac{\gamma t}{2\lambda}\right)\right] \frac{\sin(y\xi)}{\xi}\, d\xi, \quad (10)$$

where

$$r_{3,4} = \frac{-1 \pm \sqrt{1 - 4v\lambda\xi^2}}{2\lambda} \text{ and } \gamma = \sqrt{4v\lambda\xi^2 - 1}. \quad (11)$$

The values of velocity as a function of time and position were obtained from eqn (10) using the Wolfram Mathematica software, and then they was compared obtained from the simulation. Results of the comparison are illustrated in Figs 4–6.

As Figs 4–6 show, the simulations in ANSYS for a UCM model is very similar to the analytical solution in [10], that demonstrate that is possible to continue with the next topic of the research, the evaluation of the polymer with the oil at different geometries.

4 RESULTS

In this study, the simulations of the oil–water, oil–polymer and oil–polymer UDF systems were compared. The purposes of the simulation were analyzed if at the end of the transient simulation the oil trapped on the corner of expansion or contraction geometry could move to the outlet.

Two of the most important characteristics in an oil reservoir are the wettability and the surface tension between the fluid. The wettability is the preferences to adhere to the rock a drop of oil or other fluid. For that reason, this paper present different values of surface tension (0.01, 0.005 and 0.001 n/m) and wettability (120, 90 and 60 grades) for the same systems to analyze the effect of this characteristic in the oil displacement. The properties of the fluid in the simulation are in Table 2.

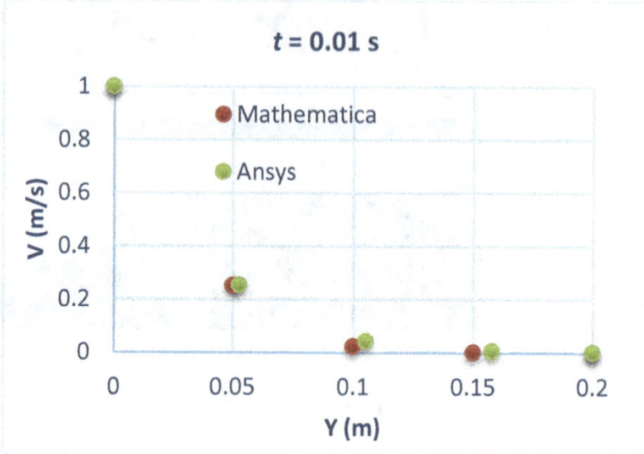

Figure 4: Velocities at different position at $t = 0.01$ s.

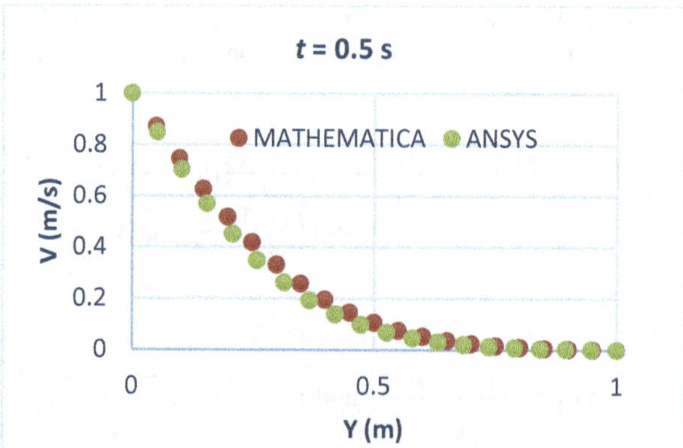

Figure 5: Velocities at different position at $t = 0.5$ s.

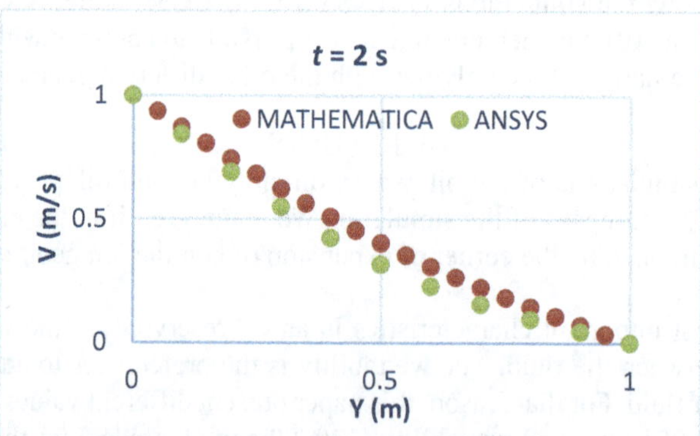

Figure 6: Velocities at different position at $t = 2$ s.

Table 2: Properties of the fluids simulated.

Properties of oil	Value
Viscosity (kg/m-s)	0.1
Density (kg/m^3)	900
Properties of water	Value
Viscosity (kg/m-s)	0.001
Density (kg/m^3)	998.2
Properties of polymer	Value
Viscosity (kg/m-s)	0.01
Density (kg/m^3)	900

In this research we consider two simple geometries: contraction and expansion. Additionally, a small part of the volume is an oil drop (red region), the rest of the volume is the other fluid (blue region) that could be water or polymer, as Fig. 7 shows.

(a) (b)

Figure 7: Oil drop at pore channel. (a) Contraction; and (b) Expansion.

4.1 Contraction geometry

For the contraction geometry, Fig. 8 shows that the most critical parameter that affects the oil movement is the wettability, represented in the contact angle. In the angle of 120 grades, in each one of the three values of surface tension evaluated, all the drop of oil moved out of the geometry, however for the other two angles, the oil drop is divided but not total evacuated.

(a) (b) (c)

Figure 8: Oil drop at different angles of wettability. (a) 120 grades; (b) 90 grades; and (c) 60 grades.

Fig. 9 shows results for the interfacial tension aspect, at least for the values of 0.01, 0.005 and 0.001 n/m does not represent change on the oil drop in comparison of the effect of the contact angle, because for these three values the volume of the oil drop that is removed remains the same as in the previous case.

Additionally, the polymer with UDF source only represents a little movement in the oil drop, as Fig. 10 illustrates. The polymer with the viscoelastic behavior moves more oil than the polymer without these characteristics in the cases of 90 and 60 grades. For these two simulations, even the oil drops look smaller in comparison with the simulation without the viscoelastic model.

4.2 Expansion geometry

For the expansion geometry, investigating streamlines between viscoelastic fluids and Newtonian fluids, there is an increase in the vortex of the corner as Fig. 11 shows. That behaviour probably will increase the oil that stay on the corner of the geometry.

(a) (b) (c)

Figure 9: Oil drop at different interfacial tension. (a) 0.01 n/m; (b) 0.005 n/m; and (c) 0.001 n/m.

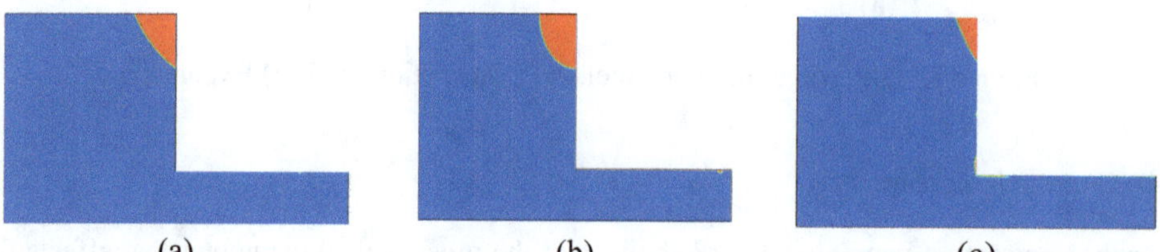

(a) (b) (c)

Figure 10: Oil drop at different systems. (a) Oil–water; (b) Oil–polymer; and (c) Oil–polymer UDF.

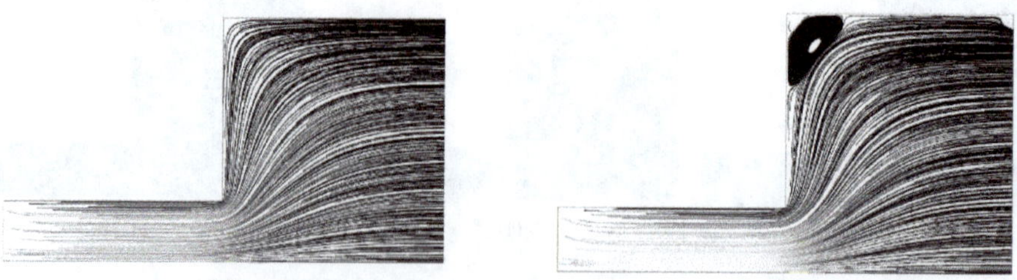

(a) (b)

Figure 11: Streamlines in ANSYS Fluent. (a) Newtonian fluid; and (b) Viscoelastic fluid.

In the case of two fluids, all the simulations have the same oil trapped on the corner, even with the changes of the system, interfacial tension, and wettability the oil drop stay on the corner. Fig. 12 shows the different shapes of the drop, but it was not possible at least in the conditions of this work to move out the oil of the geometry. The main reason is that the fluid could arrive at that corner but not with enough forces to drag the oil drop.

(a) (b) (c)

Figure 12: Oil drop at different systems at the same wettability (60 grades) and interfacial tension (0.001 n/m). (a) Oil–water; (b) Oil–polymer; and (c) Oil–polymer UDF.

ACKNOWLEDGEMENTS

This work was supported by Universidad Industrial de Santander, and the research groups: FIRST (Fenomenos Interfaciasles, Reologia y Simulacion de Transporte) and GRM (Grupo de Recobro Mejorado).

REFERENCES

[1] Wreath, D.G., A study of polymerflooding and residual oil saturation. Dissertation, 1989.
[2] Wang, D. et al., Viscous-elastic fluids can mobilize oil remaining after water-flood by force parallel to the oil–water interface. *SPE Asia Pacific Improved Oil Recovery Conference*, Society of Petroleum Engineers, 2001.
[3] Urbissinova, T., Trivedi, J.J. & Kuru, E., Effect of elasticity during viscoelastic polymer flooding: A possible mechanism of increasing the sweep efficiency. *SPE Western Regional Meeting*, Society of Petroleum Engineers, 2010.
[4] Seright, R.S., Use of Polymers to Recover Viscous Oil from Unconventional Reservoirs. Final Report, Contract No. DE-NT0006555, US Department of Energy, 2011.
[5] Qi, P. et al., Reduction of residual oil saturation in sandstone cores by use of viscoelastic polymers. *SPE Journal,* **22**(2), pp. 447–458, 2017.
[6] Yin, H., Wang, D. & Zhong, H., Study on flow behaviours of viscoelastic polymer solution in micropore with dead end. *SPE Annual Technical Conference and Exhibition*, Society of Petroleum Engineers, 2006.
[7] Afsharpoor, A. & Balhoff, M., Static and dynamic CFD modelling of viscoelastic polymer: Trapped oil displacement and deformation at the pore-level. *SPE Annual Technical Conference and Exhibition*, Society of Petroleum Engineers, 2013.
[8] Fan, J.W. et al., Hydrodynamics of residual oil droplet displaced by polymer solution in micro-channels of lipophilic rocks. *International Journal of Heat and Technology*, **35**(1), pp. 611–618, 2017.
[9] Ansys Inc., *ANSYS Fluent UDF Manual*, 2013.
[10] Fetecau, C. & Fetecau, C., The first problem of Stokes for an Oldroyd-B fluid. *International Journal of Non-Linear Mechanics*, **38**(10), pp. 1539–1544, 2003.

SECTION 4
FLOW IN POROUS MEDIA

SWELLING DURING PYROLYSIS OF FIBRE–RESIN COMPOSITES WHEN HEATED ABOVE NORMAL OPERATING TEMPERATURES

BRENT C. HOUCHENS[1], SARAH N. SCOTT[1], VICTOR E. BRUNINI[1], ELIZABETH M. C. JONES[2], MICHAEL M. MONTOYA[2], WENDY FLORES-BRITO[2] & KATHRYN N. G. HOFFMEISTER[2]
[1]Sandia National Laboratories, California, USA
[2]Sandia National Laboratories, New Mexico, USA

ABSTRACT

It is experimentally observed that multilayer fibre–resin composites can soften and swell significantly when heated above their designed operating temperatures. This swelling is expected to further accelerate the pyrolysis, releasing volatile components which can ignite in an oxygenated environment if exposed to a spark, flame or sufficiently elevated temperature. Here the intumescent behaviour of resin-infused carbon-fibre is investigated. Preliminary experiments and simulations are compared for a carbon-fibre sample radiatively heated on the top side and insulated on the bottom. Simulations consider coupled thermal and porous media flow.
Keywords: pyrolysis, carbon-fibre, porous media flow.

1 INTRODUCTION

In an accident leading to a fire, heated fibre–resin composites can be an additional fuel source. For example, modern passenger aircraft can have airframes with well above 50% composite construction, by volume and weight [1]. This added fuel source could significantly exacerbate any aircraft fire scenario, such as the recent auxiliary power unit battery fire [2]. Even without direct exposure to flame, composites heated above their designed operating ranges can first begin to soften and swell as they move through their glass transition temperature. Further heating leads to the onset of pyrolysis where liquid and gaseous products can be produced, leading to further expansion. These gases can be ignited by a spark or other sufficiently hot heat source. For carbon-fibre, the fibre itself is an additional source of fuel for combustion [3].

Here multilayer carbon-fibre composites were heated to above their designed operating temperatures but below ignition temperatures leading to pyrolysis. Swelling of the samples was measured through both visible light edge detection and digital image correlation (DIC) on the top surface. Thermographic phosphors were selected for their ability to remotely monitor the temperature on the surface with minimal invasiveness [4]. Additionally, thermographic phosphors were included between several of the carbon-fibre layers before the samples were cured, following techniques developed in previous studies [5]. Pyrolyzed gases produced in the interior of the sample flow through the porous media matrix of carbon-fibre, resin and char to the boundaries of the sample. Some phosphors are transported along with these gases and escape from the top, bottom, and edge of the sample. Laser excitation of these phosphors provides an indication of the flow of these pyrolyzed gases from the interior of the sample through the porous media.

Thermocouples were used to measure both the heater temperature and the backside temperature of the sample. An infrared (IR) camera was focused on the top of the sample to determine surface temperatures during the heating. Additionally, video and still images of the burn were recorded throughout. The goal of this paper is to experimentally measure the

WIT Transactions on Engineering Sciences, Vol 123, © 2019 WIT Press
www.witpress.com, ISSN 1743-3533 (on-line)
doi:10.2495/MPF190171

thickness change of the composite and demonstrate a model into which this thickness change can be incorporated.

2 HEATING CONFIGURATION

Experiments and computations were performed on disks of 6-layer carbon-fibre composites consisting of 33.8% prepreg epoxy by weight before curing. The samples were initially 12.7 cm in diameter and 2.5 mm thick. The geometric configuration of the sample under the heating element is shown in Fig. 1.

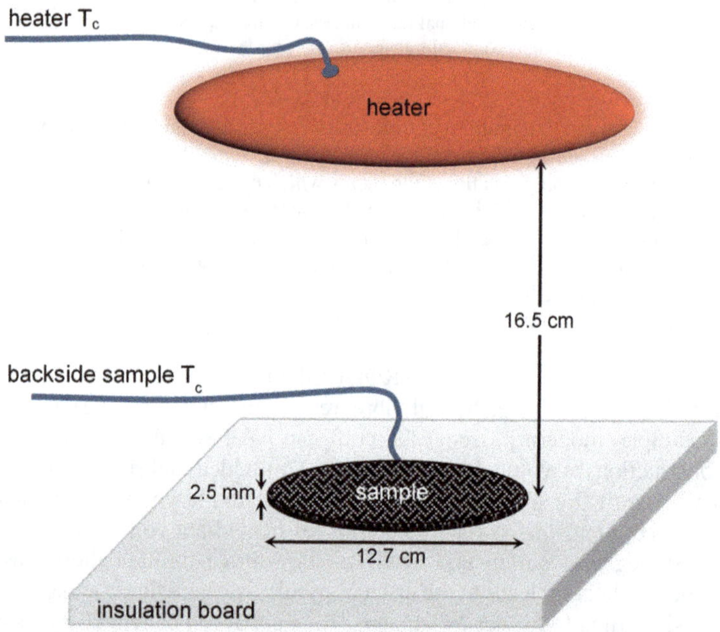

Figure 1: Carbon-fibre (6 layer) sample in the heating configuration with the backside sample thermocouple between the sample and insulation board.

The samples were heated directly on insulation board. The temperatures of both the heater and backside of the sample were measured with thermocouples. The incident heat flux was characterized by repeating the experiment with the sample replaced by an Inconel plate with known properties.

2.1 Decomposition mechanism

Previous experimental studies using thermogravimetric analysis (TGA) have suggested four-step and five-step mechanisms for the complete decomposition of composites of carbon-fibre with epoxy resin [6], [7], where complete reaction requires temperatures over 800°C. Here only data up to a temperature of approximately 280°C on the backside of the sample are considered as this is the range for which reliable expansion was measurable. TGA data from these previous studies suggests that the first reaction,

$$epoxy\ resin \rightarrow char_A + fuel, \tag{1a}$$

begins around 250°C and can occur without oxygen. This reaction is assumed independent of the environment – it proceeds independently if oxygen is or is not present. When oxygen is present, the second reaction occurs shortly after or nearly simultaneously for some heating rates, and is given by:

$$epoxy\ resin + O_2 \rightarrow char_B + fuel, \tag{1b}$$

where the potential for a different char species to be created is allowed.

At hotter temperatures, the previously developed mechanism includes the additional reactions

$$char_A + O_2 \rightarrow residue + CO, \tag{1c}$$

$$char_B + O_2 \rightarrow residue + CO, \tag{1d}$$

$$carbon\text{-}fibre + O_2 \rightarrow residue + CO_2, \tag{1e}$$

which allow the samples to decompose almost completely to ash residue.

Because the threshold temperature for eqns (1a) and (1b) is only just achieved in the experiments reported here, this mechanism suggests that most of expansion measured up to 280°C is purely thermal expansion leading through the glass transition temperature of the epoxy. Only the initial onset of decomposition is reached, as confirmed by simulations.

3 EXPERIMENTS

The experimental setup is shown in Figs 2 and 3. Independent lasers were used for digital image correlation (DIC) and thermographic phosphor measurements. Additionally, visible light cameras were used to measure the thickness change in time on the edge of the sample and record video of the experiment. An IR camera was used to measure the temperature on the top of the sample. Thermocouples measured the heater and sample (backside) temperatures.

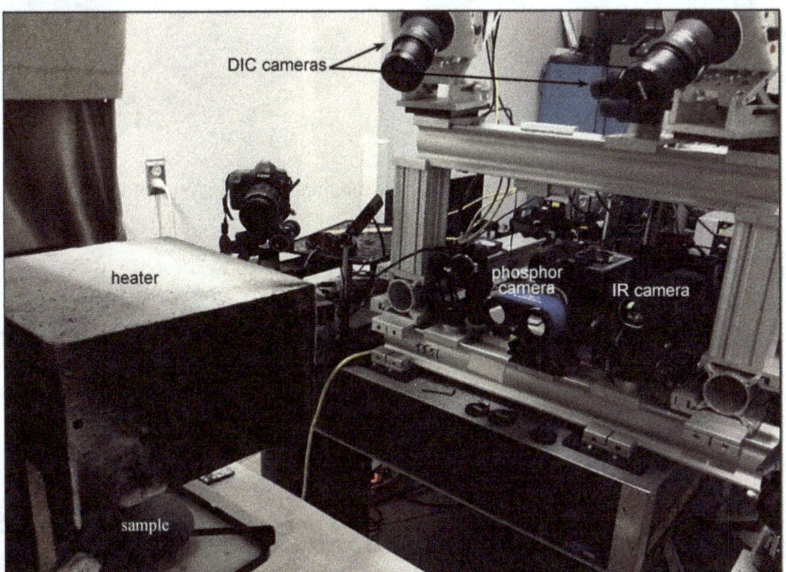

Figure 2: View with the sample in the foreground, looking back at the camera array.

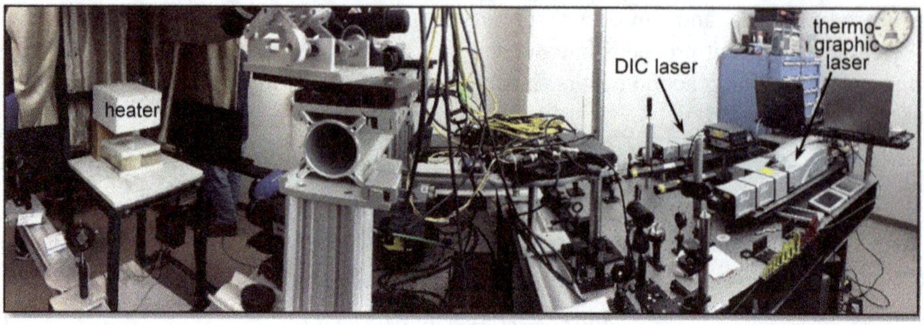

Figure 3: Panoramic view of the burn setup including lasers (far right) for DIC and thermographic phosphor measurements, camera array (centre, where the IR and visible camera are obscured), and heater and sample stage (far left).

4 COMPUTATIONAL MODELING

A disk of diameter 12.7 cm and thickness 2.5 mm was meshed with three refinements of hexahedral elements, the coarsest and finest of which are shown in Fig. 4. The coarsest, intermediate and finest meshes had 2, 3 and 4 elements through the thickness, respectively. These correspond to elements with edge lengths of approximately 1.25, 0.83 and 0.625 mm, respectively. A heat flux was applied to the top surface and the bottom surface was insulated. The thermal and porous media flow was solved using the finite element code Aria, part of the Sierra suite [8].

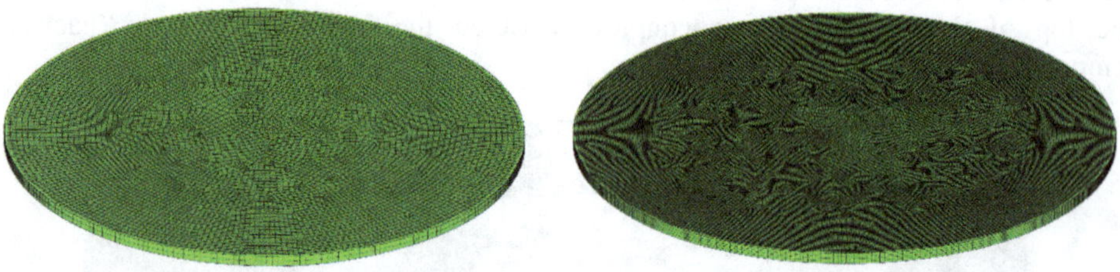

Figure 4: Images of the coarsest and finest hex meshes used for computational simulations.

The multilayer carbon-fibre composite was treated as homogeneous material until the decomposition governed by eqn (1) began. The material properties used are given in Table 1. The emissivity on the top surface was taken to be 0.91, consistent with previous studies [6].

The reaction rate k for eqn (1) is assumed to have an Arrhenius rate dependence, given by:

$$k = Ae^{-E_a/k_B T}, \tag{2}$$

where A is the pre-exponential factor, E_a is the activation energy, k_B is Boltzmann's constant, and T is the temperature in absolute units. For the only reaction initiated, eqn (1a), $A = 3.33\times10^{11}$ s^{-1} and $E_a = 1.47\times10^8$ J/kmol. The stoichiometric coefficient of the solid phase $char_A$ product on a mass basis is 0.5, and the heat of reaction is taken to be zero based on past differential scanning calorimetry (DSC) studies [6].

Table 1: Thermal parameters for the carbon-fibre epoxy sample from Scott et al. [6], with the exception of the temperature dependent density of epoxy.

		Material		
Property	**Units**	Epoxy resin	Carbon-fibre	Char
Thermal conductivity	W/m-K	0.145	$0.335 \ln(T)$-1.8257	0.029
Density	kg/m^3	$1300 + aT$	1779	650
Specific heat	J/kg-K	866	$4.0997T - 369.12$	936
Permeability	m^2	2.42×10^{-15}	2.42×10^{-14}	2.83×10^{-12}

4.1 Mesh dependency

Three mesh resolutions with 2, 3 and 4 elements through the thickness were investigated. These had corresponding element counts of 14,440, 43,249 and 141,944. Because the heat flux applied in the experiments was relatively low, the transients were very slow. Thus it was found that this relatively course resolution was sufficient to provide nearly mesh-independent results. Over the 1700 s of interest, the temperature simulated at the bottom-centre of the sample with the finest resolution mesh was within 3.2% and 3.5% of the values from the intermediate and coarse meshes, respectively, for all time steps.

5 RESULTS

Initial experimental and computational results are shown here indicating that significant thermal expansion occurs as the carbon-fibre epoxy moves through its glass transition temperature. This continues with the onset of thermal decomposition. Qualitative observations not discussed here show that expansion increases more dramatically as more gas products are generated, causing the fibre layers to separate.

5.1 Experimental results

A compiled set of data is shown in Fig. 5 near the end time of the experimental data of interest, at 25 min and 23 s. In Fig. 5(a) the thermocouple temperatures at the heater and below the sample are shown. Fig. 5(b) shows the IR camera image at this same time. Note that this image assumes a single emissivity, and thus is most accurate on the top surface and requires other rescaling on other surfaces, particularly the disk edge which appears unrealistically hot. It also shows the DIC markers as rows of hot spots near the centre of the disk. Finally, Fig. 5(c) shows a visible snapshot taken simultaneously to a laser firing for the internal phosphors. The DIC dots show up as a thick white band from front to back due to their excitement and overexposure by the visible light camera.

Detailed thermocouple data is shown in Fig. 6 with the corresponding heater power shown as a percentage of the maximum possible power. Based on the backside temperature, it is expected that the expansion measured in the first 1200 s should be due purely to thermal expansion. In later times, TGA data suggests that decomposition of the epoxy adhering to the mechanism in eqn (1) should begin between 1400 and 1600 s. The heater power was specifically increased in steps to move into this regime.

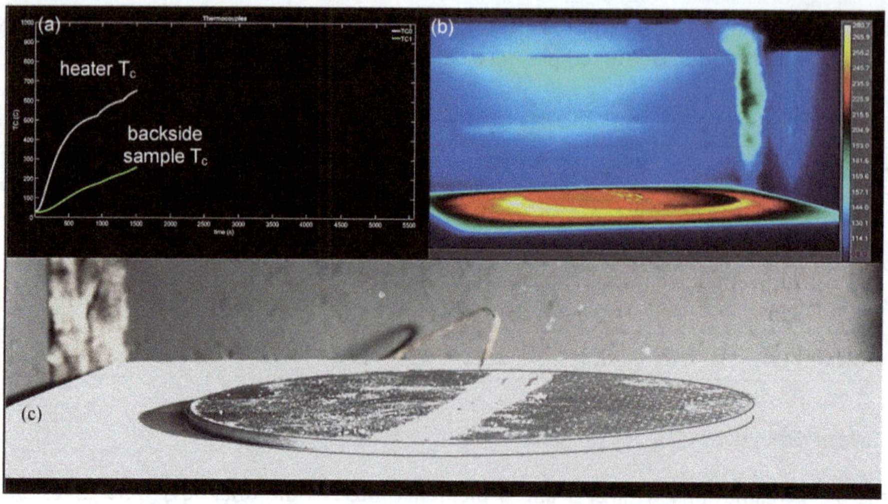

Figure 5: (a) Thermocouple temperatures at the heater and backside of sample; (b) Infrared image of the sample; and (c) Visible image at an instant illuminated by a laser.

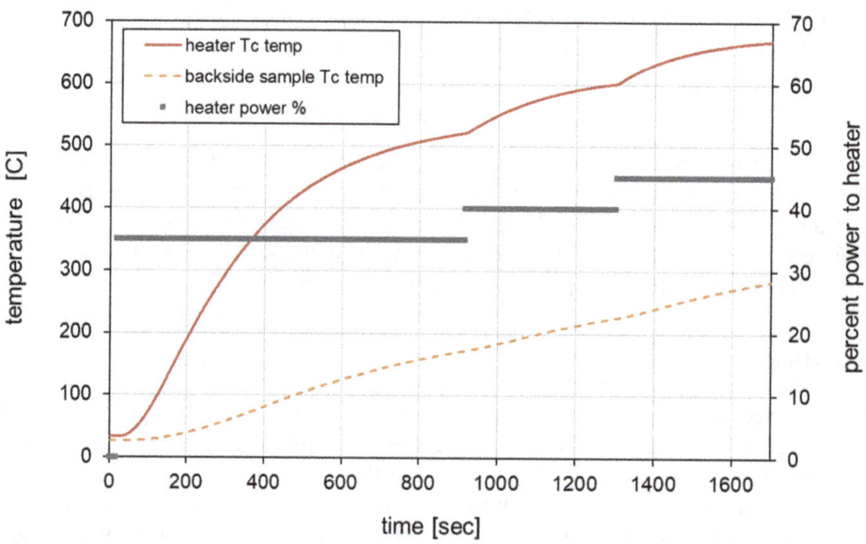

Figure 6: Temperatures (left axis) of the heater and sample backside and percent power to the heater (right axis).

Fig. 7 shows the increase of thickness near the front edge of the sample versus time. This measurement was performed by fitting a smooth surface to the entire DIC measurement field, and extrapolating that to the visible front edge measurement. There is no obvious increase in the thickness change associated with the onset of the epoxy decomposition, after the initial thermal expansion. Nevertheless, in increasing the temperature to just under 300°C on the backside, a 10% increase in sample thickness is observed.

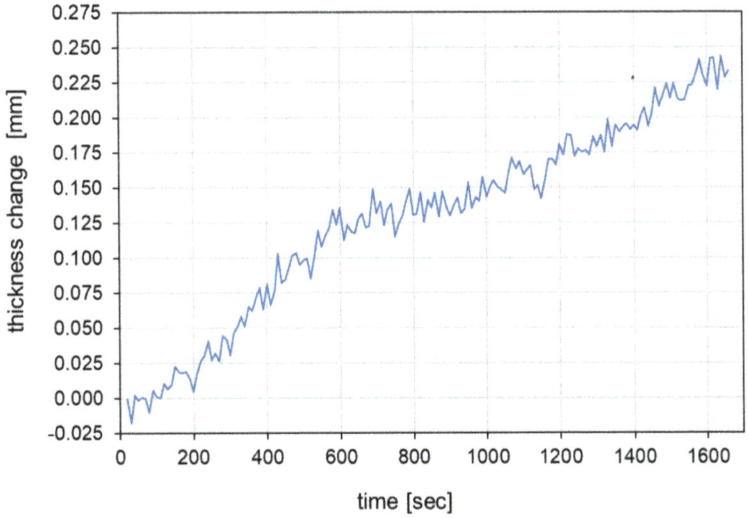

Figure 7: Average thickness change by combining visible edge measurements and a smooth fit to the DIC measurements, extrapolated to the front edge.

5.2 Modelling and simulation results

Model results shown in Fig. 8 demonstrate a similar temperature rise to experiments, though quantitatively the values diverge at later times. Previous research showed strongest sensitivities to the thermal properties including density, specific heat and conductivity of the bulk and/or carbon fibre, with generally less sensitivity to the permeability of the porous media flow [6].

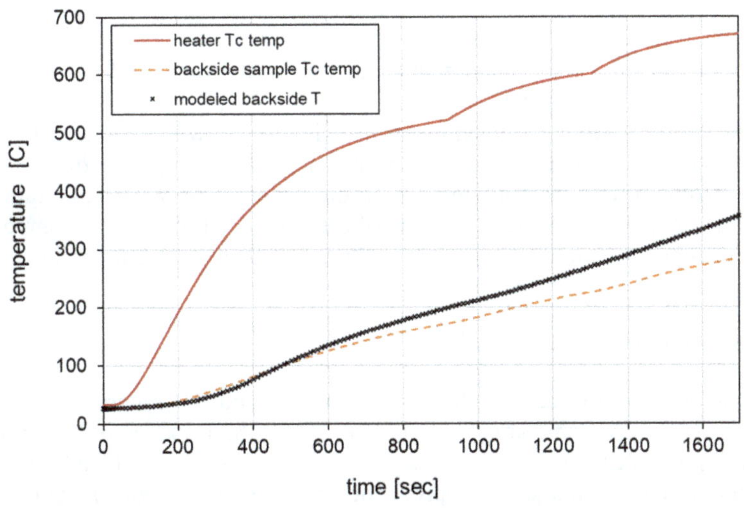

Figure 8: Modelled increase in temperature versus time at the bottom, centre of the sample, compared to the experimentally measured temperature.

Fig. 9 shows the mass fractions of the various constituents in the composite and further confirm that reactions are not expected until the very end of this experiment and model. As the epoxy begins to break down, char and fuel, the latter modelled as methane (CH_4), begin to form as in eqn (1a). Given the overshoot of temperature of the model in Fig. 8, it is likely that even these early reactions have not initiated in the lower temperature experiment. Thus all expansion measured in Fig. 7 is likely due almost purely to thermal expansion.

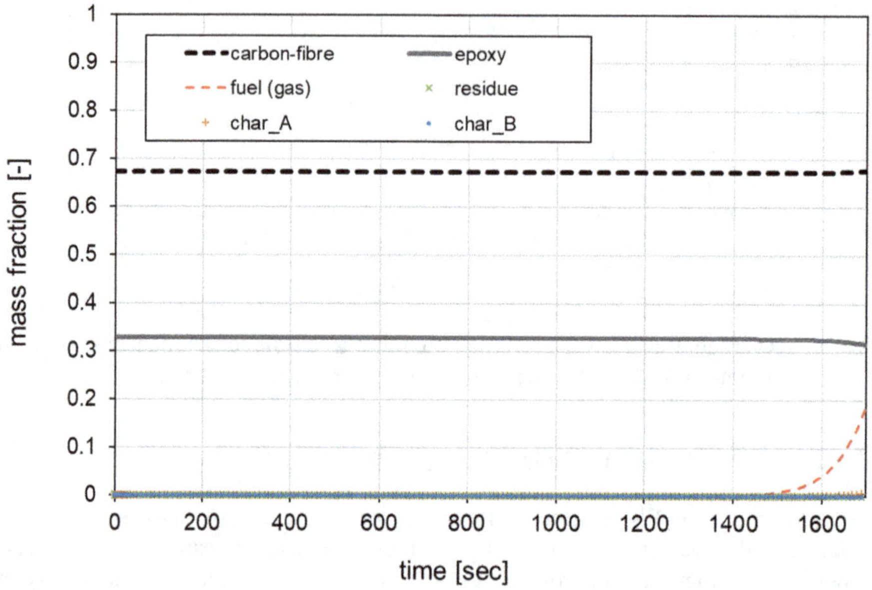

Figure 9: Simulated mass fractions of carbon-fibre, epoxy and the products of the reaction mechanisms given in eqn (1) on the backside of the sample.

By using Figs 6 and 7 in conjunction, it is possible to fit a value of the thermal expansion coefficient *a*. However, based on recent experiments at different heating rates, this has proven to not be universal. Interestingly, for experiments with very slow heating rates, it has recently been observed that it is possible to initiate decomposition that completely offsets the thermal expansion, leading to a monotonic decrease in the thickness of the sample, rather than an increase. Thus, it appears that the coefficient of thermal expansion is dependent on not only the absolute temperature of the carbon fibre sample, but also on the temporal gradient of the heating.

5.3 Simulations beyond experimental results

For completeness, the extended simulation results beyond the range for which reliable experimental data is available are shown in Fig. 10 for the backside of the sample. After 1700 s the epoxy degrades quickly into flammable, gaseous fuel and $char_A$. At the temperatures in these simulations, the production of $char_B$ and residue remains negligible. Around 2500 s the epoxy is depleted and production of fuel ceases. Furthermore, the temperatures needed to initiate the decompositions of $char_A$ to residue and carbon-fibre to residue are not attained. This is consistent with the lack of ignition observed experimentally.

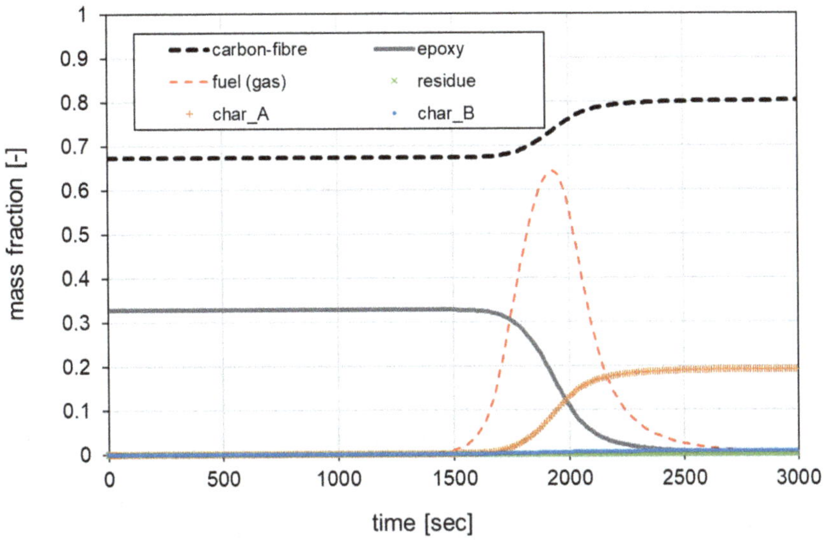

Figure 10: Simulated mass fractions of carbon-fibre, epoxy and the products of the reaction mechanisms given in eqn (1), for times beyond those in the experiment.

6 CONCLUSIONS

Swelling of up to 10% can be expected from heated composite flat panels, mostly due to thermal expansion even before decomposition mechanisms begin. Once decompositions begins, this effect is generally observed to be further magnified between the composite layers, eventually allowing gases to escape between the fibre layers at a much faster rate than the porous media flow through the layers in the normal direction. However, even this general finding is not universally observed as sufficiently slow heating rates appear to allow the porous media flow to offset the thermal expansion. This suggests that the decomposition reaction may occur at lower temperatures than originally believed, but further investigation is required before any definitive statement can be made.

The goal of future experiments and modelling is to quantify expansion (or contraction) and anisotropy in permeability for all the pyrolysis reaction mechanisms. Ongoing modelling efforts suggest that better agreement can be achieved by ramping the heater at a slow constant rate of power increase, and then holding the power constant for sufficient time to characterize the various decomposition mechanisms that occur, and then again ramping at constant rate. This may allow better association of each mechanism with TGA and DSC findings. Additionally, the effect of number of layers of composite is of interest. Experiments to investigate both effects are planned.

ACKNOWLEDGEMENTS

The authors would like to acknowledge the contributions of Brian McKay for laying up the composite samples and Vince Valdez for constructing and characterizing the heater.

Sandia National Laboratories is a multi-mission laboratory managed and operated by National Technology and Engineering Solutions of Sandia, LLC, a wholly owned subsidiary of Honeywell International Inc., for the U.S. Department of Energy's National Nuclear Security Administration under contract DE-NA0003525.

REFERENCES

[1] Teresko, J., Boeing 787: A matter of materials – Special report: Anatomy of a supply chain. *Industry Week*, 15 Nov. 2007.

[2] National Transportation Safety Board, Auxiliary power unit battery fire, Japan Airlines Boeing 787-8, JA829J, Washington, DC, 2014.

[3] McKinnon, M.B., Ding, Y., Stoliarov, S.I., Crowley, S. & Lyon, R.E., Pyrolysis model for a carbon fiber/epoxy structural aerospace composite. *J. Fire Sci.*, **35**(1), pp. 36–61, 2017.

[4] Brubach, J., Pflitsch, C., Dreizler, A. & Atakan, B., On surface temperature measurements with thermographic phosphors: A review. *Prog. Energy Combustion Sci.*, **39**, 2013.

[5] Flores-Brito, W., Westphal, E., Wilburn, B.R. & Hoffmeister, K.N.G., Study of sensitivity vs. excitation time of LED excited thermographic phosphor. *Proc. of the AIAA Science and Tech. Forum*, 2019.

[6] Scott, S.N. et al., Validation and uncertainty estimation of carbon fiber epoxy composite model. *Proceedings of the 11th US National Combustion Meeting*, 2019.

[7] Scott, S.N. et al., Modeling the decomposition behaviour of carbon fiber epoxy composite. *Proceedings of the 11th Mediterranean Combustion Symposium*, 2019.

[8] Notz, P.K., Subia, S.R., Hopkins, M.M., Moffat, H.K., Noble, D.R. & Okusanya, T.O., *SIERRA Multimechanics Module: Aria User Manual*, 2016.

GEOMETRIC VERSUS KINETIC MODELLING APPROACH FOR CHARACTERIZING POROUS METAL FOAMS

ESMARI MARÉ & SONIA WOUDBERG
Applied Mathematics Department, Stellenbosch University, South Africa

ABSTRACT

Knowledge of the geometric and kinematic parameters of porous foams are of great importance since it is used in a wide variety of industrial multiphase flow applications that require optimal functionality, e.g. gas filters, heat exchangers and catalyst supports. The large external surface area and high porosity of metal foams provide good chemical resistance, enhanced heat and mass transfer properties and low pressure drops. Four generic geometric models will be considered to characterize the metal foam geometry, namely the cubic unit cell, tetrakaidecahedron, dodecahedron and rectangular representative unit cell (RUC) models, as well as three kinetic approaches from the literature in order to predict the specific surface area (SSA). Two sets of experimental data from the literature will then be compared to the SSA model predictions of the geometric approach and to the SSA values obtained from the kinetic approach. A comparative analysis reveals that the most geometrically complex tetrakaidecahedron model indeed provides the best correspondence with the experimental data for the SSAs, followed by the geometrically simplest RUC model. The latter model, in addition, provides accurate results for the kinetic approach. The advantage of the RUC model is that it is the only geometric model that provides both a geometric and kinetic approach, and, as a result of its relatively simple geometry it is geometrically adaptable towards anisotropy. The Klinkenberg effect will also be considered to determine the influence on the predictions of the SSAs dependency on the permeability coefficients for different fluid phases.

Keywords: metal foams, specific surface area, Klinkenberg effect, permeability, porous media, porosity, Knudsen, flow.

1 INTRODUCTION

Metal foams have a vast range of industrial applications. They are used in, for example, catalytic beds as compact heat sinks for power electronics, in air cooled condensers and multifunctional heat exchangers [1], in the design of aircraft wing structures in the aerospace industry [2], etc. The properties that make foams so attractive to processes requiring gas filters, heat exchangers and catalyst supports are that they exhibit a high mechanical strength, a large external surface area and a high porosity. These physical properties in turn lead to the metal foams having good chemical resistance, enhanced heat and mass transfer properties and low pressure drops [3].

Several authors in the literature have studied multiphase flow in foams, e.g. Topin et al. [4] and Edouard et al. [5]. In multiphase flow applications (in specific), metal foams have become more popular over packed beds in, for example, the petrochemical industry due to its large specific surface area and the resulting low pressure drops. This is because in industries such as the petrochemical industry, catalytic reactions between gas, liquid and solid are essential and the rates of mass transfer increases by increasing the gas and liquid contact surfaces, as well as the turbulence within the fluid phase. Furthermore, low pressure drops at high flow rates are favourable so that the productivity and selectivity of the processes can be maintained [5]. Metal foams are thus ideal for such processes. Consequently, for design and optimization purposes, the parameters of metal foams and knowledge thereof are of great importance.

WIT Transactions on Engineering Sciences, Vol 123, © 2019 WIT Press
www.witpress.com, ISSN 1743-3533 (on-line)
doi:10.2495/MPF190181

To characterize foams, morphological parameters are needed, such as the pore diameter, the strut diameter and the specific surface area (SSA). Due to the intricate nature of foams and their complex structure, it is difficult to find reliable correlations among the morphological parameters that affect transport phenomena. The micro-structure of metal foams is approximated by mainly four different geometric models, as illustrated in Fig. 1. These are the cubic unit cell model [6], the tetrakaidecahedron model [3], the dodecahedron model [7] and the rectangular representative unit cell (RUC) model [8].

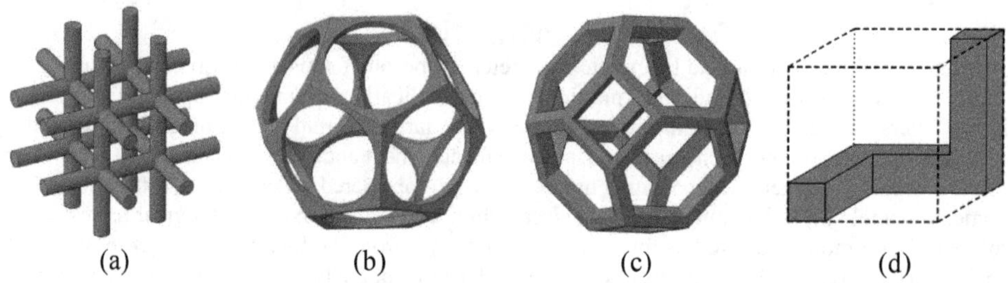

Figure 1: Geometric models for foams. (a) Cubic unit cell; (b) Dodecahedron; (c) Tetrakaidecahedron; and (d) RUC.

The objective of this paper is to investigate specifically the correlations among the morphological parameters of each of the respective model geometries by giving equations for the prediction of the SSA. Thereafter, another approach will be considered, referred to as the kinetic approach, where the transport properties of metal foams are used to determine the SSA, and are compared to the results obtained from the geometric approach. Furthermore, experimental data available in the literature for the pore-scale linear dimensions and SSA, will be compared to the model predictions of the geometric approach and to the values obtained from the kinetic approach. It will also be determined if the more complex geometries of the tetrakaidecahedron and dodecahedron models provide more accurate predictions than the simpler cubic unit cell and RUC models. Finally the influence of different fluid phases on the predicted values for the SSA will be investigated due to the Klinkenberg effect.

2 GEOMETRIC APPROACH

As mentioned in Section 1, the geometric models currently used in the literature to approximate the microstructure of foams are the cubic unit cell model [6], the tetrakaidecahedron model [3], the dodecahedron model [7] and the RUC model [8]. Furthermore, the specific surface area, denoted by S_v, is defined as:

$$S_v = \frac{A_s}{V_o}, \tag{1}$$

where A_s is the total surface area of the foam presented to the flow and V_o is the total volume of the bed.

2.1 Cubic unit cell model

The cubic unit cell resembles a cube with cylindrical fibres on the edges [6]. One approach in finding expressions for the parameters of the cubic unit cell model was presented by Giani et al. [6]. The cell volume was computed by using the volume of a cube and the pore diameter, yielding

$$V_o = d_p^3. \tag{2}$$

The SSA obtained in this approach in terms of the porosity, denoted by ϵ, and the pore diameter is then given by

$$S_v = \frac{2}{d_p}\sqrt{3\pi(1-\epsilon)}. \tag{3}$$

Another approach in finding expressions for the parameters of the cubic unit cell model was given by Lacroix et al. [9]. In this method the volume of the cube was computed using the face diameter, defined as [10],

$$D_p = d_p + d_{s,} \tag{4}$$

with d_s denoting the strut diameter, which is the thickness in the middle of the strut and usually the thinnest part between the nodes [10]. The equation for expressing the cell volume in this approach is therefore given by

$$V_o = D_p^3, \tag{5}$$

which leads to

$$S_v = \frac{2}{d_p}\left[\sqrt{3\pi(1-\epsilon)} - 2(1-\epsilon)\right], \tag{6}$$

for the SSA.

2.2 Tetrakaidecahedron model

The tetrakaidecahedron model resembles a truncated octahedron, having six square faces and eight hexagonal faces [11]. For this model, triangular strut geometry is used with an edge width d_s and side length l_s. Similarly to the cubic unit cell model, there are two approaches to determine the SSA of the tetrakaidecahedron model. The one approach makes use of the face diameter D_p [11] and the other the pore diameter d_p [12].

The approach in which D_p is used is based on the work of Gibson and Ashby [13] as used by Buciuman and Kraushaar-Czarnetzki [11]. The equation for the SSA in terms of the porosity and the pore diameter in this approach is found to be

$$S_v = \frac{4.82\sqrt{1-\epsilon}}{\left(\frac{31}{19}\sqrt{1-\epsilon}+1\right)d_p}. \tag{7}$$

For the second approach, the SSA is determined in terms of the pore diameter d_p as was done by Richardson et al. [12] and results in

$$S_v = \frac{5.621\left[1-0.971\sqrt{1-\epsilon}\right]\sqrt{1-\epsilon}}{d_p}. \tag{8}$$

In the expression obtained for S_v by Richardson et al. [12], however, the coefficient is 12.979 instead of 5.621 and it is therefore possible that there was a mistake made in the calculation of the coefficient in their equation for the SSA.

2.3 Dodecahedron model

The dodecahedral shape is formed by 12 pentagonal faces [7]. As observed by Bhattacharya et al. [2], the strut shape depends on the porosity of the foam, that is, the struts tend towards triangular prisms in the case of high porosity and towards cylindrical shape in the case of lower porosity. We will consider both cases of the slim dodecahedron model using the approaches followed by Huu et al. [7] to determine the SSA of the dodecahedron model at both high porosity ($\epsilon > 0.9$) with triangular strut geometry, and lower porosity ($\epsilon < 0.9$) with cylindrical strut geometry.

2.3.1 Triangular strut geometry

The SSA for this version of the dodecahedron model can be determined by making use of the golden ratio (~1.61803), denoted by $\emptyset$, and the following equation

$$\frac{k^2\sqrt{15}}{\emptyset^4} - \frac{k^3\sqrt{10}}{3\emptyset^4} - (1-\epsilon) = 0, \tag{9}$$

where k is defined as d_s/c and c is the length of the struts added to two times the distance between the strut and the centre mass point of the nodes [7].

The equation for the SSA in terms of the pore diameter and k is then given by

$$S_v = \frac{60k}{\sqrt{5\emptyset^3\sqrt{3-\emptyset}}d_p}\left(1 - \frac{1}{2}\sqrt{\frac{2}{3}}k\right)^2, \tag{10}$$

where k can be solved from eqn (9).

2.3.2 Cylindrical strut geometry

In this version for determining the SSA, the equation for k, defined again as d_s/c, in terms of the porosity becomes [7]

$$\frac{\sqrt{5}\pi}{\emptyset^4}k^2\left(1 - \frac{k}{2}\sqrt{\frac{2}{3}}\right) + \frac{2\sqrt{10}}{12\emptyset^4}k^3 - (1-\epsilon) = 0. \tag{11}$$

The SSA can then be given in terms of the pore diameter and k, i.e.

$$S_v = \frac{20\pi k}{\sqrt{5\emptyset^3\sqrt{3-\emptyset}}d_p}\left(1 - \frac{1}{2}\sqrt{\frac{2}{3}}k\right)^2, \tag{12}$$

where k can be solved from eqn (11).

2.4 RUC model

The RUC model represents the average geometry of the complex network of interconnected struts of actual porous foams and consists of three mutually perpendicular square duct sections [8]. In RUC notation, U_f denotes the fluid volume, $U_{||}$ the total streamwise fluid volume and U_t the fluid volumes that are not bordered by solid surfaces [8]. The geometric factor ψ can then be defined in terms of these parameters as [8]

$$\psi = \frac{U_f}{U_{||} + U_t}. \tag{13}$$

The geometric factor can also be obtained in terms of the porosity, yielding

$$\frac{\psi}{\epsilon} = \frac{4}{(3 - \psi)^2}. \tag{14}$$

Eqn (14) can be rewritten as an explicit function of porosity by using the Cardanic method of solving a cubic polynomial, yielding [14]

$$\psi = 2 + 2\cos\left[\frac{4\pi}{3} + \frac{1}{3}\cos^{-1}(2\epsilon - 1)\right]. \tag{15}$$

The SSA can be determined in terms of the pore diameter and the geometric factor, given by

$$S_v = \frac{3}{2d_p}(3 - \psi)^2(\psi - 1), \tag{16}$$

where ψ can be solved using eqn (15).

3 KINETIC APPROACH

The onset of determining the SSA of porous foams using a kinetic approach is the Darcy–Forchheimer equation, i.e. [15]:

$$\frac{\Delta p}{L} = \frac{\mu}{K_1}q + \frac{\rho}{K_2}q^2, \tag{17}$$

where $\Delta p/L$ is the pressure gradient, μ is the dynamic viscosity, ρ is the density of the fluid, K_1 and K_2 are the permeability coefficients of the Darcy and Forchheimer regimes, respectively, and q is the superficial velocity. Eqn (17) is used to determine K_1 and K_2 from experimental data for the pressure drop. These permeability coefficients are then used to find an expression for the specific surface area. To determine K_1 and K_2, three approaches will be considered. The one approach will be based on the work of Dietrich et al. [15], another approach will be based on the work done by Huu et al. [7], and the final approach will be based on the RUC modelling approach of Woudberg and Du Plessis [8].

3.1 Dietrich formulation

Dietrich et al. [15] compared the Ergun equation with eqn (17), based on the assumption that the Ergun equation is also applicable to describe the pressure drop of metal foams. The Ergun equation is given by

$$\frac{\Delta p}{L} = A\frac{\mu}{\epsilon D_h^2}q + B\frac{\rho}{\epsilon^2 D_h}q^2, \tag{18}$$

where D_h is the hydraulic diameter and $A = 110$ and $B = 1.45$ are empirical constants obtained for the metal foams [15]. An expression for the SSA in terms of the permeability coefficient K_1 can then be determined, given by

$$S_v = \sqrt{\frac{16\epsilon^3}{110K_1}}. \quad (19)$$

Similarly the SSA can be determined in terms of K_2, resulting in

$$S_v = \frac{4\epsilon^3}{1.45K_2}. \quad (20)$$

3.2 Huu formulation

The approach followed by Huu et al. [7] is similar to that of Dietrich et al. [15], but instead of using the Ergun equation defined in terms of D_h they defined it in terms of d_p, that is

$$\frac{\Delta p}{L} = E_1 \frac{\mu(1-\epsilon)^2}{\epsilon^3 d_p^2} q + E_2 \frac{\rho(1-\epsilon)}{\epsilon^3 d_p} q^2, \quad (21)$$

where $E_1 = 150$ and $E_2 = 1.75$. To avoid using a geometric approach as Huu et al. did in defining the pore diameter, it was decided to rather let $d_p = D_h$, as was done in the Dietrich approach. The SSA was then determined in terms of the permeability coefficients, yielding

$$S_v = \sqrt{\frac{16\epsilon^5}{150K_1(1-\epsilon)^2}}, \quad (22)$$

which is the SSA in terms of K_1, and

$$S_v = \frac{4\epsilon^4}{1.75K_2(1-\epsilon)}, \quad (23)$$

in terms of K_2.

3.3 RUC formulation

In the RUC approach, expressions for K_1 and K_2 are determined as shown in the study of Woudberg and Du Plessis [8]. An Ergun-type equation for porous foams is determined from the RUC model and simplified to give the following expression for the pressure gradient:

$$\frac{\Delta p}{L} = \frac{9\psi^2(\psi-1)(3-\psi)^2}{\epsilon^2 d_p^2}\mu q + \frac{(\psi-1)(3-\psi)^2\psi^2 c_d}{8\epsilon^3 d_p}\rho q^2, \quad (24)$$

where ψ is as defined in eqn (15) and c_d is the interstitial form drag coefficient with a value of $c_d = 2$. Defining d_p to be equal to D_h, as in the previous approaches, once again leads to an equation for the SSA in terms of K_1, i.e.

$$S_v = \frac{4\epsilon}{3\psi(3-\psi)\sqrt{K_1(\psi-1)}}, \quad (25)$$

and K_2, i.e.

$$S_v = \frac{16\epsilon^4}{K_2(\psi-1)(3-\psi)^2\psi^2}. \quad (26)$$

4 GEOMETRIC AND KINETIC APPROACHES COMPARED TO EXPERIMENTAL DATA

The influence of the porosity on the SSA equations obtained in the geometric approach in dimensionless form, that is $S_v d_p$, is shown in Fig. 2. The dimensionless SSA predictions obtained using the geometric approach is compared to the experimental data of Dietrich et al. [15] and Incera Garrido et al. [10]. It can be seen that all of the equations predict that the SSA decreases as the porosity increases, which makes physical sense. The SSA predicted by the dodecahedron model were split into two porosity ranges, 0.7 to 0.9 and 0.9 to 1 for the cylindrical and triangular strut geometry, respectively. Although Fig. 2 shows that the most accurate predictions obtained when compared to this experimental data are those from the tetrakaidecahedron model, which is considered to be the most complicated of the geometric models, the second most accurate model is the prediction determined from the RUC model, which is considered to be the simplest geometric model of the four considered. The dodecahedron model is the third most accurate (although the results are very similar to that of the RUC) and the cubic unit cell the fourth. It should be emphasized that these findings are purely based on the experimental data considered in this study. Comparison with other data may produce different results.

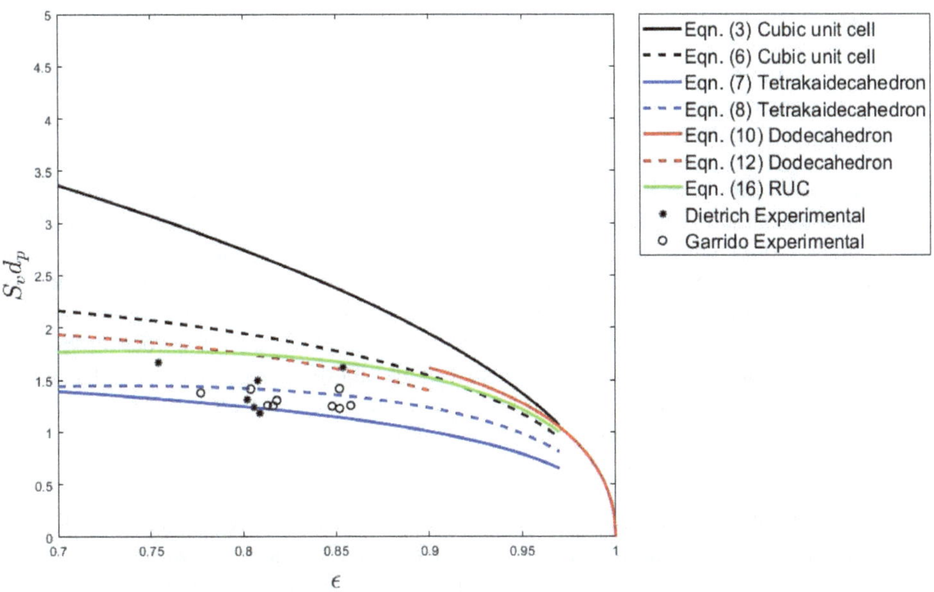

Figure 2: Dimensionless SSA versus porosity predicted by all the Geometric approach models compared to experimental data from the literature.

To evaluate the equations for the SSA obtained from the kinetic approaches, two sets of experimental data for K_1 and K_2 were considered. One set was that of Richardson et al. [12] and the other that of Liu et al. [16], shown in Fig. 3. The experimental data of Dietrich and Garrido correspond well with the values of the dimensionless SSA obtained from the RUC and Dietrich approaches. The values obtained from the Huu approach does, however, not appear to be a good approximation when compared to the experimental data.

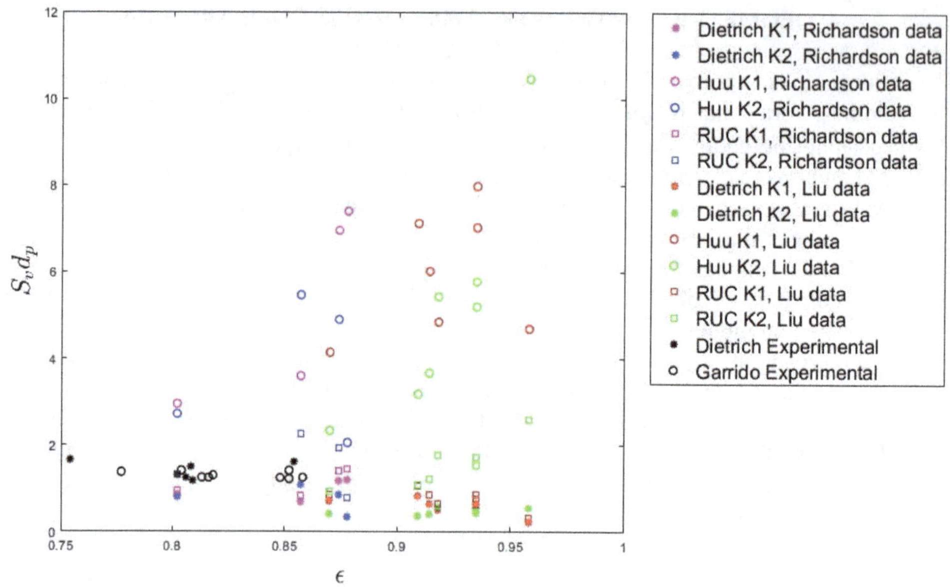

Figure 3: Dimensionless SSA values versus porosity obtained from all the Kinetic approaches compared to experimental data from the literature.

To address the question of whether the more complex geometries of the tetrakaidecahedron and dodecahedron models provide more accurate predictions than the simpler cubic unit cell and RUC models, the comparisons of the geometric models with the experimental data are considered. If the tetrakaidecahedron and dodecahedron models are considered together as the more complex models, the complex models provide more accurate predictions than the simpler RUC and cubic unit cell models. If the four models are considered individually where the complexity of the models in decreasing order are the tetrakaidecahedron, dodecahedron, RUC and the cubic unit cell model, the more complex models do not provide more accurate predictions due to the RUC model providing slightly more accurate predictions than the dodecahedron model.

5 FLUID PHASE DEPENDENCE DUE TO THE KLINKENBERG EFFECT

Gas and liquid flow in porous media may produce different permeability values. Reasons for these differences include the fact that gas is much more compressible than liquid, adsorption occurring in porous media of low porosity and the Klinkenberg effect [17]. In this study only the Klinkenberg effect will be considered and how it influences the flow of gas through porous media when compared to liquid flow.

The Klinkenberg effect is related to the slip condition of the flow at the walls of the pore. The Klinkenberg effect becomes significant at low or near atmospheric pressure, when the permeabilities of the flow are low and when the molecular mean free path, denoted by λ, is comparable to the dimensions of the pore [17].

Due to this effect, the flow deviates from Darcy's law, which in turn leads to an adjustment of the Navier–Stokes equation to implement a slip-boundary condition. To determine when the boundary condition should be implemented, a ratio is used, known as the Knudsen number and, given by [17]

$$K_n = \frac{\lambda}{D_h}. \tag{27}$$

When $K_n \leq 10^{-3}$ the continuum assumption is valid in the Navier–Stokes equation and a no slip-boundary condition can still be implemented. When $10^{-3} < K_n < 10^{-1}$ the continuum approach is still valid, but a slip-boundary condition at the pore walls needs to be incorporated in the Navier–Stokes equation and when $K_n \geq 10^{-1}$ the Navier–Stokes equation cannot be used since the gas would then exhibit non-continuum effects [17].

An expression relating the liquid and gas permeabilities involving the Knudsen number is [17]

$$K_g = K_l\left[1 + 12\frac{(2-\sigma)}{\sigma}K_n\right], \tag{28}$$

where σ is the tangential momentum accommodation coefficient (the fraction of molecules which undergo diffuse reflection at the pore walls [17]).

In the Kinetic approach in Section 4, the experimental results obtained from the Richardson et al. and Lui et al. data were from measuring gas flow through foams. The permeability coefficients in the Kinetic approach was that of liquid and need to be expressed in terms of the permeability coefficient of gas to determine whether the Klinkenberg effect affects the SSA results, or alternatively, whether different fluid phases would lead to different SSA results when determined as in the Kinetic approach. Rearranging eqn (28) therefore leads to an expression of K_l in terms of K_g, i.e.

$$K_l = \frac{K_g}{1 + 12\frac{(2-\sigma)}{\sigma}K_n}. \tag{29}$$

The tangential momentum accommodation coefficient σ ranges mostly between 0.2 and 1 and tends towards 0 for a smooth surface and towards 1 for a rough surface, since it depends on the properties of the wall. The SSA versus porosity values for the case of gas flow plotted in Fig. 4 are those obtained from substituting eqn (29) into the Dietrich and RUC formulations, i.e. eqns (19) and (25), respectively. Due to the results of the Huu formulation not performing well when compared to the Dietrich and RUC formulations, the Huu formulation was not added in this investigation of fluid phase effects on the SSA results. For each formulation σ equalling 0.2, 0.5 and 1.0 were plotted to determine the effect of the surface smoothness on the results. The Knudsen number was taken to be equal to 10^{-3} to determine the significance of the Klinkenberg effect on the SSA values in the case of the lower bound of estimated slip-flow occurrence.

It can be observed in Fig. 4 that the predictions for the dimensionless SSA decreases with increasing σ-values. All the Dietrich formulation values, obtained using the Richardson data are pink, blue and green, in decreasing order, corresponding to σ -values of 0.2, 0.5 and 1, respectively, for each porosity value in the data set. Physically this also makes sense, since it would be expected that the slip-flow at the boundaries would decrease for surfaces that are less smooth and as a result the Klinkenberg effect would decrease. The closest prediction to the experimental values is provided by the cases where $\sigma = 0.2$. This can be observed more clearly when the predictions for the tetrakaidecahedron geometric model, which was estimated to represent the data most accurately among the geometric models, is compared to the SSA estimations.

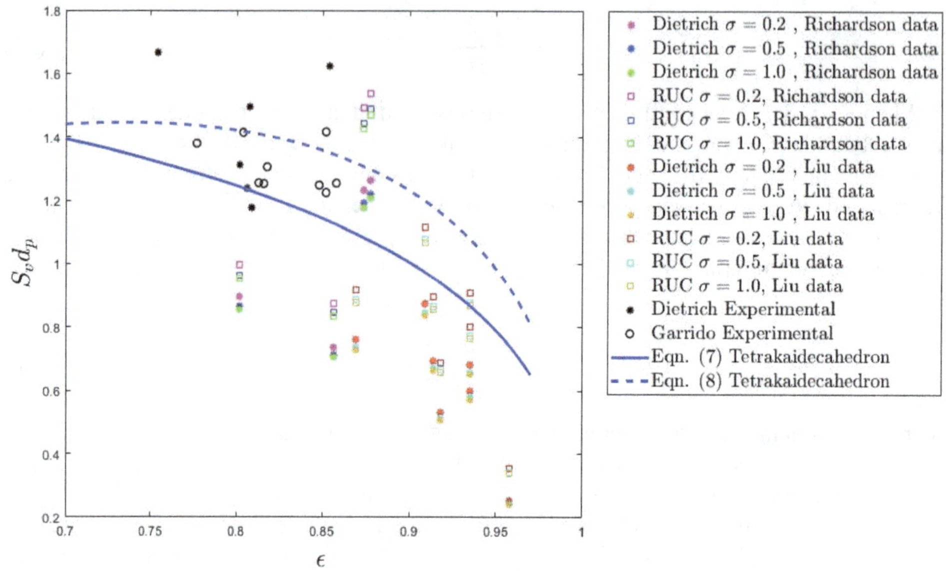

Figure 4: Dimensionless SSA values versus porosity obtained from the Kinetic approaches and implementation of the Klinkenberg effect compared to experimental data from the literature and the tetrakaidecahedron model predictions for $K_n = 10^{-3}$.

The predicted values for the SSA shown in Fig. 4 has a relative percentage difference range of 0.6% to 5.3% for all three σ-values when compared to the predicted values obtained in the Kinetic approach. It furthermore shows a better correspondence with the trend of the experimental data when K_n is incorporated. When the upper bound of estimated slip-flow occurrence along with a continuum approach is considered, i.e. when $K_n = 10^{-1}$, the percentage difference range is 48.3% to 243.5%, which is significantly higher, as expected.

In conclusion, the Klinkenberg effect can have an influence if the Knudsen number is high enough. The question would now, however, be whether the Knudsen number can reach the required interval for it to have an effect. In Miguel and Serrenho [18] the influence of gas properties on the permeability coefficient in porous media was investigated. They determined that air has a mean free path of $\lambda = 70$ nm and for helium $\lambda = 200$ nm. Miguel and Serrenho [18] then calculated K_n for sand and fibrous material, by using the latter λ-values and eqn (27). Once again letting D_h be equivalent to d_p, we can determine the approximate order of Knudsen number in the data used in this study. For the data of Richardson et al. [12] and Lui et al. [16] the order of magnitude of d_p is 10^{-3}. The Knudsen number for air flow would therefore be of an order of approximately 10^{-5}, which is too small to truly have an effect. Similarly, the Knudsen number for helium flow through the metal foams would be of an order of approximately 10^{-4}, which is also insignificant. Other data obtained for stainless steel fibres by Manzo et al. [19] has an order of magnitude of 10^{-6} for d_p, which results in the Knudsen number for air and helium flow being of the order of magnitude 10^{-2} and 10^{-1}, respectively. Since foams are considered to fall under the category of fibres and not granular media [20], it can therefore be concluded that the Klinkenberg effect can affect the SSA results for foams of small enough diameter, i.e. for pore diameters of the order of 10^{-6} or less.

6 CONCLUSIONS

The geometric model that compared most favourably to the experimental data for the specific surface area in the geometric approach was the geometrically most complex tetrakaidecahedron model. In the kinetic approach the RUC and Dietrich predictions were the most accurate. The Klinkenberg effect did have a measure of influence in the data considered in this study, but it was regarded as negligible due to the relatively large pore sizes. For other finer materials, i.e. for pore diameters of the order of 10^{-6} or less., however, the effect is expected to produce more significant deviations from liquid flow.

ACKNOWLEDGEMENTS

The support of the South African DST-NRF Centre of Excellence in Mathematical and Statistical Sciences (CoE-MaSS) towards this research is hereby acknowledged. Opinions expressed and conclusions arrived at, are those of the authors and are not necessarily to be attributed to the CoE.

REFERENCES

[1] Mancin, S., Zilio, C., Cavallini, A. & Rossetto, L., Pressure drop during air flow in aluminum foams. *International Journal of Heat and Mass Transfer*, **53**(15–16), pp. 3121–3130, 2010.

[2] Bhattacharya, A., Calmidi, V. & Mahajan, R., Thermophysical properties of high porosity metal foams. *International Journal of Heat and Mass Transfer*, **45**(5), pp. 1017–1031, 2002.

[3] Inayat, A., Freund, H., Zeiser, T. & Schwieger, W., Determining the specific surface area of ceramic foams: The tetrakaidecahedra model revisited. *Chemical Engineering Science*, **66**(6), pp. 1179–1188, 2011.

[4] Topin, F., Bonnet, J., Madani, B. & Tadrist, L., Experimental analysis of multiphase flow in metallic foam: Flow laws, heat transfer and convective boiling. *Advanced Engineering Materials*, **8**(9), pp. 890–899, 2006.

[5] Edouard, D., Lacroix, M., Pham, C., Mbodji, M. & Pham-Huu, C., Experimental measurements and multiphase flow models in solid SiC foam beds. *American Institute of Chemical Engineers,* **54**(11), pp. 2823–2832, 2008.

[6] Giani, L., Groppi, G. & Tronconi, E., Mass-transfer characterization of metallic foams as supports for structured catalysts. *Industrial and Engineering Chemistry Research,* **44**(14), pp. 4993–5002, 2005.

[7] Huu, T., Lacroix, M., Pham Huu, C., Schweich, D. & Edouard, D., Towards a more realistic modeling of solid foam: Use of the pentagonal dodecahedron geometry. *Chemical Engineering Science*, **64**(24), pp. 5131–5142, 2009.

[8] Woudberg, S. & Du Plessis, J., An analytical Ergun-type equation for porous foams. *Chemical Engineering Science*, **148**, pp. 44–54, 2016.

[9] Lacroix, M., Nguyen, P., Schweich, D., Pham Huu, C., Savin-Poncet, S. & Edouard, D. Pressure drop measurements and modeling on SiC foams. *Chemical Engineering Science*, **62**(12), pp. 3259–3267, 2007.

[10] Incera Garrido, G., Patcas, F., Lang, S. & Kraushaar-Czarnetzki, B., Mass transfer and pressure drop in ceramic foams: A description for different pore sizes and porosities. *Chemical Engineering Science,* **63**(21), pp. 5202–5217, 2008.

[11] Buciuman, F. & Kraushaar-Czarnetzki, B., Ceramic foam monoliths as catalyst carriers. 1. Adjustment and description of the morphology. *Industrial and Engineering Chemistry Research*, **42**(9), pp. 1863–1869, 2003.

[12] Richardson, J., Peng, Y. & Remue, D., Properties of ceramic foam catalyst supports: pressure drop. *Applied Catalysis A: General*, **204**(1), pp. 19–32, 2000.

[13] Gibson, L.L. & Ashby, M.F. (eds), *Cellular Solids, Structure and Properties*, Pergamon Press: Oxford, 1988.

[14] Wilms, J.M., On the hydrodynamic permeability of foamlike media. MSc Eng. thesis, Stellenbosch University, South Africa, 2006.

[15] Dietrich, B., Schabel, W., Kind, M. & Martin, H., Pressure drop measurements of ceramic sponges—determining the hydraulic diameter. *Chemical Engineering Science*, **64**(16), pp. 3633–3640, 2009.

[16] Liu, J., Wu, W., Chiu, W. & Hsieh, W., Measurement and correlation of friction characteristic of flow through foam matrixes. *Experimental Thermal and Fluid Science*, **30**(4), pp. 329–336, 2006.

[17] Woudberg, S. & Du Plessis, J.P., Predicting the permeability of very low porosity sandstones. *Transport in Porous Media*, **73**, pp. 39–55, 2008.

[18] Miguel, A.F. & Serrenho, A., On the experimental evaluation of permeability in porous media using a gas flow method. *Journal of Physics D: Applied Physics*, **40**, pp. 6824–6828, 2007.

[19] Manzo, G., Wu, Y., Chase, G. & Goux, A., Comparison of nonwoven glass and stainless steel microfiber media in aerosol coalescence filtration. *Separation and Purification Technology*, **162**, pp. 14–19, 2016.

[20] Jackson, G.W. & James, D.F., The permeability of fibrous porous media. *The Canadian Journal of Chemical Engineering*, **64**, pp. 364–374, 1986.

BED EXPANSION CHARACTERISTICS IN MICROJET AND VIBRATION ASSISTED FLUIDIZATION OF TIO_2 NANOPARTICLES

KEJU AN[1] & JEAN M. ANDINO[2,3]
[1]Mechanical Engineering, Arizona State University, USA
[2]Chemical Engineering, Arizona State University, USA
[3]Civil, Environmental and Sustainable Engineering, Arizona State University, USA

ABSTRACT

A microjet and vibration assisted (MVA) fluidized bed was designed and tested to improve the fluidization quality of nanosized particles. Previous work relied on the addition of alcohol support in order to minimize electrostatic forces, thereby improving fluidization. The MVA method increased the fluidized bed height smoothly without the need for alcohol support. Thus, the MVA system is more versatile for use where the addition of a chemical may impact the process (e.g. environmental remediation). Commercial, nanosized TiO_2 particles (P25 from Evonik Degussa) were analyzed in the MVA system. It was operated at its resonance frequency (50Hz), under different vibrational amplitude levels. The microjet was operated using a flow of nitrogen at 135.8, 170.3, 204.7, 239.2, 273.7 and 308.2 kPa resulting in a downward stream through the micronozzle (500 µm). The additional vibrational support enabled nanoparticle fluidization under lower inlet pressure and velocity conditions compared to systems where only a microjet was used. In the MVA system, the micronozzle velocity coupled with a vibrational intensity of 1.6 showed 5 times higher fluidized bed height compared to the packed bed. Optimal operating/processing conditions of the MVA system are suggested based on the experimental results. The results presented provide valuable parameters to develop computer simulations of large commercial and pilot scale fluidized bed systems.

Keywords: fluidization, nondimensional-height, microjet, vibration.

1 INTRODUCTION

The fluidization of nanoparticles in a gas has been broadly used in various industrial applications due to its exceptional ability of continuous mixing, large gas–solid contact surface area per unit mass, and high heat and mass transfer rate. Nanosized particles spontaneously lead to the formation of large agglomerates due to the large interparticle forces. Particularly with agglomerate bubbling fluidization (ABF) types of particles like TiO_2 P25, fluidization is described by highly restricted bed expansion and large bubble formation when only using the superficial velocity to fluidize the particles [1] This is the reason why previous researchers applied assistance methods such as vibration or a microjet to obtain smooth fluidization behavior [2]–[4].

We have studied the behaviour of TiO_2 nanoparticles in a microjet and vibration assisted (MVA) fluidized bed. This new experimental method aims to enhance the fluidization process in the absence of alcohol support. The nondimensional bed height was measured as a parameter of fluidization quality and the Richardson and Zaki's model [5] is used to examine the advantages of the MVA system. The theoretical Richardson and Zaki equation may be used to characterize the fluid-like regime that exists when gas is used to fluidize nanoparticles. By plotting the void fraction against the superficial gas velocity in the Richardson and Zaki form, the value of the exponent, n, which is a measure of bed expansion, can be determined.

WIT Transactions on Engineering Sciences, Vol 123, © 2019 WIT Press
www.witpress.com, ISSN 1743-3533 (on-line)
doi:10.2495/MPF190191

This work compares the MVA system to a vibrating fluidized bed (VFB) to show the advantages. Ultimately, the goal is to use the results of the current work to enhance the design of a photocatalytic chemical reactor for environmental remediation purposes.

2 EXPERIMENTAL SETUP

The experiments were performed in a circular column reactor made of clear cast acrylic with an internal diameter of 7.62 cm and a height of 1 m. The bottom part of the reactor consists of a gas distributor which is a sintered, 2 mm thick quartz plate having a pore size of 20 microns with 40% opening area. The nitrogen leaving the compressed gas cylinder passed through a water bubbler that connected to the gas distributor at the bottom of the reactor. The exhaust gas through the reactor was filtered by a HEPA filter to remove any elutriated nanoparticles as shown in Fig. 1.

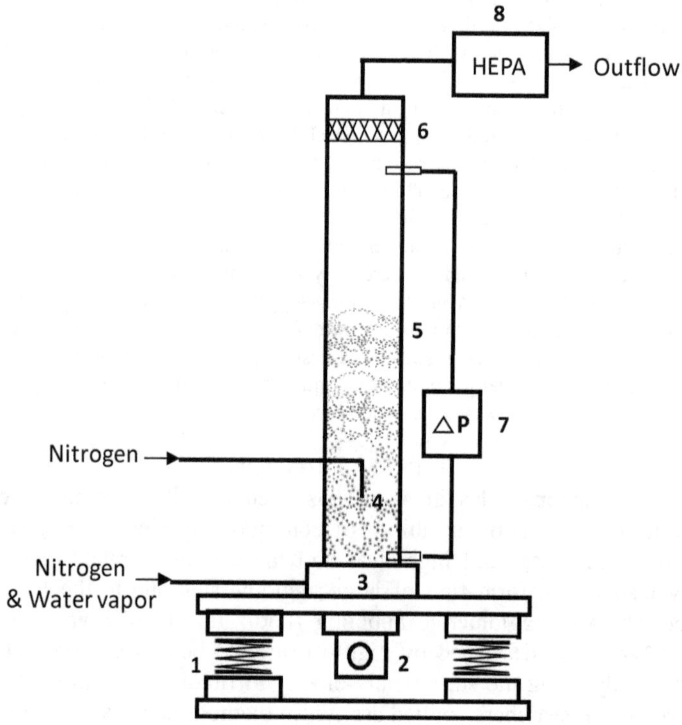

Figure 1: Schematic diagram of experimental setup. 1: Vibration isolator; 2: Magnetic vibrator; 3: Gas distributor; 4: Microjet; 5: Fluidization reactor; 6: Pre-filter; 7: Pressure manometer; 8: HEPA filter.

The electromagnetic vibrator was mounted below the wooden table that supported the column reactor. Vibration was controlled by a signal generator which provided vertical sinusoidal motions with controlled amplitude and frequency.

Dry nitrogen was supplied by a compressed nitrogen gas cylinder, and a mass flow controller (MFC) was used to regulate the gas flow rate through the bottom gas distributor. This flow was considered as the primary gas flow. The secondary gas flow passed through the downward pointing micro-sized nozzle at a height of about 10 cm above the distributor.

2.1 Materials

The commercial nanopowder, TiO_2 P25, manufactured by Degussa Evonik was used for this work. The nanopowder was sieved using a 500 μm sieve placed on a vibration shaker in order to remove large agglomerates that may have been gradually enlarged during packing, storage, and transportation. The properties of TiO_2 P25 are: primary particle size 21 nm, surface area 50 m^2/g and tap density 130 kg/m^3 [6].

2.2 Operating conditions

Nanosized powder was fluidized with nitrogen gas at a superficial velocity range from 0.005 to 0.035 m/s, supplied to the fluidized bed through a gas distributor at the bottom of the reactor. The resonance frequency of 50 Hz and vibrational amplitude were monitored using a vibration meter attached to the reactor. The vibrator was turned on before the microjet was started in order to avoid large bubble and channel formation. The fluidization behaviour was investigated by increasing the superficial gas velocity in small steps (0.2 cm/s), starting from 0.5 cm/s. Additional experimental conditions appear in Table 1.

Table 1: Experimental condition.

Powder mass (g)	50
Vibrational frequency (Hz)	50
Vibrational intensity	1, 1.2, 1.4, 1.6, 1.8, 2
Superficial gas velocity (m/s)	0.005–0.033
Microjet upstream pressure (kPa)	135.8, 170.3, 204.7, 239.2, 273.7, 308.2

3 RESULTS AND DISCUSSION

3.1 Bed expansion behavior

Fig. 2 shows the variation in nondimensional height (H/H_0) as a function of the gas velocity for the VFB and MVA fluidization systems. The nondimensional heights are obtained by dividing the fluidized bed height, H by the initial bed height, H_0 (=0.12 m). The MVA system's nondimensional height was 5 when gas velocity reached 0.02 m/s, which is a larger nondimensional height as compared to the VFB system's nondimensional height of 2.5. In both cases, the fluidized bed height increased as the gas velocity increased.

The Richardson and Zaki equation (shown in eqn (1) [7]) correlates the superficial gas velocity and the void fraction of the fluidized bed, and our results have been shown to be consistent with the trend that the fluidized bed expands uniformly as the superficial gas velocity is increased

$$U_g = U_{pt}\, \varepsilon_g^{\,n}, \tag{1}$$

where U_g = superficial gas velocity; U_{pt} = terminal velocity; ε_g = void fraction of fluidized bed; n = Richardson and Zaki exponent.

Eqn (2) shows the overall mass balance of the powder in the bed where ρ_a and ε_g indicate the bulk density of the powder and the void fraction of fluidized bed, and ρ_{a0} and ε_{g0} indicate the initial powder density and the initial void fraction of fluidized bed, respectively. The density of the powder does not vary under the experimental conditions imposed by the gas flow and the additional vibration [8]. Thus, eqn (3) applies

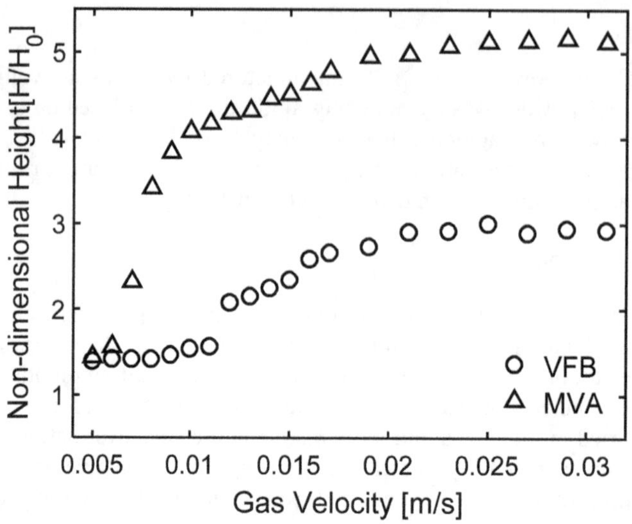

Figure 2: Nondimensional height of TiO_2 P25 as a function of superficial gas velocity.

$$\rho_a(1 - \varepsilon_g)HA = \rho_{a0}(1 - \varepsilon_{g0})H_0A, \tag{2}$$

where ρ_a = density of agglomerate; ρ_{a0} = initial density of agglomerate; ε_{g0} = void fraction of packed bed; A = cross sectional area of column

$$\rho_a = \rho_{a0}. \tag{3}$$

Eqns (2) and (3) can be combined to generate eqn (4), which shows that the void fraction of the fluidized bed can be calculated from the nondimensional height and initial void fraction of the packed bed

$$\varepsilon_g = 1 - (H_0 / H)(1 - \varepsilon_{g0}). \tag{4}$$

The relationship between U_g and ε_g, eqn (1) can be written as a linear equation (eqn (5))

$$\ln U_g = \ln U_{pt} + n \ln\varepsilon_g. \tag{5}$$

From the nondimensional height, we can simply calculate the void fraction of fluidized bed at each gas velocity using eqn (4). We used the gas velocity range up to 0.02 m/s (i.e. before the plateau region) where nondimensional height is increasing continuously to be able to employ the Richardson and Zaki equation which describes the bed expansion as the gas flow is increased. Fig. 3 shows the data plotted according to eqn (5). The lines in Fig. 3 are the linear fits of the data points, and the slope represents the Richardson Zaki exponent "n", is an empirical parameter that can be directly related to Re_t, the Reynolds number of the particle, under terminal settling conditions. Rowe [9] fitted their experimental data by the relation of $(4.7\text{-}n)/(n\text{-}2.35)=0.1Re_t^{3/4}$ and Garside and Al-Dibouni [10] proposed the empirical relation $(5.1\text{-}n)/(n\text{-}2.7)=0.1Re_t^{0.9}$.

In this work, we determined the exponent (n) values for TiO_2 nanoparticle fluidization in the VFB and MVA system; these were 3.01 and 2.34, respectively. Kramer et al. [11] stated that the Richardson Zaki index, n, tends to approach 2.3 for the inertial regime and 4.6 for the viscous regime. Based on the values that were obtained, the MVA system is operating in the inertial regime, corresponding to high particle Reynolds number, suggesting that the

MVA system has improved fluidization as compared to the VFB system. The MVA system resulted in a fluidized bed that expanded smoothly and was stabilized by a vibrational intensity of 1.6.

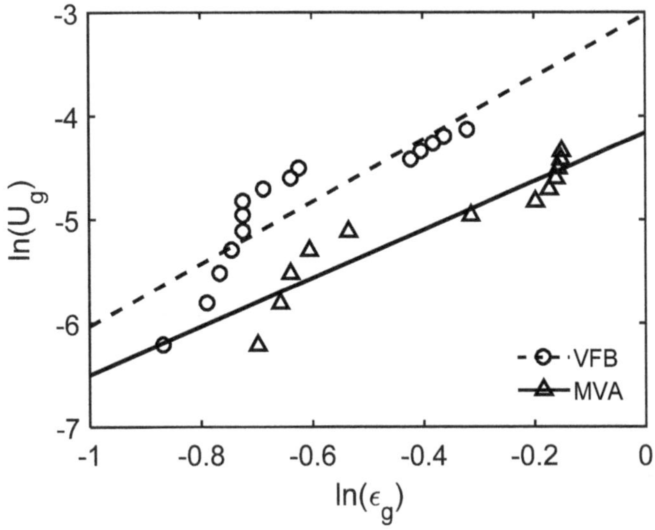

Figure 3: Logarithmic plot of superficial gas velocity against bed voidage derived theoretically. The points represent the theoretical data, while the lines are the linear fit.

3.2 Influence of microjet

In this work, the pressure upstream of the 500 μm micronozzle was varied between 135.8 and 308.2 kPa to show the influence of microjet assistance on the bed height. Theoretically, the flow through the micronozzle is said to be choked at pressures greater than 204.7 kPa upstream, with a surrounding atmospheric pressure of 101.3 kPa. The resulting flow through the micronozzle becomes sonic if we assume adiabatic, frictionless, compressible flow of nitrogen [12]. We performed the experiments under the conditions as shown in Table 1. The vibrational intensity, Γ, is a nondimensional value of the ratio of vibrational acceleration to gravitational acceleration, as indicated in eqn (6) [3]. The parameter f is the frequency in Hz, as set with the signal controller

$$\Gamma = (Aw^2)/g, \tag{6}$$

where A = amplitude of vibration, as adjusted and measured by vibration meter; w = angular velocity, $w = 2\pi f$; f = the frequency in Hz; g = gravitational acceleration, 9.8 m/s^2.

Fig. 4 shows the nondimensional height profile as a function of vibrational intensity at different upstream pressures of nitrogen that are fed into the micronozzle. When the gas velocity and the vibrational frequency were, respectively, 0.02 m/s and 1.6, the nondimensional height hit the maximum value. When the frequency was varied from the resonant frequency of 50 Hz to the antiresonant frequency of 60 Hz, the non-dimensional bed height decreased. It should be noted that at higher pressures, variation in the bed height occurred due to the high flowrate from the microjet. This is illustrated in the error bars that are included with the data points in Fig. 4. The results indicate that the MVA system achieved higher nondimensional heights using lower upstream pressures (and hence lower micronozzle flow rates). However, when the micronozzle flow was completely turned off, the nondimensional height went down to 2.5, as shown in Fig. 2. Quevedo et al. [4] employed a similarly sized micronozzle as in the current work, but employed an alcohol solution and a nitrogen gas pressure through the micronozzle of 239.2 kPa (i.e. sonic flow conditions); they achieved stable fluidization with bed expansion of up to six times the original bed height. As seen in Fig. 4, in the MVA system, subsonic velocity through the micronozzle was sufficient to achieve fluidization (and preferred in terms of enhancing the bed expansion in the absence of alcohol support) due to the added vibrational assistance. This result is particularly significant in terms of enabling more applications that would benefit from nanoparticle fluidization but where relatively pristine chemical conditions are required.

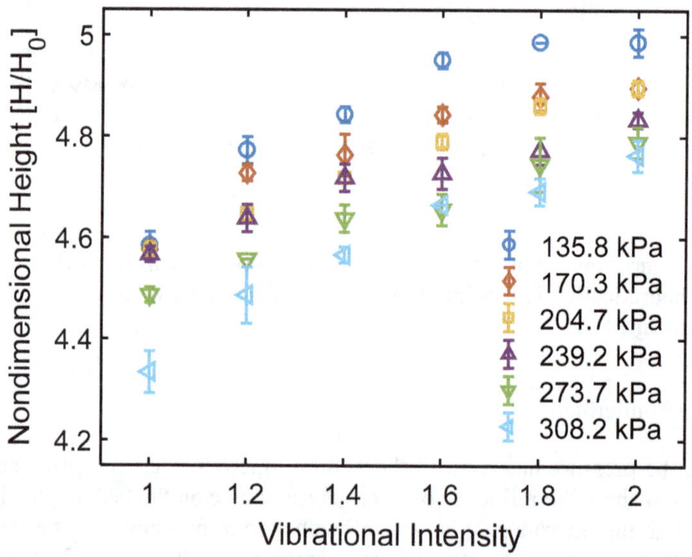

Figure 4: Nondimensional height as a function of vibrational intensity at different upstream pressures through the microjet (frequency: 50 Hz).

ACKNOWLEDGEMENT

The authors gratefully acknowledge Arizona State University, Ira A. Fulton Schools of Engineering for research funding and graduate student support.

REFERENCES

[1] Wang, Y., Gu, G., Wei, F. & Wu, J., Fluidization and agglomerate structure of SiO_2 nanoparticles. *Powder Technology,* **124**(1–2), pp. 152–159, 2002.

[2] Mawatari, Y., Ikegami, T., Tatemoto, Y. & Noda, K., Prediction of agglomerate size for fine particles in a vibro-fluidized bed. *J. Chem. Eng. Jpn.,* **36**(3), pp. 277–283, 2013.

[3] Nam, C.H., Pfeffer, R., Dave, R.N. & Sundaresan S., Aerated vibrofluidization of silica nanoparticles. *AIChE J.*, **50**(8), pp. 1776–1785, 2004.

[4] Quevedo, J.A., Omosebi, A. & Pfeffer, R., Fluidization enhancement of agglomerates of metal oxide nanopowders by microjets. *AIChE J.*, **56**(6), pp. 1456–1468, 2010.
[5] Richardson, J.F. & Zaki, W.N., Sedimentation and fluidisation. Part 1. *Trans. Inst. Chem. Eng.*, **32**, pp. 35–53, 1954.
[6] Evonik Industries, *AEROXIDE®, AERODISP® and AEROPERL® Titanium Dioxide as Photocatalyst*, pp. 5–6, 2018.
[7] Wang, H., Zhou, T., Yang, J.S., Wang, J.J., Hiroyuki, K. & Mawatari, Y., Model for calculation of agglomerate sizes of nanoparticles in a vibro-fluidized bed. *Chem. Eng. Technology*, **33**(3), pp. 388–394, 2010.
[8] Wang, Y., Gu, G., Wei, F. & Wu, J., Fluidization and agglomerate structure of SiO_2 nanoparticles. *Powder Technology,* **124**, pp. 152–159, 2002.
[9] Rowe, P.N., A convenient empirical equation for estimation of the Richardson and Zaki exponent. *Chem. Eng. Sci.*, **42**(11), pp. 2795–2796, 1987.
[10] Garside, J. & Al-Dibouni, M.R., Velocity-voidage relationships for fluidization and sedimentation in solid-liquid systems. *Ind. Eng. Chem. Process Des. Develop.*, **16**(2), pp. 206–214, 1977.
[11] Kramer, O.J.I., Mdeol, P.J. & Van der Heok, J.P., Improvement of the Richardson–Zaki liquid–solid fluidization model on the basis of hydraulics. *Powder Technology,* **343**, pp. 465–478, 2019.
[12] Pope, S.B., *Turbulent Flows: Free Shear Flows*, Cambridge University Press: Cambridge, 2000.

EXPANSION PROCESSES AND STRUCTURES OF TWO-PHASE FLOWS IN AERATED-LIQUID JETS DISCHARGED INTO A QUIESCENT ENVIRONMENT

KUO-CHENG LIN[1], ALAN KASTENGREN[2] & CAMPBELL CARTER[3]
[1]Taitech, Inc., USA
[2]Argonne National Laboratory, USA
[3]Air Force Research Laboratory, USA

ABSTRACT
Expansion processes and near-field structures in the two-phase flow within an aerated-liquid jet injected into a quiescent environment were experimentally explored with the confocal x-ray fluorescence technique available at Argonne National Laboratory. Quantitative time-averaged liquid and gas density distributions within the aerated-liquid jet were spatially resolved simultaneously. For the present injection condition, the observation of a dome-shaped gas plume head and a high gas density gradient within a short distance from the nozzle exit clearly indicates that the initial gas expansion processes across the nozzle exit are highly similar to those within a typical under-expanded gaseous jet. With the assistance of compressible gas expansion near the nozzle exit, the liquid plume exhibits a plume width larger than that of the aerating gas plume, leading to a separation between liquid and gas plumes at the downstream location. Expansion of the gas plume creates a region of low gas density, followed by a region of gas density recovery in the near field. The low gas density region resembles the region with over-expanded gas in front of a Mach disk inside a typical under-expanded gaseous jet. The density variation within the low density region, however, is small, indicating a low level of over expansion and a weak Mach disk. The present analysis of Mach disk location within the discharged aerating gas plume shows that, while the location can be generally identified from gas density measurements, the use of correlations for location identification may be unreasonable, due to the lack of direct measurement data on characteristic pressures.
Keywords: two-phase flow, aerated-liquid jet, confocal x-ray fluorescence, under-expanded gaseous jet.

1 INTRODUCTION

Two-phase flow generation and handling are of great interest in applications in the petroleum, civil, chemical, nuclear, propulsion, and power industries. For instance, liquid (or effervescent, or barbotage) aeration is a plausible scheme to enhance liquid atomization for painting, agriculture sprays, or propulsion/power systems [1]. By creating a two-phase mixture inside an aerated-liquid injector using a small amount of gas mixed with the liquid fuel, the liquid atomization processes can be greatly accelerated, for the generation of a well-dispersed plume for effective mixing with the ambient air and, therefore, for efficient combustion [2], [3]. Successful characterization of structures and expansion processes in the two-phase flow discharged from an aerated-liquid injector or other device can greatly advance these applications.

In order to better characterize the discharged two-phase flow, which is compressible and can be optically dense, capable in-situ diagnostic techniques are required. Recently, two-phase flow structures of aerated-liquid jets have been successfully explored with various x-ray diagnostic techniques, including high-speed x-ray phase contrast imaging (PCI), x-ray radiography, and x-ray fluorescence [4]–[7]. In particular, the study of Lin et al. [7] utilized the x-ray fluorescence technique to quantitatively characterize the line-of-sight (*LOS*) liquid and gas mass distributions within the discharged aerated-liquid jets. For the injection

WIT Transactions on Engineering Sciences, Vol 123, © 2019 WIT Press
www.witpress.com, ISSN 1743-3533 (on-line)
doi:10.2495/MPF190201

condition with a high injection pressure, the discharged two-phase flows exhibit gas-jet-like expansion processes in the aerating gas. In addition, the gas plume is narrower than the liquid plume. The expansion processes, however, were not further explored.

The capability of x-ray diagnostics to characterize the structures of an under-expanded argon gas jet was demonstrated in the study of Scarcelli et al. [8]. The Mach disk location was clearly identified from the *LOS* gas density contours, using x-ray radiography. This approach to identify the possible shock structure within a two-phase jet can, however, be problematic, due to the liquid, which has high x-ray absorbance within the line of sight.

The confocal x-ray fluorescence technique was recently applied by Lin et al. [9] to spatially resolve both liquid and gas mass distributions within discharged aerated-liquid jets. Quantitative cross-sectional mass distribution contours were obtained at several axial locations in a quiescent environment. The phenomenon of a narrower gas plume persisting inside the liquid plume was confirmed. No efforts were made, however, to spatially resolve shock structures within the aerated-liquid jet.

The objective of the present study is to use the confocal x-ray fluorescence technique to spatially resolve both liquid and gas mass distributions, with emphasis on exploring the expansion processes within the near field of an aerated-liquid jet.

2 EXPERIMENTAL METHODS

2.1 Injector assembly

The experiment was conducted at the 7-BM beamline of the Advanced Photon Source (APS) at Argonne National Laboratory. For the present study, the discharged spray was placed in the path of a focused x-ray beam. Liquid and aerating gas were supplied into the aerated-liquid injector at the desired flow rates to form a two-phase mixture inside the injector before discharge into a quiescent environment. The aerated-liquid injector features the *outside-in* aeration configuration, with the aerating gas flowing through an annular passage before entering the central liquid stream for mixing. Fig. 1(a) shows a schematic of the injection assembly, which has a stainless steel injector body, an aerating tube, and an exchangeable beryllium nozzle. The aerating tube has an internal diameter of 2.2 mm and features 32 small orifices located in 16 rows with two 0.18 mm diameter orifices in each row, as shown in Fig. 1(b). Orifices in each row are offset by 90 degrees from the orifices in adjacent rows,

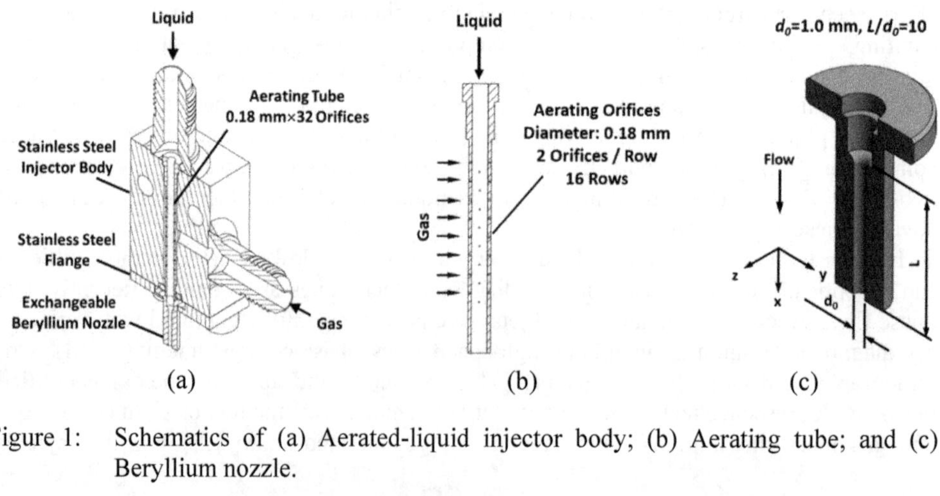

Figure 1: Schematics of (a) Aerated-liquid injector body; (b) Aerating tube; and (c) Beryllium nozzle.

resulting in an orifice distribution pattern in four columns on the aerating tube. The beryllium nozzle used for the present study features a contoured entrance, followed by a cylindrical passage with a constant diameter (d_0) of 1.0 mm and passage length (L) of 10 mm, and thus an L/d_0 of 10. The nozzle connects smoothly with the aerating tube via a short plenum section with a diameter of 2.2 mm. A schematic of this nozzle is shown in Fig. 1(c).

The aerated-liquid jet was discharged vertically into a collecting bucket with a small opening on the cap to minimize the probability of stray droplets entering the beam path. The distance between the nozzle exit and the bucket cap was kept around 5 mm, in order to allow probing of the discharged jet and at the same time avoid splashing. Both the aerated-liquid injector and the collecting bucket were rigidly mounted on a traversing table, which provided movement normal to the x-ray beam.

For the present study, the injection condition consists of a liquid flow rate, m_L, of 18.2 g/s and a gas-to-liquid mass ratio, GLR, of 6.72%. The injection pressures (P_{inj}) and temperatures (T_{inj}) measured just upstream of the injector assembly were 3.69 MPa and 294 K, respectively. The ambient is at atmospheric pressure. At this injection condition with a relatively large P_{inj}/P_∞ of 36.5, the discharged plume is highly dispersed with a fairly large plume width for x-ray probing.

2.2 X-ray measurements

The 7-BM beamline at Argonne National Laboratory is dedicated to ultrafast x-ray radiography, fluorescence, and tomography experiments in fuel sprays and associated phenomena. The x-ray source for the beamline is a synchrotron bending magnet, which produces nearly collimated, broadband x-ray emission. The beamline consists of two radiation enclosures. The first enclosure (7-BM-A) houses a pair of slits to limit the x-ray beam size and a double multilayer monochromator (1.2–4.3% ΔE/E) to filter the beam. The resulting monochromatic beam then passes into the second radiation enclosure (7-BM-B), which houses the experimental equipment. More information regarding the beamline performance can be found in [10].

Fluorescence of x-rays from tracer elements in the liquid and gas is caused by the relaxation of atoms to a more stable state after the ionization of a core-shell electron due to absorption of an incident x-ray photon. The energy of the emitted photons is characteristic of the atomic number of the atoms emitting the photons, and the emitted fluorescence flux is proportional to the amount of fluorescent material in the beam. X-ray fluorescence is commonly used for nondestructive elemental analysis. More detail of the x-ray fluorescence process is given in [11] and [12].

Here, x-ray fluorescence was used to simultaneously measure gas and liquid concentrations in the spray. X-ray fluorescence from the liquid and gas phases can be used to determine the distributions of liquid and gas independently, by doping the working fluids with bromine (*Br*) and krypton (*Kr*) atoms, respectively. Sodium bromide (*NaBr*) was dissolved into the water at a concentration of 0.42% by mass. This concentration was monitored during the experiment by measuring the electrical conductivity of the water, which was approximately 4.5 mS/cm. The gas phase was pre-mixed (by the gas vendor) to a concentration of 6% *Kr* (and 94% N_2) by volume. For the current x-ray fluorescence experiments, the mean x-ray beam photon energy was set at 15 keV; a photon energy of at least 14.3 keV is required to excite both *Br* and *Kr*. The beam was focused using a pair of 300 mm long Kirkpatrick–Baez focusing mirrors. The beam focus is approximately 5 × 6 μm FWHM V × H, located approximately 500 mm from the center of the horizontal focusing mirror. The effective size of the beam for the current sprays (which are several mm wide) is somewhat greater than this minimum focus size. The experimental setup at the 7-BM beamline is shown schematically in Fig. 2.

The fluorescence photons emitting from the media of interest within the focused probe volume were collected by a polycapillary x-ray optic and then measured by a silicon drift diode detector. Fig. 3 shows the present x-ray fluorescence test section setup. The polycapillary x-ray optic has a focal length of 50 mm and a spatial resolution of approximately 200 μm at the Br and Kr fluorescence photon energies, and it was placed in front of the fluorescence detector to filter out all x-rays but those originating from the focal region. The integrated detector assembly was then positioned at 90° relative to the x-ray beam in the horizontal plane to minimize the scattering detected by the detector. The effective probe volume for the present measurements is defined by the focused beam size and the focal spot size of the polycapillary optic: 5 × 6 × 200 μm. With this small probe volume, distributions of liquid and gas mass can be resolved in three dimensions by moving the spray normal to the x-ray beam and the detector and polycapillary together along the x-ray beam.

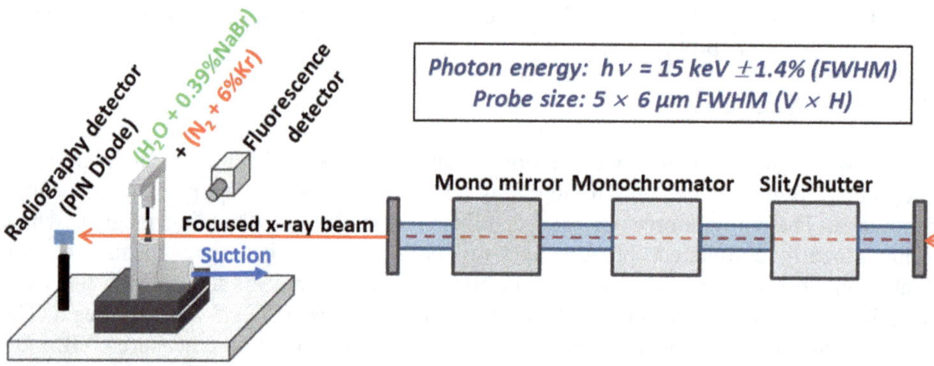

Figure 2: X-ray and injection stand setup at the 7-BM beamline at the Argonne National Laboratory.

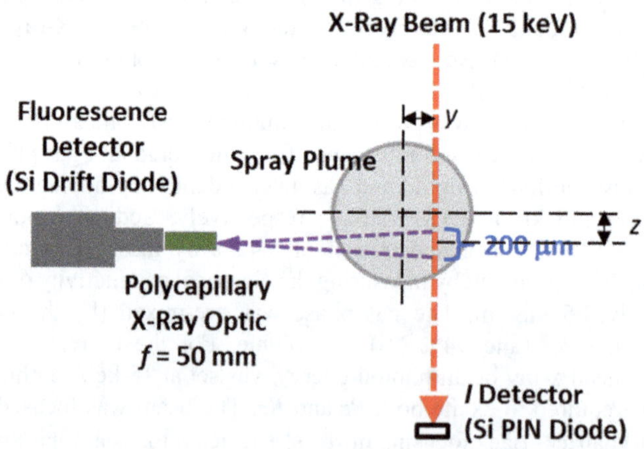

Figure 3: Confocal x-ray fluorescence setup.

The measurement was carried out within the near field of the discharged plume (less than 3.0 mm from the nozzle exit plane, i.e., $x < 3.0$ mm; $x = 0$ at the nozzle exit) with prescribed probing locations within a center sliced plane (x-y plane, y in the transverse direction relative to the injector axis). Calibration scans at known liquid and gas densities were carried out to quantitatively correlate the measured signal to local densities.

3 RESULTS AND DISCUSSION

3.1 Two-phase flow structures

Fig. 4 shows the cross-sectional liquid and gas density contours within the near field of the discharged plume. Please note that in contrast to the previous x-ray fluorescence measurements in the study of Lin et al. [7], the current data are not pathlength integrated. Instead, these data truly represent a cross-sectional slice of the fluid distribution. The liquid density contours exhibit an annular-like liquid distribution pattern near the nozzle exit plane, with the most upstream measurement location being $x = 0.1$ mm. This observation clearly indicates that more liquid stays close to the nozzle passage wall before discharge for the present injection condition. A smaller amount of liquid is distributed in the core region of the two-phase mixture within the nozzle passage and the discharged plume at $x < 0.6$ mm. The discharged liquid is quickly dispersed as the flow progresses downstream, with no distinct annular liquid distribution pattern in the downstream locations (x > 0.6 mm).

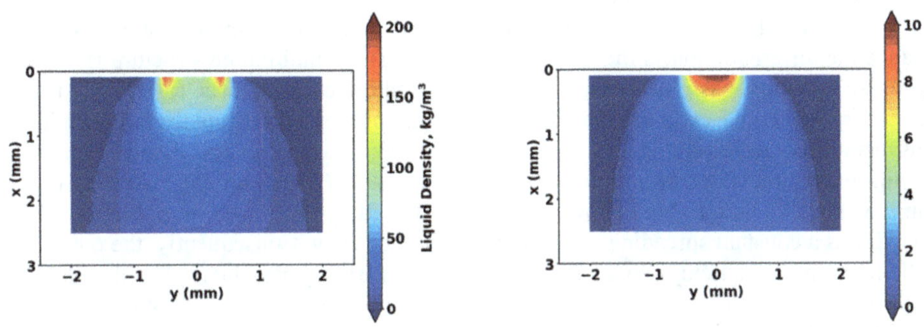

(a) (b)

Figure 4: Cross-sectional (a) Liquid; and (b) Gas density contours in linear scale within the near field of the discharged plume. Injector exit plane at $x = 0$ and injector axis at $y = 0$.

The discharged aerating gas exhibits a dome-shaped region of high density within the near field. The high-density gas quickly expands to a low value within a short distance of $x < 1.2$ mm from the nozzle exit. It appears that the present injection condition with a relatively large (P_{inj}/P_{∞}) of 36.5 exhibits a gas plume structure similar to that of a highly under-expanded gaseous jet. Unlike in the study of Scarcelli et al. [8], however, the typical features of an under-expanded gaseous jet, such as the presence of barrel shock and Mach disk, cannot be observed in the present planar gas density contours.

A gaseous jet injected from a sonic nozzle into a quiescent environment can be classified as subsonic, moderate under-expanded, or highly under-expanded jet, depending on the (P_0/P_{∞}) value of the injection condition, where P_0 is the stagnation pressure in the plenum of the nozzle. Based on the study of Donaldson and Snedeker [13], the range of (P_0/P_∞) values for each type of discharged gaseous jet for air is listed in Table 1. Also listed in the table is the range of (P_{exit}/P_∞) values for each type of discharge jet, where P_{exit} is the local pressure at the nozzle exit. The large (P_{inj}/P_∞) value of 36.5 for the present injection condition indicates that the aerating gas jet is highly under-expanded.

Table 1: Ranges of (P_0/P_∞) and (P_{exit}/P_∞) values for three major variations of gaseous jets discharged from a sonic nozzle into a quiescent environment. *(Source: Donaldson and Snedeker [13].)*

Subsonic jet	Moderate under-expanded jet	Highly under-expanded jet
$1 < P_0/P_\infty < 1.89$	$2.08 < P_0/P_\infty \leq 3.85$	$P_0/P_\infty > 3.85$
$P_{exit}/P_\infty = 1$	$1.1 < P_{exit}/P_\infty \leq 2$	$P_{exit}/P_\infty > 2$

It should be noted that the use of the measured injection pressure, P_{inj}, to present the stagnation pressure, P_0, for the purposes of jet structure classification may be problematic, due to pressure losses during the mixing of liquid and aerating gas within the injector. In addition, the presence of liquid within the gas plume of an aerated-liquid jet may affect the gas expansion processes. Nonetheless, the aerating gas jet seen here exhibits a certain degree of under-expansion, as evidenced by the dome-shaped plume head and a rapid decrease in gas density within a short distance.

Fig. 5 shows the density contours from Fig. 4 plotted in logarithmic scale, in order to enhance the differentiation in the mass distributions. The liquid plume structure is affected by the compressible gas plume. The most pronounced effect takes place right at the nozzle exit, where the initial expansion of the aerating gas pushes liquid rapidly outward in the transverse direction, resulting in a large initial spreading angle for the liquid plume. The aerating gas plume exhibits a dome-shaped plume head to constrain the spreading rate in plume width. The liquid plume, however, is not affected by the change in gas plume shape and exhibits a constant spreading rate within the field of view. Consequently, the outer edge of the liquid plume quickly separates from that of the aerating gas plume. The liquid plume width is larger than that of the aerating gas plume. The narrow gas plume stays within the liquid plume at the end of the probing area.

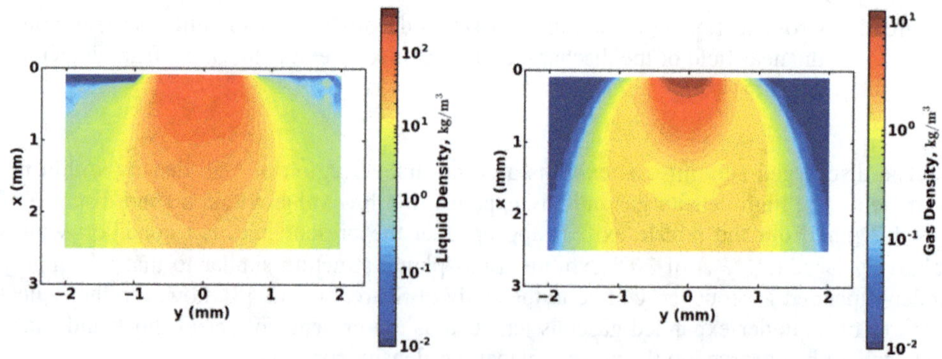

(a) (b)

Figure 5: Cross-sectional (a) Liquid; and (b) Gas density contours in logarithmic scale within the near field of the discharged plume.

In the gas density contours, a region with a noticeably low gas density can be observed between $x = 1.4$ and 2.0 mm, indicating a small degree of over expansion within the aerating gas plume. The slightly over-expanded gas density recovers within a short distance. This observation is similar to the phenomenon associated with pressure rise across a Mach disk in a typical under-expanded gaseous jet.

Fig. 6 shows the radial distribution profiles of liquid and gas densities at selected axial locations. The liquid distribution exhibits an annular-like profile at both $x = 0.1$ and 0.4 mm locations. The annular liquid distribution pattern quickly transitions to a Gaussian distribution pattern around $x = 0.5$ mm. The radial distribution profiles of gas density exhibit a quick reduction within a short distance ($x < 1.2$ mm), followed by a fairly constant distribution pattern with a gradual increase in plume width. A region with a noticeably low gas density (on a linear scale) can be observed around the $x = 1.5$ mm location, indicating a slight gas over expansion, followed by a recovery, as described earlier.

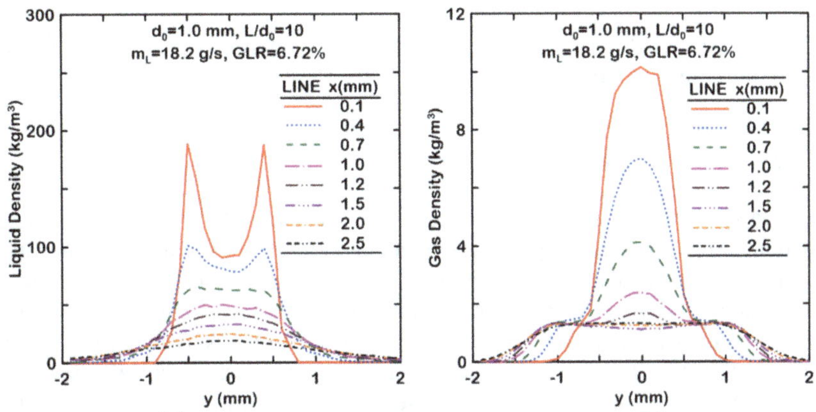

(a) (b)

Figure 6: Radial distribution profiles for (a) Liquid; and (b) Gas densities at selected axial locations.

Fig. 7 shows the axial distribution profiles for liquid and gas densities in linear and logarithmic scales, respectively. The axial liquid density profile decreases continuously from the nozzle exit within the present probing range. The axial gas density profile also exhibits a rapid decrease initially and, with the use of logarithmic scale, reaches a density lower than the ambient gas density of around 1.18 kg/m^3 (as shown in Fig. 7(b)), followed by a gradual recovery to a gas density of 1.32 kg/m^3 at the end of the present probing range. This gas density recovery is not as dramatic as the recovery processes across a Mach disk within a typical under-expanded gaseous jet. It seems that a weak Mach disk or compression wave(s) exists within the present discharged aerating gas jet; the unique features of gas over-expansion and density recovery indicate that the discharged aerating gas jet in the present injection condition weakly resembles an under-expanded gaseous jet.

The presence of liquid phase within the aerating gas plume of the present aerated-liquid jet may greatly affect the expansion processes, retarding the gas expansion processes and limiting gas density variation. The liquid phase may also disrupt shock wave formation at a particular spatial location, rendering the identification of a shock wave in the present time-averaged measurements uncertain or even impossible.

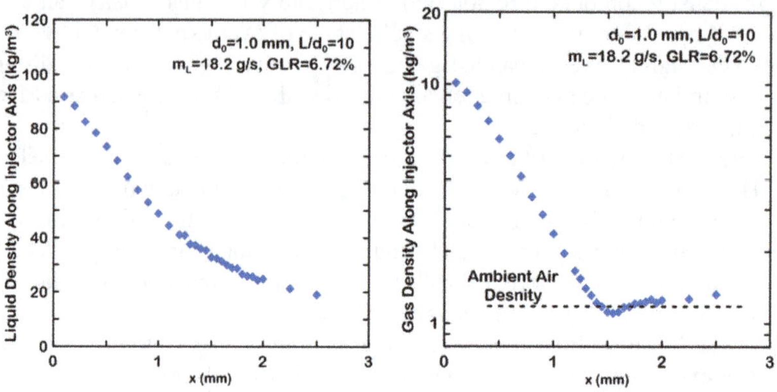

(a) (b)

Figure 7: Axial distribution profiles for (a) Liquid (in linear scale); and (b) Gas (in logarithmic scale) densities.

3.2 Further examination of gas plume structures

The characteristics of a discharged aerated-liquid jet, such as mass distribution profiles, plume width, and plume spreading angle, depend on the two-phase flow structures inside the injector and the subsequent expansion processes of the near-field plume. The present measurements have shown that the initial expansion processes of the discharged liquid are greatly affected by the expansion of the compressible aerating gas. In particular, the aerating gas plume in the near field exhibits features highly similar to a typical under-expanded gaseous jet. If a Mach disk does exist inside the discharged aerating gas plume (even though not identified in the present measurements), it is possible that its location could be predicted from correlations that were developed for predicting Mach disk location within an under-expanded gaseous jet.

Two correlations for the prediction of Mach disk location inside a typical under-expanded gaseous jet, one by Crist et al. [14], and the other by Lewis and Carlson [15], are explored here, along with a correlation for Mach disk diameter by Addy [16]. The correlation of Crist et al. [14] is based on the (P_0/P_∞) ratio, as follows:

$$\frac{x_{MD}}{d_0} = \sqrt{\frac{\left(\frac{P_0}{P_\infty}\right)}{2.4}} = 0.6455 \times \sqrt{\frac{P_0}{P_\infty}}. \quad (1)$$

The parameter x_{MD} is the Mach disk location measured from the nozzle exit. Please note that eqn (1) is independent of gas type or the specific heat ratio (γ) and is a function of the (P_0/P_∞) ratio. It can be readily applied to an under-expanded gaseous jet undergoing an isentropic expansion process. The direct use of P_{inj} to replace P_0 in eqn (1) gives a predicted Mach disk location of x_{MD} = 3.9 mm. This predicted value does not agree with the location of gas density recovery illustrated in Fig. 7(b) and places the Mach disk downstream of the present probing range.

As noted previously, however, it is highly problematic to treat the formation of two-phase mixtures within the aerated-liquid injector as an isentropic process, given pressure losses during the mixing of liquid and aerating gas within the injector. If the Mach disk location is defined as the location where the axial gas density rise from the minimum value intersects with the levelled-off density value, the gradual density recovery in Fig. 7(b) prevents an unambiguous identification of the Mach disk location for the present aerating gas plume. For assessment purposes, therefore, the Mach disk is estimated at the x_{MD} = 1.75 mm location. With this measurement, the effective stagnation pressure, $P_{0,eff}$, which is defined as the stagnation pressure for the present gas plume to generate a Mach disk location from eqn (1) to perfectly match the measurement, is 0.72 MPa. The derived $P_{0,eff}$ is about 20% of the present P_{inj} value. The differences between P_{inj} and $P_{0,eff}$ can be treated as the pressure losses associated with the injector design and the mixing between liquid and aerating gas. Based on the classification criteria with (P_o/P_∞) in Table 1, the present $(P_{o,eff}/P_\infty)$ value of 7.1 indicates that the aerating gas jet is highly under-expanded for the present Mach disk location measurement of x_{MD} = 1.75 mm.

A correlation for Mach disk diameter prediction was proposed by Addy [16], as follows:

$$\frac{d_{MD}}{d_0} = 0.31 \times \sqrt{\frac{P_0}{P_\infty} - 5}. \tag{2}$$

The parameter d_{MD} is the Mach disk diameter. The use of the present $(P_{o,eff}/P_\infty)$ value gives d_{MD} = 0.46 mm. Fig. 8 shows the gas density contour with measured and predicted locations and diameter of the Mach disk. The Mach disk diameter predicted using this correlation is significantly smaller than the measured gas plume width.

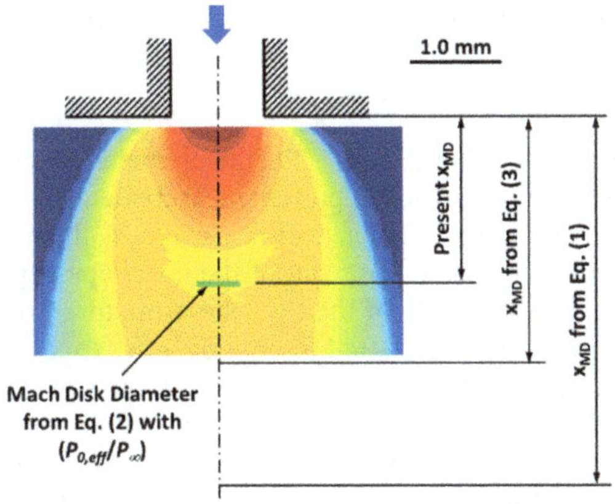

Figure 8: Measured Mach disk location and predicted Mach disk locations and diameter from [14]–[16].

The second correlation for Mach disk location prediction is based on the (P_{exit}/P_∞) ratio of Lewis and Carlson [15], as follows:

$$\frac{x_{MD}}{d_0} = 0.69 \times M_{exit} \times \sqrt{\gamma \times \frac{P_{exit}}{P_\infty}}. \tag{3}$$

The parameter M_{exit} is the Mach number at the nozzle exit. In the present study, P_{exit} was not measured. Instead, the local pressures derived from x-ray fluorescence measurements in the study of Lin et al. [17] inside the same beryllium nozzle with the same injector assembly were used to estimate the P_{exit} value. Fig. 9 (from [17]) shows the local pressure profiles in the axial direction inside the nozzle for the present injection condition. The value of P_{exit} was extrapolated from the third-order polynomial fitting of the derived local pressures. The study of Lin et al. [17] has also shown that the two-phase flows are choked at the nozzle exit, that is, $M_{exit} = 1.0$, mainly due to the reduction in the speed of sound for compressible two-phase mixtures. The present krypton/nitrogen mixture has a γ value of 1.409, which is similar to that of air (1.4).

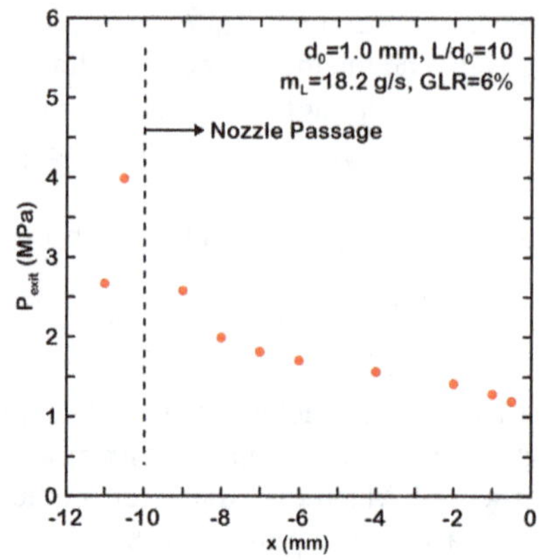

Figure 9: Local pressure within the nozzle section of an aerated-liquid injector. *(Source: Lin et al. [17].)*

With the extrapolated P_{exit} value of 1.08 MPa, the predicted Mach disk location from eqn (3) is $x_{MD} = 2.7$ mm, which is closer to the present measurements but still outside the present probing range, as is shown in Fig. 8. The use of P_{exit} excludes the pressure losses associated with the liquid/gas mixing processes upstream of the nozzle passage and may therefore be more representative of the local state of the two-phase mixture prior to discharge into the ambient. Eqn (3) can therefore be expected to provide a better prediction of the Mach disk location. The use of the derived P_{exit} value may, of course, carry over uncertainties from [17], and should be replaced by directly measured P_{exit} in the future. An effective exit pressure, $P_{exit,eff}$ – which is defined as the exit pressure for the present gas plume to generate a Mach disk location from eqn (3) to perfectly match the measurement – is 0.45 MPa. Table 2 summarizes the ratios between characteristic pressures and the ambient pressure. Based on the classification criteria with (P_{exit}/P_∞) in Table 1, both (P_{exit}/P_∞) value of 10.7 and ($P_{exit,eff}/P_\infty$) value of 4.4 indicate that the aerating gas jet is also highly under-expanded.

Table 2: Measured and derived pressures and ratios with ambient pressure.

P_{inj}	$P_{0,eff}$	P_{exit}	$P_{exit,eff}$	P_{inj}/P_∞	$P_{0,eff}/P_\infty$	P_{exit}/P_∞	$P_{exit,eff}/P_\infty$
3.69 MPa	0.72 MPa	1.08 MPa	0.46 MPa	36.5	7.1	10.7	4.4

4 SUMMARY

Expansion processes and near-field structures of an aerated-liquid jet injected into a quiescent environment were experimentally explored with the confocal x-ray fluorescence technique available at Argonne National Laboratory. Quantitative time-averaged liquid and gas density distributions within the aerated-liquid jet were spatially resolved simultaneously. The major conclusions of the present study are, as follows:

1. For the present injection condition, the observation of a dome-shaped gas plume head and a high gas density gradient within a short distance from the nozzle exit clearly indicate that the initial gas expansion processes across the nozzle exit are highly similar to those within a typical under-expanded gaseous jet.
2. With the assistance of compressible gas expansion near the nozzle exit, the liquid plume exhibits a plume width larger than that of the aerating gas plume, leading to a separation between liquid and gas plumes at the downstream location.
3. Expansion of the gas plume creates a region of low gas density, followed by a region of gas density recovery within the near field. The low gas density region resembles the region with over-expanded gas in front of a Mach disk inside a typical under-expanded gaseous jet. The density variation within the low density region, however, is small, indicating a low level of over expansion and a weak Mach disk.
4. The present analysis of Mach disk location within the discharged aerating gas plume shows that the location can be approximately identified from the present gas density measurements. The use of correlations from the literature to further pinpoint the location poses a significant challenge, due to the lack of direct measurements of characteristic pressures.

Accurate and relevant pressure measurements inside an aerated-liquid injector would greatly improve the understanding of the structures of aerated-liquid jets.

ACKNOWLEDGEMENTS

This work was sponsored by the AFRL/Aerospace Systems Directorate under contract number FA8560-14-D-2316 (Contract monitor: Stephen Smith) and by the Air Force Office of Scientific Research (AFOSR). Use of the Advanced Photon Source at Argonne National Laboratory was supported by the U.S. Department of Energy, Office of Science, Office of Basic Energy Sciences, under Contract No. DE-AC02-06CH11357. The authors would like to thank Travis Tidball, Steve Enneking, Justin Stewart, and Andrew Baron of Taitech, Inc. and Lt. Jonathon Hill of AFRL for their assistance in hardware design, setup, data acquisition, and data reduction.

REFERENCES

[1] Lefebvre, A.H., Wang, X.F. & Martin, C.A., Spray characteristics of aerated-liquid pressure atomizers. *J. Propulsion Power*, **4**, pp. 293–298, 1988.

[2] Sovani, S.D., Sojka, P.E. & Lefebvre, A.H., Effervescent atomization. *Progress in Energy and Combustion Science*, **27**, pp. 483–521, 2001.

[3] Mathur, T. et al., Liquid JP-7 combustion in a scramjet combustor. *AIAA Paper 2000-3581*, 2000.

[4] Lin, K.-C., Rajnicek, C., McCall, J., Carter, C. & Fezzaa, K., Investigation of pure- and aerated-liquid jets using x-ray phase contrast imaging technique. *Nuclear Instruments and Methods in Physics Research Section A*, **649**(1), pp. 194–196, 2011.

[5] Lin, K.-C., Carter, C., Smith, S. & Kastengren, A., Exploration of aerated-liquid jets using x-ray radiography. *AIAA Paper 2012-0347*, 2012.

[6] Lin, K.-C., Carter, C., Smith, S. & Kastengren, A., Characterization of aerated-liquid jets using simultaneous x-ray radiography and x-ray fluorescence measurements. *Proceedings of the 26th Annual Conference on Liquid Atomization and Spray Systems*, 2014.

[7] Lin. K.-C., Carter, C., Kastengren, A. & Peltier, S., Exploration of gas phase properties in aerated-liquid jets using x-ray fluorescence. *AIAA Paper 2015-0165*, 2015.

[8] Scarcelli, R., Kastengren, A., Powell, C., Wallner, T. & Matthias, N., High-pressure gaseous injection: a comprehensive analysis of gas dynamics and mixing effects. *ASME Paper No. ICEF2012-92137*, pp. 793–801, 2012

[9] Lin, K.-C., Kastengren, A. & Carter, C., Exploration of temporal and time-averaged two-phase flow structures using x-ray diagnostics. *Proceedings of the 29th Annual Conference on Liquid Atomization and Spray Systems*, 2017.

[10] Kastengren, A.L., Powell, C.F., Arms, D., Dufresne, E. & Wang, J., Spray diagnostics at the advanced photon source 7-bm beamline. *Proceedings of the 22nd Annual Conference on Liquid Atomization and Spray Systems*, 2010.

[11] Kastengren, A.L. & Powell, C.F., Synchrotron x-ray techniques for fluid dynamics. *Experiments in Fluids*, **55**(3), article 1686, 2014.

[12] Kastengren, A.L., X-ray fluorescence as a gas-phase mixing diagnostic. *Proceedings of the Central States Section of the Combustion Institute Spring Technical Meeting*, 2014.

[13] Donaldson, C.D. & Snedeker, R.S., A study of free jet impingement. Part 1. Mean properties of free and impinging jets. *J. of Fluid Mechanics*, **45**(2), pp. 281–319, 1971.

[14] Crist, S., Sherman, P.M. & Glass, D.R., Study of the highly underexpanded sonic jet. *AIAA Journal*, **4**, pp. 68–71, 1966.

[15] Lewis, C.H. Jr & Carlson, D.J., Normal shock location in underexpanded gas and gas-particle jets. *AIAA Journal*, **2**(4), pp. 776–777, 1964.

[16] Addy, A.L., Effects of axisymmetric sonic nozzle geometry on Mach disk characteristics. *AIAA Journal*, **19**(1), pp. 121–122, 1981.

[17] Lin, K.-C., Bornhoft, B., Kastengren, A. & Carter, C., Derivations of averaged two-phase flow properties using x-ray fluorescence measurements. *AIAA Paper 2018-1547*, 2018.

PROTON RADIOGRAPHY OF EXPLOSIVELY DISPERSED METAL PARTICLES WITH VARYING CARRIER FLUID

KYLE HUGHES[1], KATHY PRESTRIDGE[1], NAM-HO KIM[2],
RAPHAEL HAFTKA[2] & SIVARAMAKRISHNAN BALACHANDAR[2]
[1]Physics Division, Los Alamos National Laboratory, USA
[2]Department of Mechanical and Aerospace Engineering, University of Florida, USA

ABSTRACT

A series of experiments performed at Los Alamos National Laboratory aimed to provide validation data for numerical simulations performed at the Centre of Compressible Multiphase Turbulence at the University of Florida. Five explosive tests were performed. Approximately 2.8 g of PBX-9501 was initiated by a RP-80 detonator. A 13 x 6 mm cylindrical packet of 100 μm steel particles was dispersed by the explosive. Three ambient carrier fluids were investigated: air, xenon, and SF_6. To penetrate the optically opaque explosive products at the early time and track the dispersed particle cloud, radiography was required. Proton radiography performed at the Los Alamos Neutron Science Centre allowed extraction of 21 radiograph images at 2 μs temporal resolution. Upstream and downstream particle fronts were extracted from the transmission radiographs. Centreline particle fronts for the three carrier phases demonstrated close agreement, suggesting the shock traversing the bed of particles provided little additional impulse to the dispersed particles in this regime. An additional shot was performed in vacuum and showed close agreement with the carrier phase shots, furthering this conclusion.

Keywords: multiphase flow, particle bed, explosive dispersal, proton radiography.

1 INTRODUCTION

The experiments considered in this investigation are to demonstrate the predictive ability of the code developed by the Centre of Compressible Multiphase Turbulence (CCMT) at the University of Florida. Compressible multiphase turbulent flows are a challenging and rich topic within the fluid mechanics community. Such a flow may arise in the natural world in the case of dispersal of particulate from a volcanic eruption or in engineering applications such as dispersal of particles by ordinance. Large-scale, exploratory experiments have been performed by Frost and collaborators [1]–[4] examining the behaviour of a bed of particles subjected to a detonation wave. The detonation replaces the central explosive with a high-density, high-temperature gas that propagates outward. A shock wave travels through the particle bed, followed by a contact discontinuity between the shocked air and the explosive products. In the early times, the detonation wave causes many of the particles to collide and compact with each other. As the particles disperse, the experiments observe a series of large-scale jets or fingers that form in the advancing particle front.

Small scale experiments by Sun, Tanno, and collaborators [5]–[7] have measured the forces on a single, stationary particle subjected to a moderate shock. When comparing the experimental results to drag relations proposed as compressible generalizations of the Maxey–Riley–Gatignol equation [8], [9], the generalized force formula captures the magnitude and the timing of the peak force quite well [10]–[12]. Shock-tube experiments examining the behaviour of moving particles have been performed by variety of researchers [13]–[16]. In these experiments, the position of the particle is observed after being subjected to a shock and the time-varying drag force extracted. Together these small scale experiments provide a wealth of validation data in the modest Mach number regime for a single particle for the proposed drag force relations. Hughes et al. [17] attempted to extend the validation

WIT Transactions on Engineering Sciences, Vol 123, © 2019 WIT Press
www.witpress.com, ISSN 1743-3533 (on-line)
doi:10.2495/MPF190211

data for a single particle, where the particles are subjected both to a high Mach number shock and a strong contact discontinuity.

Similar to the investigations of a single particle, other investigations have been performed at where many particles are subjected to a moderate Mach number shock. In this regime, not only are the individual drag forces important, but the particles compact and collide with each other as well. However, the flows still lack the strong contact discontinuity present in detonation conditions. Dilute suspensions of particles ($\varphi < 1\%$) subjected to a shock have been studied extensively by researchers such as Rudinger [18]. Less common is examination of multiphase flows where moderate to large volume fractions are present ($\varphi < 5\%$). Wagner et al. [19] examined shock wave impingement on a free-falling particle curtain of approximately 20% volume fraction by tracking the upstream and downstream particle fronts through high-speed Schlieren photography. Under very similar experimental conditions, Kellenberger et al. [20] examined the behaviour of stationary particle wafers of approximately 48% volume fraction subjected to a shock wave through high-speed photography. Note in these studies the volume fraction is less than the 60% commonly found in particle beds typical for the explosive dispersal of particles [21].

Examining the current level of experiments present in the literature, it was desirable to produce a set of experiments that contained all the physics of the large-scale explosive tests, but much reduced in the number of particles. The large-scale experiments contain O(10^7–10^9) particles, whereas the new experiments are estimated to contain O(10^5) particles. The reduction of particles is a significant reduction in computational burden while still providing validation data within the high volume fraction and high Mach number flow.

2 EXPERIMENTAL SETUP

The experiments were performed at the Los Alamos Neutron Science Centre Proton Radiography facility. The proton radiography apparatus utilizes 800 MeV protons provided by a linear particle accelerator [22]. A brief schematic is shown in Fig. 1. A beam of collimated protons is passed through the object of interest. The protons are guided by a series of magnetic lenses to focus the protons on the image plane, where a scintillator converts the protons to photons. The photons are turned out of the beam path with a mirror and passed to the camera group.

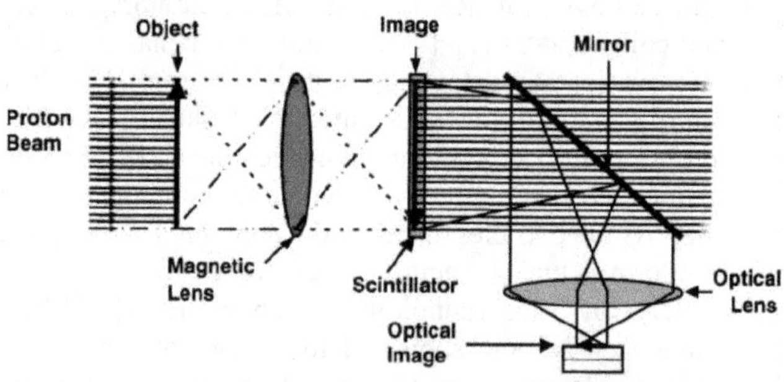

Figure 1: Proton radiography schematic showing the collimated beam of protons, optical system to capture the radiographs, and scintillator [22].

The protons pierce the opaque explosive products produced during detonation to provide 21 time-resolved images of the particle fronts. Images are spaced 2 μs apart with a resolution of 844x853 pixels. The x3 magnifier is used to provide a 45 mm by 45 mm field of view with a resolution of 50 µm/pixel. Tests are performed within a 1.83 m diameter spherical containment vessel that is pumped down to vacuum.

Details of the test article are shown in Fig. 2. A 10.2 cm outer diameter steel casing (AISI 4340) was used with a central bore of 13.1 mm nominal diameter and a depth of 19.1 mm. The casing showed minor deformation post-test, but no fragmentation was observed. Examination of post-detonation photographs shows the bore diameter increased from 13.1 mm to 15±3 mm, an increase of approximately 15%. A pellet of PBX-9501 (12.7 mm diameter with 12.7 mm length) is initiated by a Teledyne RISI RP-80 exploding bridgewire detonator. When examining Fig. 2 a slight step is visible along the length of the centre bore. The step was necessary to locate the pellet of PBX-9501 exactly and provide a consistent particle bed depth. The PBX-9501 pellet was epoxied to this step using Barcobond to prevent movement of the explosive as the detonator is brought into contact with the explosive using the adapter nut. Time is measured from detonator initiation, determined from a current viewing resister, with approximately ±0.1 µs uncertainty.

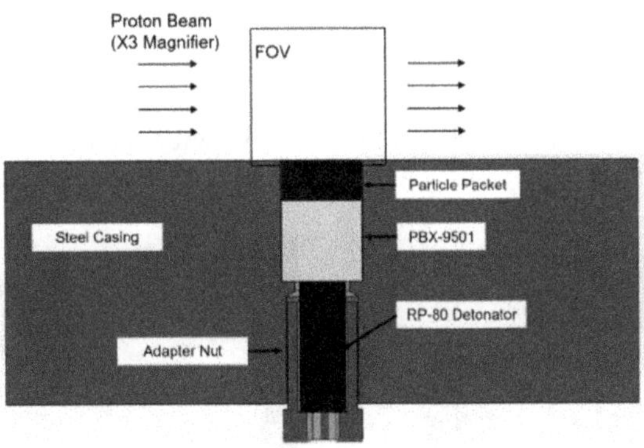

Figure 2: Cross-sectional view showing the particle packet and explosive train. Note the small step used to locate the pellet of PBX-9501.

After setting the explosive, a 13.1 mm diameter pocket with a depth of 6.35 mm was then filled with a steel particle packet and covered with tape. The tape was necessary to secure the particle bed during loading of the test article into the test vessel due to the significant jostling present in the loading process. The steel particles were 410 alloy, sieved between 75–125 µm, and obtained from Sandvik Osprey. The steel particles were divided into three layers of 1.233±0.001 g by paper dividers (3.697±0.002 g total). The particles were packed to approximately 56±1% volume fraction and no binder was used to constrain their movement.

The test articles were placed in the 1.83 m diameter pressure vessel, oriented vertically so the particles were held in place by gravity before detonation as well as by the tape. Wrapping the test article was a 6.35 mm thick aluminium cylinder to contain the various carrier phases. A gas handling system was used to ensure each cylinder was brought to 100±2 kPa pressure and the temperature recorded to determine the initial thermodynamic state of the gas. Xenon and SF_6 were chosen for their large nuclear cross-sections in attempt to capture the shock with the proton radiography. A series of traverses were used to align the centre of the test article with the proton beam. Table 1 contains a shot log detailing the gas used in each shot and the measured pressure and temperature.

Table 1: Carrier fluid used in each shot with measured temperature and pressure.

Shot #	Gas	Temperature, K	Pressure, kPa
1	Air	298.6	101.67
2	Xenon	297.3	100.13
3	SF_6	298.5	100.22
4	Air	299.2	100.32
5	-	-	-

3 RESULTS

Sample transmission radiographs of the four different test conditions 40 μs after detonation initiation are shown in Fig. 3. Note that an aluminium ring was placed around the lip of the steel casing to secure it and is visible in the radiographs as a thin layer of less dense material above the dark steel casing. Two screws used to secure the restraining ring are also visible at the edges of the radiographs. The particle packet shows a piston-like behaviour as it is expelled by the explosive. The three initial layers maintain their structure as they are expelled from the barrel and exhibit minimal spreading. At late times, some slow intermixing of the layers is exhibited. Though small differences in structure and spreading are apparent, the particles are remarkably close to each other considering the varying initial conditions.

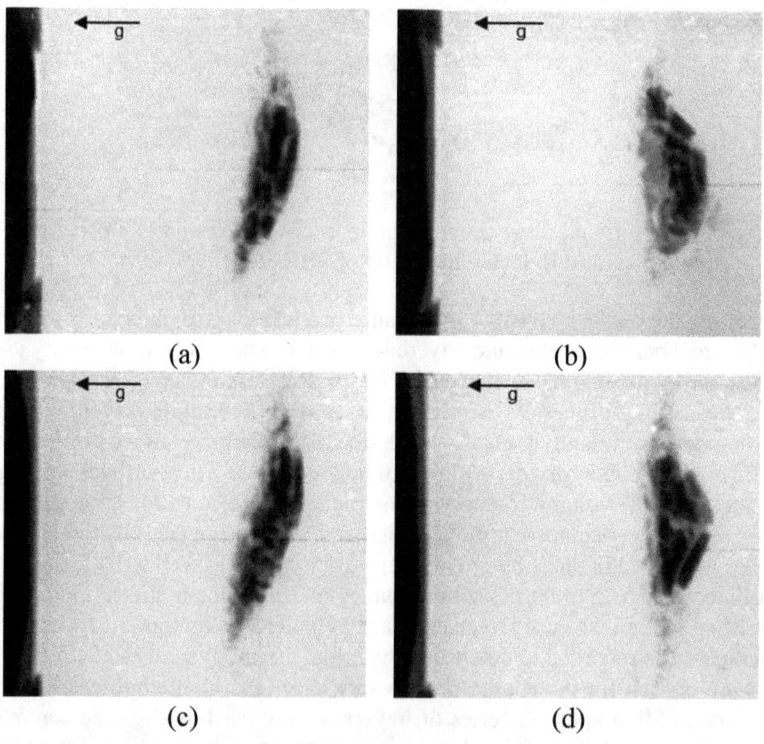

Figure 3: Transmission radiographs showing the dispersal of the particles 40 μs after detonator initiation for the four different test configurations: (a) Air; (b) Xenon; (c) SF6; (d) Vacuum. Flow is from left to right.

Particle fronts are determined from thresholding of the transmission intensity. The intensity is filtered with a moving average with a 17 pixel window size. A background threshold is determined for each image line (see Fig. 4(a)). Casing location is determined from the first frame as the image location that drops below 40% of the background threshold. The 40% threshold was necessary to capture the location of the steel casing while ignoring the less dense aluminium retaining ring. In subsequent frames, the downstream front is determined as the image location where the intensity drops below 90% of the background and the upstream front is where the intensity rises again above the 90% threshold. Greater threshold values, such as 95% and 98%, were attempted but produced too many spurious front locations. Sample particle front results are overlaid on the radiograph in Fig. 4(b). The seam down the centre of the image is a scintillator tile artefact. The reader will note that the seam is of sufficient contrast that it trips the particle front algorithm but is a spurious reading.

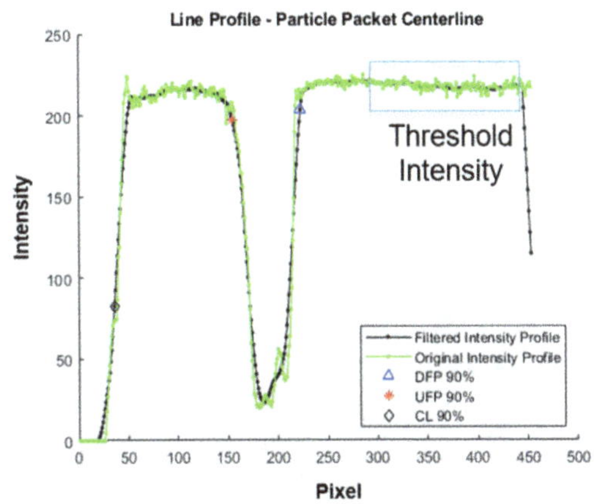

(a)

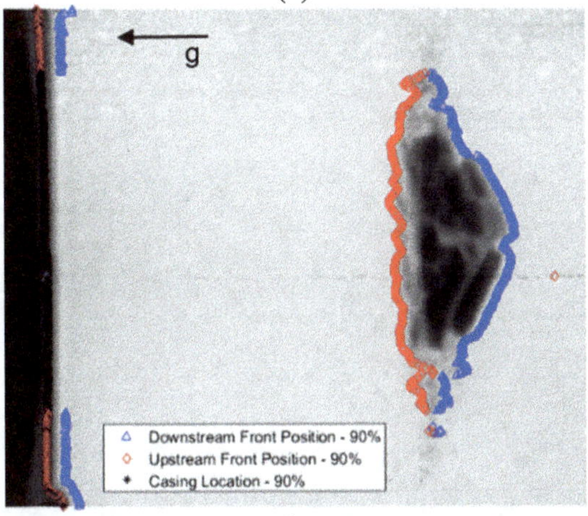

(b)

Figure 4: Threshold operations are used to identify the particle fronts. (a) Sample line profile showing the filtered intensity and the identification of the downstream front (blue triangle), upstream front (red star), and casing location (black diamond); (b) Results overlaid on the radiograph. Flow is from left to right.

The behaviour of the particle packets at the centreline was extracted and plotted in Fig. 5. The initial time is from the detonation initiation. An approximately 4 μs delay is present for the detonator to activate and for the explosive to completely react. The repeated air tests show close agreement on the centreline. One may note that the centreline results demonstrate a highly linear behaviour during this early time suggesting that the particles are impulsively acted upon by the contact interface and the shock. Furthermore, comparing the results of the four gas shots with the vacuum shot, we note that the results are within close agreement. The vacuum shot lacks an ambient medium for a shock to form, suggesting that the contact interface is almost solely responsible for the delivered impulse.

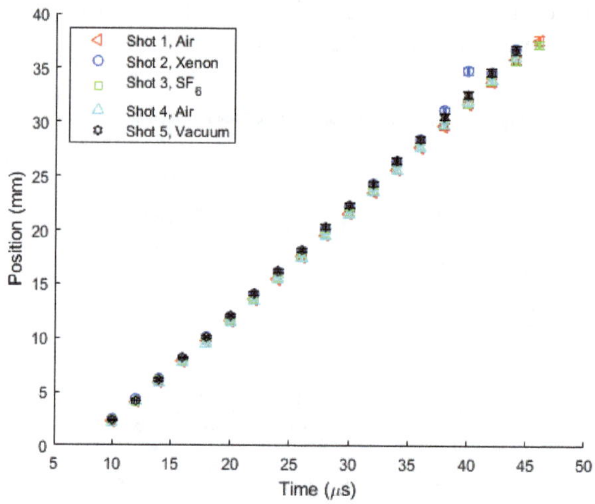

(a)

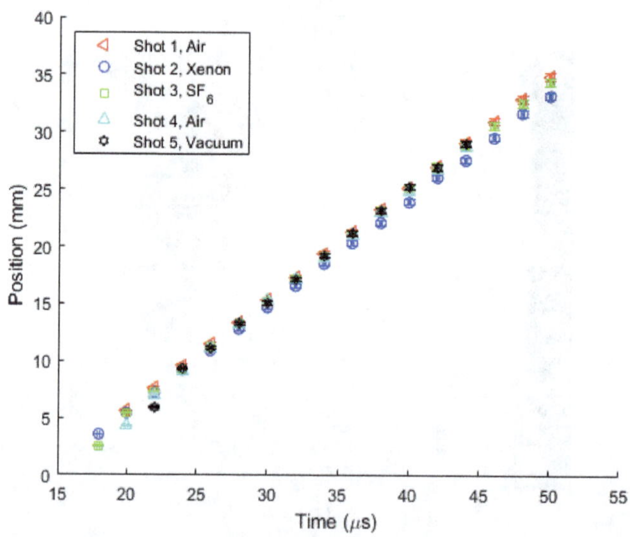

(b)

Figure 5: Centreline particle fronts obtained from the radiographs. (a) Downstream particle position; (b) Upstream particle position.

4 CONCLUSIONS

A series of experiments performed at Los Alamos National Laboratory was aimed to provide validation data for numerical simulations performed at the Centre of Compressible Multiphase Turbulence at the University of Florida. The experiments were designed to be small scale, containing O(10^5) particles, to reduce the computational burden while maintaining the full physics of the explosive dispersal of particles. Five explosive tests were performed. A 13 x 6 mm cylindrical packet of 100 μm steel particles was dispersed by an explosive into three different carrier fluids: air, xenon, and SF_6. A vacuum test was also performed to benchmark the relative strength of the shock compared to the contact interface. To penetrate the optically opaque explosive products at the early time and track the dispersed particle cloud, proton radiography was used. Twenty-one radiograph images at 2 μs temporal resolution were obtained. However, as this study only presents the motion of the particle cloud, additional tests are needed to validate the accompanying particle-fluid interaction and explosive model. Tests of similar configuration, but that offer information on the behaviour of the shock and contact interface at early time, are needed to provide these necessary prediction metrics.

Centreline particle fronts for the three carrier phases demonstrate close agreement, suggesting the ambient carrier phase has little impact on the dispersal of the particles. Rather the forces exerted on the particle bed by the dense, propagating explosive products are dominant. Furthermore, the results demonstrate a highly linear behaviour during this early time suggesting that the particles are impulsively acted upon by the contact interface and the shock. Finally, comparing the results of the four gas shots with the vacuum shot, we note that the results are within close agreement. The vacuum shot lacks an ambient medium for a shock to form, suggesting that the contact interface is almost solely responsible for the delivered impulse.

REFERENCES

[1] Zhang, F., Frost, D.L., Thibault, P.A. & Murray, S.B., Explosive dispersal of solid particles. *Shock Waves*, **10**(6), pp. 431–443, 2001.

[2] Frost, D.L., Zarei, Z. & Zhang, F., Instability of combustion products interface from detonation of heterogeneous explosives. *International Colloquium on the Dynamics of Explosions and Reactive Systems*, pp. 1–6, 2005.

[3] Frost, D.L., Ornthanalai, C., Zarei, Z., Tanguay, V. & Zhang, F., Particle momentum effects from the detonation of heterogeneous explosives. *Journal of Applied Physics*, **101**(11), p. 113529, 2007.

[4] Frost, D.L., Grégoire, Y., Petel, O., Goroshin, S. & Zhang, F., Particle jet formation during explosive dispersal of solid particles. *Physics of Fluids,* **24**(9), p. 091109, 2012.

[5] Sun, M., Saito, T., Takayama, K. & Tanno, H., Unsteady drag on a sphere by shock wave loading. *Shock Waves*, **14**(1–2), pp. 3–9, 2005.

[6] Tanno, H., Itoh, K., Saito, T., Abe, A. & Takayama, K., Shock wave interaction with a sphere in a shock tube. *Shock Waves*, **13**, pp. 483–497, 2004.

[7] Bredin, M.S. & Skews, B.W., Drag measurement in unsteady compressible flow part 1: An unsteady flow facility and stress wave drag balance. *R & D Journal of the South African Institution of Mechanical Engineering*, **23**(1), pp. 1–10, 2007.

[8] Maxey, M.R. & Riley, J.J., Equation of motion for a small rigid sphere in a nonuniform flow. *Physics of Fluids*, **26**(4) p. 883, 1983.

[9] Gatignol, R., The Faxén formulas for a rigid particle in an unsteady non-uniform Stokes-flow. *Journal de Mécanique Théorique Appliquée*, **2**(2), pp. 143–160, 1983.

[10] Parmar, M., Haselbacher, A. & Balachandar, S., Generalized Basset–Boussinesq–Oseen equation for unsteady forces on a sphere in a compressible Flow. *Physical Review Letters*, **106**(8), pp. 2–5, 2011.

[11] Parmar, M. & Haselbacher, A., Balachandar, S., Equation of motion for a sphere in non-uniform compressible flows. *Journal of Fluid Mechanics*, **699**(5), pp. 352–375, 2012.

[12] Annamalai, S. & Balachandar, S., Faxén form of time-domain force on a sphere in unsteady spatially varying viscous compressible flows. *Journal of Fluid Mechanics*, **816**, pp. 381–411, 2017.

[13] Britan, A., Elperin, T., Igra, O. & Jiang, J.P., Acceleration of a sphere behind planar shock waves. *Experiments in Fluids*, **20**(2), pp. 84–90, 1995.

[14] Jourdan, G., Houas, L., Igra, O., Estivalezes, J-L., Devals, C. & Meshkov, E.E., Drag coefficient of a sphere in a non-stationary flow: New results. *Proceedings of the Royal Society A: Mathematical, Physical and Engineering Sciences*, **463**(2088), pp. 3323–3245, 2007.

[15] Wagner, J.L., Beresh, S.J., Kearney, S.P., Pruett, B.O.M. & Wright, E.K., Shock tube investigation of quasi-steady drag in shock-particle interactions. *Physics of Fluids*, **24**(12), 123301, 2012.

[16] Bordoloi, A.D., Martinez, A.A. & Prestridge, K., Relaxation drag history of shock accelerated microparticles. *Journal of Fluid Mechanics*, **823**(R4), 2017.

[17] Hughes, K. et al., Uncertainty quantification of experiments on a small number of explosively-driven particles. *55th AIAA Aerospace Sciences,* pp. 1–14, 2017.

[18] Rudinger, G., Some properties of shock relaxation in gas flows carrying small particles. *Physics of Fluids*, **7**(5), pp. 658–663, 1964.

[19] Wagner, J.L. et al., A multiphase shock tube for shock wave interactions with dense particle fields. *Experiments of Fluids*, **52**(6), pp. 1507–1517, 2012.

[20] Kellenberger, M., Johansen, C., Ciccarelli, G. & Zhang, F., Dense particle cloud dispersion by a shock wave. *Shock Waves*, **23**(5), pp. 415–430, 2013.

[21] Dullien, F.A.L., *Porous Media : Fluid Transport and Pore Structure*. 2nd ed., Elsevier Science, p. 598, 1992.

[22] King, N.S. et al., An 800-MeV proton radiography facility for dynamic experiments. *Nuclear Instruments and Methods in Physics Research Section A: Accelerators, Spectrometers, Detectors and Associated Equipment*, **424**(1), pp. 84–91, 1999.

Author index

Materials and Contact Characterisation IX

Edited by: ***S. HERNÁNDEZ****, University of A Coruña, Spain,* ***J. DE HOSSON****, University of Groningen, The Netherlands,* ***D.O. NORTHWOOD****, University of Windsor, Canada and* ***R. VILAR****, Lisbon University, Portugal*

Including papers from the 9th edition of the International Conference on Computational Methods and Experiments in Material and Contact Characterisation this volume presents the work of selected researchers on the subject. Material and contact characterisation is a rapidly advancing field and this volume contains the latest research.

Of particular interest to industry and society is the knowledge of surface treatment and contact mechanics of these materials to determine the in-service behaviour of components subject to contact conditions. Modern society requires systems that operate at conditions that use resources effectively. In terms of components durability, the understanding of surface engineering wear frictional and lubrication dynamics has never been so important.

Current research is focussed on modification technologies that can increase the surface durability of materials. The characteristics of the system reveal which surface engineering methods should be chosen and as a consequence it is essential to study the combination of surface treatment and contact mechanics.

The accurate characterisation of the physical and chemical properties of materials requires the application of both experimental techniques and computer simulation methods in order to gain a correct analysis. A very wide range of materials, starting with metals through polymers and semiconductors to composites, necessitates a whole spectrum of characteristic experimental techniques and research methods.

The papers in the book cover a number of topics, including: Experimental approaches; Recycled materials; Metallic materials; Mechanical properties; Mechanics and thermal properties; Composite materials; Materials for bioengineering applications; Performance based design materials; Numerical models.

ISBN: 978-1-78466-331-5   eISBN: 978-1-78466-332-2
Published 2019 / 286pp

www.ingramcontent.com/pod-product-compliance
Lightning Source LLC
LaVergne TN
LVHW062232200926
848484LV00006B/1525

* 9 7 8 1 7 8 4 6 6 3 2 9 2 *